W9-BHV-193

fifth symposium on

the structure of low-medium mass nuclei

October 26-28, 1972
University of Kentucky
Lexington, Kentucky

SPONSORED BY

The Graduate School and
The Department of Physics and Astronomy
of The University of Kentucky

ORGANIZING COMMITTEE

Jerry D. Brandenberger
Marcel Coz
Fletcher Gabbard
Bernard D. Kern, Chairman
Alan D. MacKellar
Marcus T. McEllistrem
Rudolph Schrills
Jesse L. Weil

fifth symposium on
the structure of low-medium mass nuclei

EDITED BY

J. P. DAVIDSON &

BERNARD D. KERN

THE UNIVERSITY PRESS OF KENTUCKY

ISBN: 0-8131-1293-1

Library of Congress Catalog Card Number: 73-77254

A statewide cooperative scholarly publishing agency
serving Berea College, Centre College of Kentucky,
Eastern Kentucky University, Georgetown College,
Kentucky Historical Society, Kentucky State University,
Morehead State University, Murray State University,
Northern Kentucky State College, Transylvania University,
University of Kentucky, University of Louisville, and
Western Kentucky University.

Editorial and Sales Offices: Lexington, Kentucky 40506

CONTENTS

EDITORS' PREFACE

A change of locale and imprint has altered neither the format nor contents of these Proceedings, the third to be published formally. All of the invited papers as well as formal comments (short papers) and the discussion following each, which were presented at the Fifth Symposium at the University of Kentucky on October 26-28, 1972, are included. The organizers at the host institution while expanding somewhat the size of the group of participants maintained the same structure which has served so well during the past decade. The only notable change from the previous two symposia was the absence of an industrial session.

The invited papers are denoted by the Roman numeral of the session followed by a capital letter indicating the order in the session. The formal comments are denoted by the session numeral followed by a lower case letter. The discussion was tape recorded, transcribed, and edited on-the-spot by the speakers involved. As before we have found that this process preserves the flavor of the spontaneous rejoinder and at the same time makes the comments an integral part of the thoughtful presentation of ideas from several quarters. A few editorial insertions have been made and are enclosed in square brackets.

Xerox copies of the completed pages were sent to each contributor for correction. This process has not delayed the appearance of these Proceedings and it has left the editors with the feeling that many typographical errors and figure permutations have been found and rectified. For their kind cooperation in supplying manuscripts within a very narrow time frame and for promptly returning the page proofs the editors wish to express their sincere appreciation.

Our grateful acknowledgment is made for the assistance given by Marie Kelleher in the typing of the discussion from the taped records, and by the several University of Kentucky graduate students who competently managed the details of slide projection and tape recording.

The support necessary to provide the camera-ready copy has come from the University of Kansas Physics Department and we wish to express our appreciation to Professors David B. Beard and Gordon G. Wiseman, Chairman and Associate Chairman respectively, for cheerfully dipping into their special funds to see this task to its completion. One of us in particular (JPD) wishes to express here his especial appreciation to Betsy Goldberg not only for the typing and retyping of the copy but also for the many hours spent agonizing with him over author's meanings, word order and spellings. The task of producing a book such as this is always entered into with great enthusiasm which in due course leads to flagging spirits that we overcame only by the greatest of mutual support. This I (JPD) received in full measure and it should not be said now that the task is done that we were gluttons for punishment--only gourmets.

The University Press of Kentucky is to be held in no degree responsible for any errors or omissions or lack of good judgment of the editors, inasmuch as in order to facilitate the rapid publication of these Proceedings, the Director has agreed to forgo the heretofore strict editorship of publications by the University Press.

J. P. Davidson
University of Kansas

Bernard D. Kern
University of Kentucky

KEY TO THE PHOTOGRAPH

1. J. D. Fox
2. H. C. Lee
3. J. F. Sharpey-Schafer
4. H. F. Glavish
5. T. K. Alexander
6. C. van der Leun
7. G. M. Temmer
8. F. B. Malik
9. S. Maripuu
10. P. Goode
11. G. E. Mitchell
12. M. Coz
13. M. J. A. de Voigt
14. P. M. Endt
15. P. Goldhammer
16. E. K. Warburton
17. R. D. Lawson
18. B. C. Robertson
19. B. D. Kern
20. S. E. Darden
21. H. A. Van Rinsvelt
22. F. D. McDaniel
23. J. D. McCullen
24. S. Tryti
25. C. R. Gould
26. T. W. Burrows
27. D. R. Tilley
28. J. F. Walker
29. G. Kanatas
30. J. E. Sherwood
31. A. D. MacKellar
32. L. A. Alexander
33. A. Mueller-Arnke
34. J. L. Weil
35. L. Brown
36. F. E. Dunnam
37. R. D. Koshel
38. F. D. Snyder
39. R. W. Krone
40. D. J. Donahue
41. T. R. Donoghue
42. F. Gabbard
43. M. W. Greene
44. J. P. Davidson
45. E. M. Diener
46. P. Taras
47. L. W. Seagondollar
48. T. T. Thwaites
49. D. M. Sheppard
50. G. I. Harris
51. W. A. Lanford
52. G. M. Crawley
53. R. Y. Cusson
54. R. E. Azuma
55. K. P. Jackson
56. D. R. Goosman
57. J. R. Risser
58. F. W. Prosser
59. C. P. Swann
60. H. L. Scott

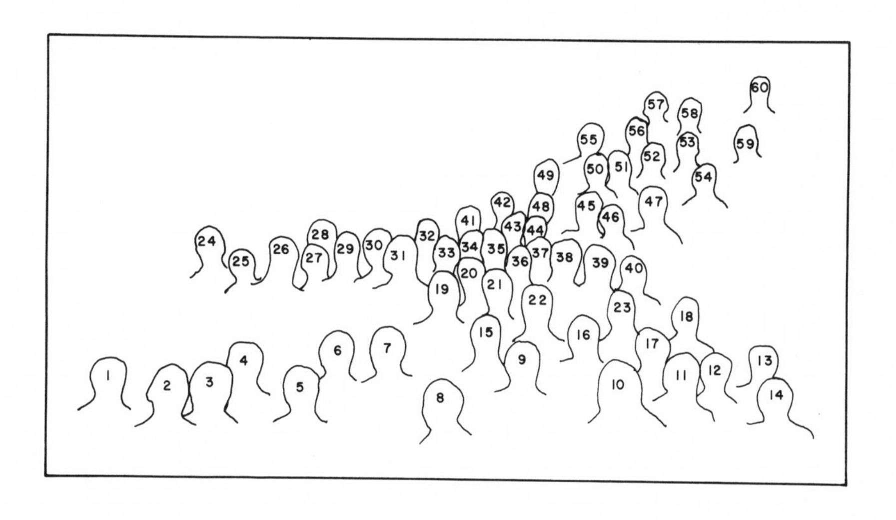
1
2
3
4
5
6
7
8
9
10
11
12
13
14
15
16
17
18
19
20
21
22
23
24
25
26
27
28
29
30
31
32
33
34
35
36
37
38
39
40
41
42
43
44
45
46
47
48
49
50
51
52
53
54
55
56
57
58
59
60

I.A. PROGRESS IN THE sd-SHELL

C. van der Leun
Fysisch Laboratorium, Rijksuniversiteit
Utrecht, Netherlands

I. TRENDS IN NUCLEAR SPECTROSCOPY

Some twenty years ago, it was possible to summarize the spectroscopic information on sd-shell nuclei in one table of a few pages. That table listed all the known levels, including resonance levels.

Today one has to apply all the tricks of condensation and of concise writing to be able to review our knowledge of the sd-shell nuclei in the scope of a book. Even then it is imperative to forget about all the data that have only historical significance, however interesting they may be. Moreover, one has to treat the practically infinite number of resonance levels in a stepmotherly fashion. The rate of progress indeed has been high.

The numbers required for a more quantitative formulation of this general statement are available in compilations. They give a summary of the results obtained at several different moments over this 20 year period. The growth of our spectroscopic knowledge can be followed by comparing the consecutive issues of these review studies. For the sd-shell (more precisely: the Z = 11-21 nuclei, and later the A = 21-44 nuclei) one can use references 1-6.

The number of nuclides observed in this mass-region as given in the reviews published in the years 1950, 1954, 1957, 1962, 1967 and 1973 is plotted against time in Figure 1. The "measuring points" are connected by a dotted line, which is merely a guide to the eye. A few selected nuclear properties have been plotted in a similar way:

the number of observed *bound states*,

the number of levels to which a *spin* has been assigned unambiguously, and

of which the *lifetime* has been measured.

Each of these curves represents a specific trend in nuclear spectroscopy. They have much in common. In all cases a slow start is followed by a period of rapid expansion. Once the interest in a certain quantity has been excited--quite often by the development of a new or refined nuclear model--it obviously takes some time to devise the methods and to design the apparatus to determine these quantities. The periods of rapid growth are clearly correlated with the introduction of new apparatus or methods of analysis.

The number of known nuclides increases rapidly both in the beginning of the fifties and seventies, with the advent of accelerators providing beams of sufficiently energetic particles or heavy ions.[7,8] The period of rapid increase in the number of observed bound states is correlated with the introduction of high-resolution magnetic spectrographs. The upward turn in the spin-curve of Figure 1 is to a large extent due to the large computers which shorten the time-consuming analysis of angular correlation measurements. The sudden jump in the number of measured lifetimes finally, is a consequence of the development of Ge(Li) γ ray detectors, which enable Doppler-shift attenuation measurements.

Figure 1 exhibits a few trends. It illustrates for example, the close interrelation between technical innovations and progress in science. It should also be stressed, however, that it only gives an incomplete picture of the enormous progress. Partly so since a number of quantities have not been plotted; similar curves could be drawn for parities, spectroscopic factors, γ ray multipolarity mixing ratios and others. Most of all, however, Figure 1 is incomplete since it reflects nothing of the growing theoretical insight in the structure of light nuclei.

Sometimes the rapid growth in our knowledge of nuclear properties is met with some distress. It gives rise to the idea that everything has already been measured. Everybody who had to go through the literature, *e.g.* to check some theoretical calculations or to obtain statistical information on certain quantities knows that this is a serious

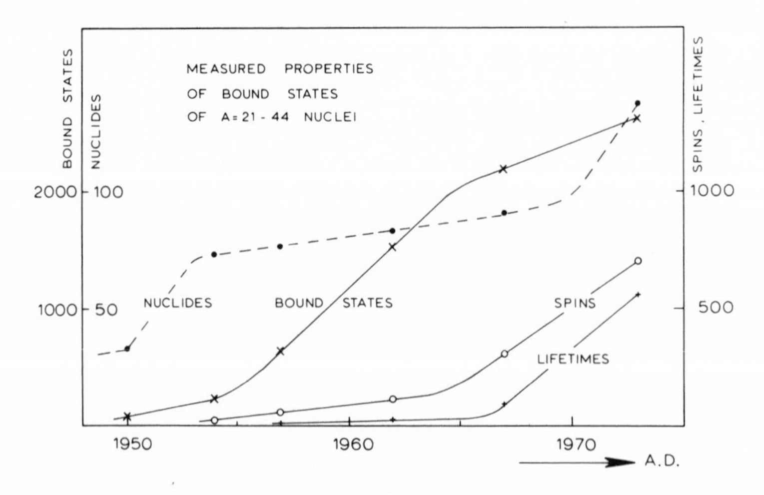

Fig. 1. Spectroscopic information on bound states of sd-shell nuclei, available over the period 1950-1972. The "measuring points" are based on the data summarized in References 1-6.

misunderstanding. More often than not one gets the impression that virtually nothing has been measured. To illustrate this one might try to trace in Figure 1 the curves for multi-polarity mixing ratios, M2, E3, etc. transition strengths, dipole and quadrupole moments of excited states, high ($J \geq 5$) spins. For most of these quantities, the abscissa adequately represents the growth-curve! The next section, on transition strengths, clearly illustrates our very limited knowledge. In the subsequent sections, which concentrate on some recent Utrecht work, it will be shown that useful contributions to this field are possible, also with small accelerators.

II. GAMMA RAY STRENGTHS

The lifetimes of about 600 bound levels of sd-shell nuclei have been measured (cf. Figure 1). If it is assumed that a level deexcites with an average of two or three branches, and that each transition with a non-zero multipolarity mixing ratio gives rise to two transition strengths, the 600 lifetimes could lead to almost 3000 transition strengths. The calculation of a transition strength, however, is possible only if the spins and parities of the initial and final state are also known, and if the mixing ratio of the transition has been measured. At present the real bottleneck in the calculations is the low number of measured mixing ratios. In this respect it is illustrative that in extensively studied nuclei like Mg^{24} and Si^{28}, the mixing ratios, δ, of only two transitions are known. In general, measurements of more δ-values would be a valuable contribution to the spectroscopy of sd-shell nuclei; especially, however, in the self-conjugated nuclei, where all $\Delta T = 0$ dipole transitions are forbidden or retarded.

This underdeveloped area of our knowledge reduces the above estimated number of roughly 3000 transition strengths to about 400. Of these 400, more than 100 strengths could be calculated since the transitions have necessarily a pure multipolarity ($J \rightarrow 0$, $0 \rightarrow J$ and $\frac{1}{2} \rightarrow \frac{1}{2}$ transitions). The assumption that $\Delta J = 0, 1$ transitions between states with opposite parity and $\Delta J = 2$ transitions between states with the same parity are (almost) pure, seems to be justified since very few non-zero M2/E1 or M3/E2 mixings have been found experimentally and because those which have been found are small. On the basis of this assumption an additional 100 E1 and E2 strengths could be calculated, bringing the total for the A = 21-40 nuclei to about 500. A number of impossible (and statistically not significant) strengths were eliminated by acceptance of the rule that only strengths with an experimental error smaller than 50% of the value should be included in the survey. These experimentally determined strengths are represented in the histograms of Figure 2. Detailed tables on which these figures are based will be published elsewhere. Comparison of the data, with those reported eight years ago at a previous meeting,[11] again shows remarkable progress notwithstanding the above mentioned limitations.

The practice of giving "average"-transition strengths has been in use for a long time. The averages, however, are to a large extent determined by the quality and sensitivity of the detectors used. The many upper limits given in published decay schemes for non-observed transitions, strongly suggest that extension of the histograms to the left simply waits for more precise measurements. These left hand wings (and thus also the averages) are determined by our instruments, the right-hand wing, however, by nature. Therefore it might be good to concentrate on the strongest transitions of each type.

An additional reason to do this is the fact that many of our spin and parity assignments, and also the choices between two possible δ-values, are based on what is considered an impossible, unacceptable, improbable, or unlikely transition strength. A large variety of limits could be quoted from the literature; they sometimes seem to be set *pour besoin de la cause*.

Especially for spin assignments this is in contrast to the usual carefulness, which is reflected by the almost general acceptance of the 99.9% probability limit for unambiguous spin assignments from angular correlation work. This very strict criterion seems to originate from the geodetic services. An equally careful set of "acceptable" transition strengths (for a suggestion see below) might do no harm to nuclear spectroscopy.

E1 Transitions

The two strongest observed E1 transitions both occur in Si^{29}. They deexcite the lowest $J^{\pi} = 3/2^-$ and $J^{\pi} = 1/2^-$ levels. The 4.93 → 0 MeV and 6.38 → 0 MeV transitions have strengths of 7 and 5 mW.u., respectively. The lifetimes of the initial levels (1.2 and 0.5 fs, respectively) have been measured by resonance scattering techniques. The strengths reflect the relatively pure single-particle character of these transitions.

M1 Transitions

The highest strength in the M1 histogram is due to the 0.66 → 0.58 MeV, $0^+ \rightarrow 1^+$ transition with $\Delta T = 1$ in Na^{22}, which has a strength of 3 W.u. The only two other strengths which exceed 1 W.u. are also $\Delta T = 1$ transitions in a self-conjugated nucleus: the 1.06 → 0.23 MeV ($1^+ \to 0^+$) and 2.07 → 1.06 MeV ($2^+ \to 1^+$) transitions in Al^{26}, with $|M|^2$ = 1.6 and 1.8 W.u., respectively. In the next highest block of eight strengths one finds four ΔT = 1 transitions in self-conjugated nuclei. All transitions of this type are shaded in Figure 2.

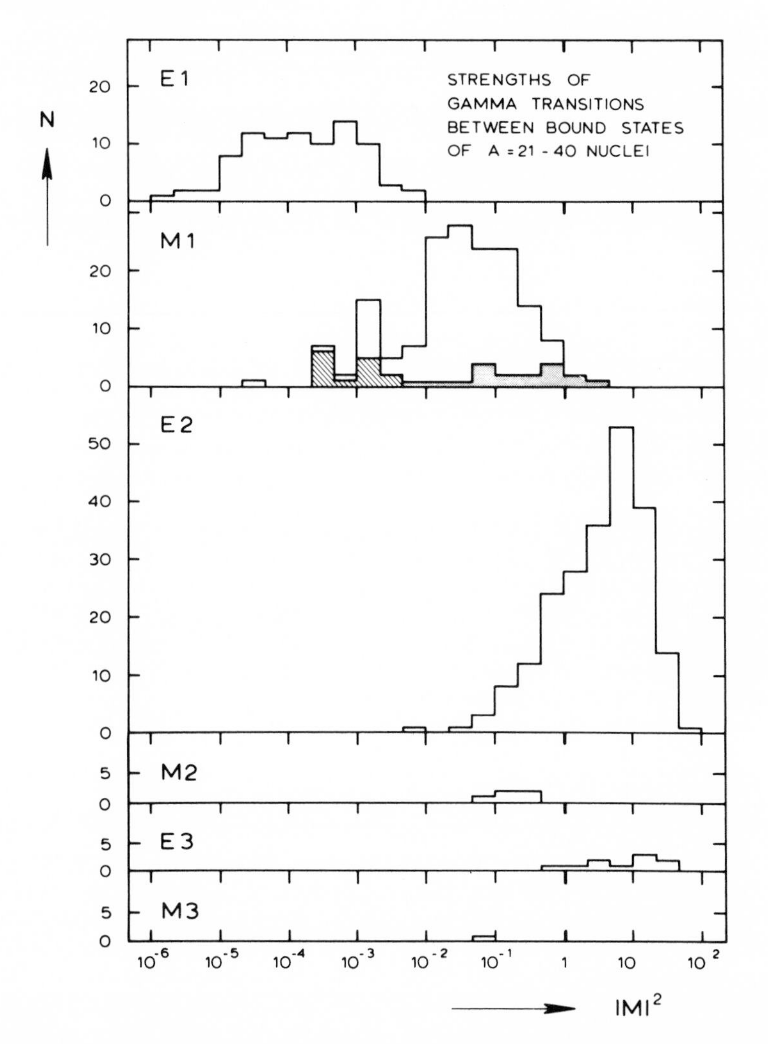

Fig. 2. Histogram of measured strengths of γ ray transitions between bound states of A = 21-40 nuclei. The hatched and shaded parts of the M1 histogram correspond to transitions in self-conjugated nuclei with ΔT = 0 and ΔT = 1, respectively.

E2 Transitions

The 5.28 →3.90 MeV, $4^+ \to 2^+$ transition in Ca^{40}, with a strength of 60 ± 20 W.u. is by far the strongest E2 transition observed in the sd-shell. The 14 E2 transitions in the next highest block have all strengths between 20 and 30 W.u.

M2 Transitions

Only five M2-transition strengths could be used. The two strongest, in Cl^{35} and K^{40}, have $|M|^2 \simeq 0.25$ W.u. It should be noted that a much stronger (6.3 ± 3.7 W.u.) M2-transition in Cl^{35} has been reported. It is not included in this review because of the precision-rule given above.

E3 Transitions

The two strongest observed E3 transitons, with $|M|^2 \simeq$ 25, occur in Ca^{40} (3.73 → 0 MeV) and S^{32} (5.01 → 0 MeV).

M3 Transition

The only example is the $3^+ \to 0^+$, 0.15 → 0 MeV transition in Cl^{34}, with $|M|^2 = 0.07$ W.u.

On the basis of the above observations, one might formulate the following criteria for "acceptable" transition strengths between bound states of A = 21-40 nuclei (due to the strong A-dependence of some "average" strengths[11], these criteria are not necessarily valid for other mass regions):

E1 : $|M|^2 < 0.01$ W.u.

M1 : $|M|^2 < 1$ W.u. (in self-conjugated nuclei, however,
< 10 W.u. if $\Delta T = 1$
< 0.01 W.u. if $\Delta T = 0$)

E2 : $|M|^2 < 50$ W.u.

M2 : $|M|^2 < 0.5$ W.u.

E3 : $|M|^2 < 50$ W.u.

These limits are rounded off by a factor of two higher than the strongest observed transition in each category. If a more precise measurement of the strength of the 5.28 → 3.90 MeV E2 transition in Ca^{40} would substantiate its present exceptionally high value, the E2 acceptability limit should be increased to 100 W.u. It should also be noted that the last two entries of the above table are based on rather few data, and thus are amenable to changes. Efforts to measure accurate values of more M2 strengths will be especially welcome.

In general, a careful remeasurement of the strongest observed transitions in each category, might be useful in establishing a generally accepted set of limits for use in further spin and parity assignments.

Isospin Selection Rules

The material given above may also be used to check the isospin selection rules for transitions in self-conjugated nuclei.

E1. Most of the observed E1 transitions are forbidden $\Delta T = 0$ transitions. Only one transition has $\Delta T = 1$ and is thus isospin allowed. Its strength, however, is just an average E1 strength. It would be interesting to have more examples of this type.

E2. Another type on which information is scarce is the inhibited $\Delta T = 1$, E2 transition. Only two cases are known, such that comparison with the allowed $\Delta T = 0$ transitions is not very useful.

M1. A statistically significant comparison can only be made for the M1 transitions. In self-conjugated nuclei the $\Delta T = 0$ transitions are expected to be weak.[9] The transitions of this type are marked (hatched) in the histogram of Figure 2. The strength reduction amounts to a factor of 100 to 1000. The latter value is applicable in a comparison with the $\Delta T = 1$ transitions in the self-conjugated nuclei (shaded), which are favored since the proton and neutron spin contributions reinforce each other.[9]

III. CAPTURE REACTIONS

The discussion of some recent Utrecht experiments should, following the historic line, start with the capture reaction investigations. The interest in the analog resonances, which have also been studied extensively in (p,γ) reactions, has during the past few years stressed an aspect that in fact is not typical for capture reaction work. The classical picture of a resonance reaction implies that the capture reaction is not particularly selective. Now the selective strong population of a few states (IAS) is typical, but the fact that, for example, a (p,γ) reaction excites almost any unbound nuclear state, if only the proton energy is sufficiently high to overcome the Coulomb and centrifugal barriers. Even at energies where proton capture just starts one finds that practically no levels are missed (as far as they are known from other reactions). Since the study of unbound states usually is not the strongest point in charged particle reaction work, it is not easy to find a good example for a comparison. In Na^{22} however, the unbound states have been investigated through four different charged-particle reactions: $Ne^{20}(t,p)Na^{22}$, $Ne^{20}(Li^6,\alpha)Na^{22}$, $Na^{23}(t,\alpha)Na^{22}$ and $Mg^{24}(d,\alpha)Na^{22}$. In Reference 6 the results are compared with those of the $Ne^{21}(p,\gamma)Na^{22}$ and $Ne^{21}(p,p'\gamma)Ne^{21}$ resonances. In the E_x = 8-9 MeV range only two of the twenty known levels are not observed as proton resonances.

Such a non-selective reaction mechanism has certain advantages in systematic spectroscopic studies. If indeed almost all unbound states of however complicated configuration are populated, it might be expected that in the γ decay of these resonances also most of the bound states are excited. A good example is the work of Alderliesten[12] on the reaction $Cl^{37}(p,\gamma)Ar^{38}$. About 240 resonances were found in the energy range of E_p = 0.65-1.78 MeV, that is up to about 200 keV above the neutron threshold. The γ ray spectra at about 15 resonances were studied in detail. It turns out that in the decay of these resonances all but one of the 40 bound states with $E_x < 7$ MeV known from the charged particle reactions $Cl^{35}(\alpha,p)Ar^{38}$, $Cl^{37}(\tau,d)Ar^{38}$, and $K^{41}(p,\alpha)Ar^{38}$ (plus a few hitherto unknown states), are excited (see Figure 3) with sufficient intensity to allow:

measurements of the excitation energies with sub-keV precision,

determination of the branching ratios,

Doppler shift attenuation measurements that yield the mean lives (or limits) for 30 levels.

These mean lives, combined with some angular distribution measurements determine the spin of seven of the bound states (and limit the possibilities for several others). Further angular correlation work might provide the remaining undetermined spins and the highly wanted (see above) multipolarity mixing ratios. Simultaneous measurements of the (p,γ) and (p,α) reactions on Cl^{37}proved to be very useful in the identification of resonances of which the J^π values were known from previous (p,α) data. The only bound state with $E_x < 7$ MeV not excited in these (p,γ) experiments, is the level at 6420 ± 20 keV found[13] in the reaction $K^{41}(p,\alpha)Ar^{38}$ at E_p = 9 and 10.5 MeV. The reason for the non-observation of this level in the $Cl^{37}(p,\gamma)Ar^{38}$ reaction might be its probably high spin value. In the $Mg^{24}(O^{16}, 2p)Ar^{38}$ reaction at $E(O^{16})$ = 38 MeV Engelbertink[14] located

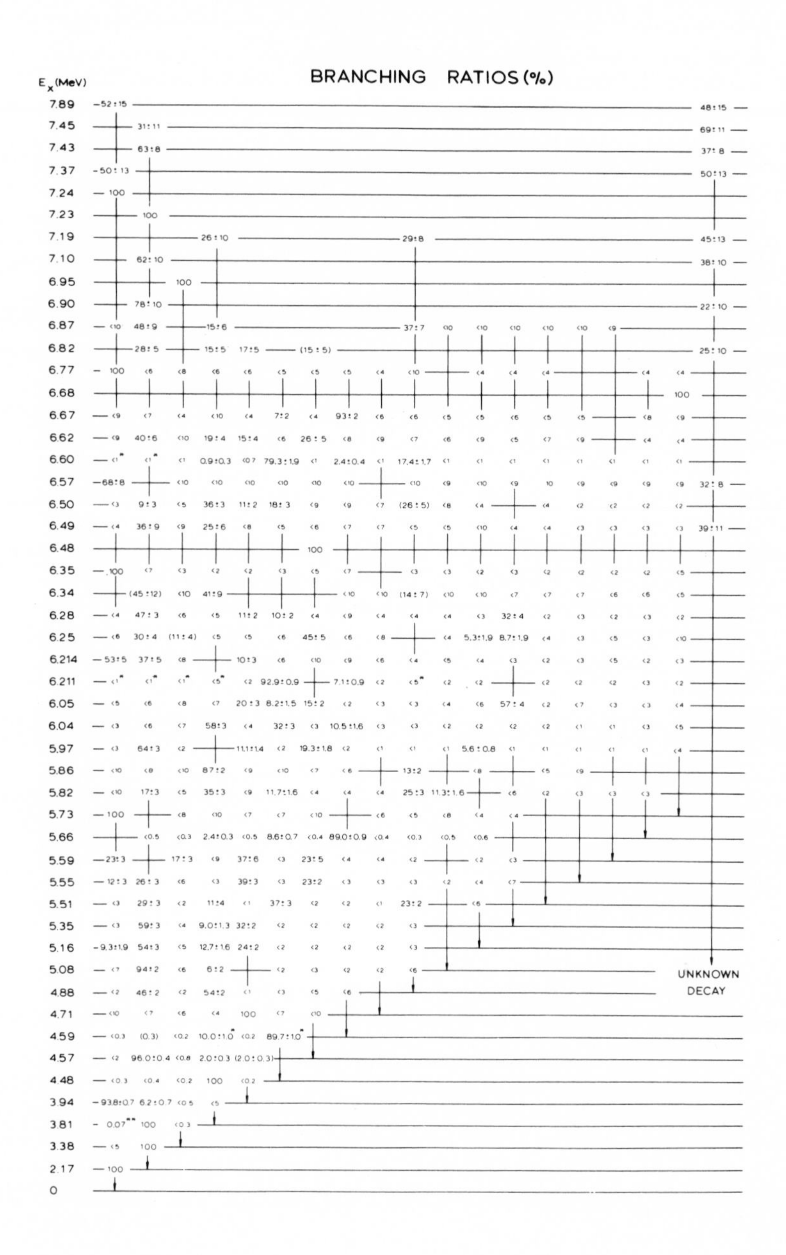

Fig. 3. Decay scheme of the bound states of Ar^{38} as deduced from the data at 15 $Cl^{37}(p,\gamma)Ar^{38}$ resonances.

a level at 6407.1 ± 1.0 keV, which decays exclusively to the 4.59 MeV level with $J^{\pi} = 5^{-}$. Since heavy ion work is not a topic at this conference, however, we proceed to α capture work.

The above mentioned problem of low-energy proton-capture experiments which imply low momentum transfer and thus low-spin resonance levels, can be partly overcome by bombardment with heavier particles, for example, α particles. In the early years of the study of (α,γ) reactions[15], the experiments were essentially limited to even-even target nuclei, where it is possible to study (α,γ) reactions without competing (α,p) or (α,n) reactions. These experiments are well known, and here we only recall the result that is relevant for the following discussion: even among the lowest energy resonances, one finds $\ell = 3$ and $\ell = 4$ capture resonances.

If one could do α capture experiments on odd-A nuclei, and start with, for example, a $J^{\pi} = 5/2^{+}$ nucleus, $\ell = 4$ capture could lead to resonance levels with J^{π} up to $13/2^{+}$. That would provide an excellent starting point for the study of high spin bound states; a welcome supplement to (p,γ) studies. Two problems, however, had to be solved before these experiments could be successfully performed. The first, the sensitivity of NaI crystals for background neutrons from the $C^{13}(\alpha,n)O^{16}$ reaction, that prohibited γ ray spectroscopy below $E_{\gamma} \simeq 7$ MeV in previous experiments, was overcome by the advent of Ge(Li) detectors. This is not only due to the excellent energy resolution of these detectors. Maybe even more important is the fact that the ratio of the efficiencies for γ ray detection and thermal neutron capture is about one order of magnitude larger for a 60 cm^3 Ge(Li) detector than for a 10 cm X 10 cm NaI scintillator. The sharper peaks and the lower background in Ge(Li) spectra facilitate the observation of weak and/or low-energy γ ray transitions. The relatively low absolute γ ray detection efficiency and the consequently long measuring times have to be taken for granted for the time being.

The second problem, the competition of (α,p) and/or (α,n) reactions, turns out to be very helpful in some cases. In Al^{27} for example, the α particle binding energy is 1.8 MeV higher than the proton binding energy. This implies that the resonances in the usual $E_{\alpha} = 2$-4 MeV range will decay via the p_0 channel. At high spin resonances, however, the large centrifugal barrier for transitions to the $J^{\pi} = 0^{+}$ Mg^{26} ground state, narrows the proton decay channel. An estimate based on average proton and radiative widths, indicated[11] that at resonances with $J \geq 9/2$, γ ray decay may compete favorably with proton emission. Such high spin resonances may readily be formed: γ capture in Na^{23}, with $J^{\pi} = 3/2^{+}$ leads to resonances with spins as high as $J^{\pi} = 11/2^{+}$. The low spin resonances will decay primarily by proton emission and the observed (α,γ) resonances are therefore expected to have mainly high spins. The competing (α,p) reaction thus facilitates the study of the (α,γ) reaction. By eliminating the many, sometimes already broad and overlapping low spin resonances, it makes the narrow high spin (α,γ) resonances stand out more clearly. The selection mechanism provided by nature turns out to be a useful sieve.

The results obtained by De Voigt *et al.* in investigations of the reactions $Na^{23}(\alpha,\gamma)Al^{27}$ and $Al^{27}(\alpha,\gamma)P^{31}$ have been published in detail[16,17] and only a few points relevant for the present discussion will be recalled here. The selection mechanism of high-spin resonances is illustrated most clearly by the fact that 17 out of the 20 resonances observed in the reaction $Na^{23}(\alpha,\gamma)Al^{27}$ for $E_\alpha < 3.3$ MeV decay to the $J^\pi = 9/2^+$ level at 3.00 MeV and/or the $J^\pi = 11/2^+$ level at 4.51 MeV and very few of them to the lower 0.84 and/or 1.01 MeV levels with $J^\pi = 1/2^+$ and $3/2^+$, respectively. Further experimentation resulted in the spins and parities indicated in Figure 4. Only spins of the lowest levels ($E_x \leq 4.51$ MeV) were known from other reactions. Another attractive possibility is the measurement of lifetimes with the DSA-method. The (α,γ) reaction is attractive for experiments of this type, since the complication of coincidence measurements is avoided in capture reactions and because of the relatively high initial velocity of the recoiling ions. In the case of $Na^{23}(\alpha,\gamma)Al^{27}$ resonance at $E_\alpha = 3$ MeV, the velocity amounts to $v/c = 0.006$. The energy shift between spectra measured at 0° and 130° amounts to $\Delta E_\gamma/E_\gamma \simeq 0.01$, that is 30-50 keV for γ rays of typical energy. One hardly needs Ge(Li) detectors to measure these large shifts.

A nucleus like Al^{27} that abounds in excellent information, to a large extent due to a combination of the results from (p,γ) and (α,γ) reactions, indeed calls for theoretical calculations. A detailed discussion of these calculations in the framework of the many particle shell model has been published.[18] The results on the electromagnetic transition rates in Al^{27} are reproduced in Figure 5; the calculated quantities are given in square brackets together with the measured quantities. In general the agreement between the two values for the lifetimes, branching and mixing ratios is remarkable, especially in view of the low number of free parameters used.

IV. GAMMA RAY ENERGIES[19]

From the early days of α capture experiments a continuous battle has been fought against carbon build up. Though the natural abundance of C^{13} is only 1%, the cross section of the $C^{13}(\alpha,n)$ reaction is so high, that without special precautions like high vacuum pumping speeds, tantalum lined cooling traps, carbon free vacuum seals, *etc.*, the (α,γ) spectra are useless.

Notwithstanding all these precautions, a peak was observed in a few of the γ ray spectra measured during the angular distribution measurements mentioned above. Since these peaks did show neither the usual Doppler broadening nor the expected Doppler shift with angle, it was concluded that they could not be due to any (α,γ) reaction. The energy of the line, 7.6 MeV, led to the assumption that it was due to $C^{13}(\alpha,n)$ background neutrons, thermalized in the material of the experimental set-up, and capture in the iron of the turning table. A verification of this assumption was easy: by placing some extra iron in the neighborhood of the Ge(Li) detector, the intensity of the background lines could

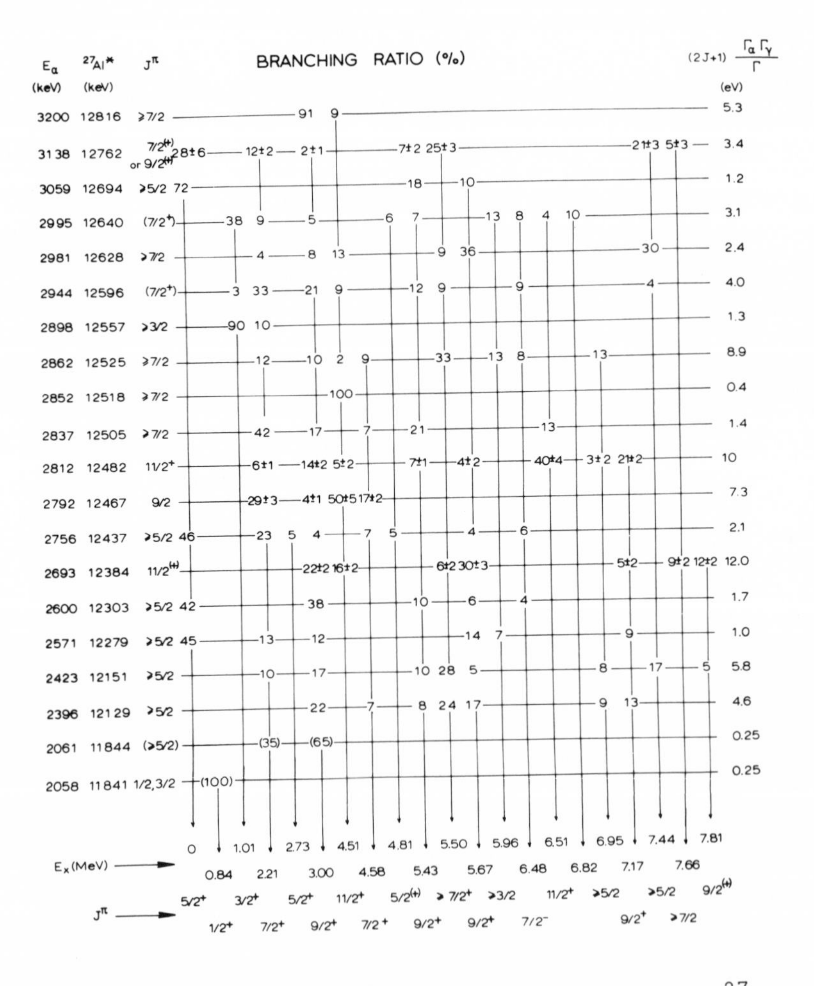

Fig. 4. Decay scheme of 20 resonance levels of Al^{27}.

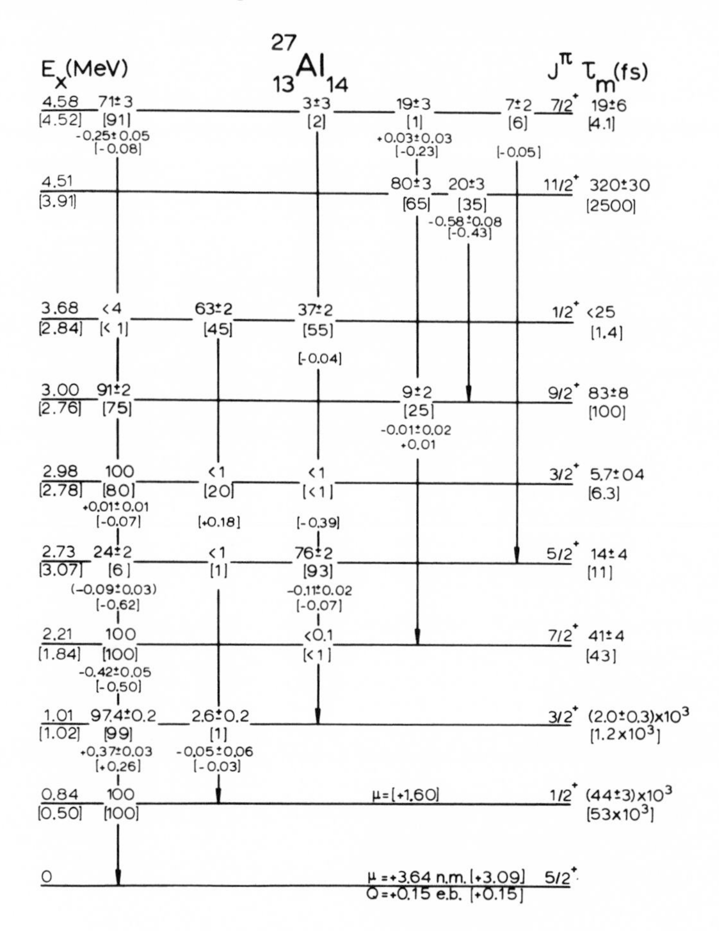

Fig. 5. Calculated [in square brackets] and experimental level and decay scheme of Al^{27}.

be increased. Such a simple production of sharp monoenergetic high energy γ lines offers an attractive possibility for the energy calibration of γ ray spectra. Usually, γ ray energies are measured relative to that of a series of well known calibration γ lines, mainly from radioactive sources. An adequate variety of sources is available[20] for the energy calibration of Ge(Li) γ ray spectra up to $E_\gamma \simeq 3.5$ MeV. In capture reaction experiments the difficulties arise at the more typical γ ray energies of $E_\gamma = 4\text{-}10$ MeV. A γ ray source, that covers this range of high energies has been constructed along the lines indicated above. Gamma rays from thermal neutron capture in many different materials can in principle be used as convenient calibration lines. It has been shown[19] that the thermal neutron flux from a 2.6×10^6 n/s Am^{241}-Be source, surrounded by a moderator, is sufficiently high to produce capture γ ray intensities comparable with the γ ray intensities produced in typical proton and α particle capture reactions.

The capturing materials most frequently used up to now are:

(i) Fe, with the 7.63-7.64 MeV doublet from the $Fe^{56}(n,\gamma)Fe^{57}$ transitions to the ground state and to the first excited state at 14.4 keV; this doublet is a well known test case for the quality of Ge(Li) spectrometers, and

(ii) Cl, not only since the thermal neutron capture cross section is high (33 b) but also since it has a γ ray spectrum with rather well isolated peaks over a broad energy range (0.52-8.58 MeV). A 5 mm thick vessel with CCl_4 was placed in front of the detector.

Of course, these and other neutron capture lines can only be fully used in the sub keV precision range, when the energies have been determined with adequate accuracy. In this process, the charged-particle capture reactions on sd-shell nuclei are useful.

The obvious advantages of (n,γ) over (p,γ) experiments, (i) a high flux of thermal neutrons and (ii) the exactly known energy of the incoming particle, are compensated by the fact that the one and only (n,γ) capturing state for each target nucleus is replaced by say 100 resonances in (p,γ). This abundance of initial states makes it possible to select resonance levels which decay strongly both in a cascade of low energy γ rays ($E_\gamma \lesssim 2.5$ MeV) and directly with a high energy γ ray to one of the lowest bound states. At these resonances it is possible to measure the excitation energy of the resonance level (typically 5-10 MeV in the sd-shell) with a precision of typically 0.2 keV (sum of the errors of three of four $\lesssim$ 2 MeV γ rays). The deexciting high energy γ rays are for practical reasons unsuitable as standard calibration lines, but they can be used to determine a precise value for a few convenient (n,γ) lines. In these measurements the E_p = 316, 390, 592, 724, 1044, 1106 and 1165 keV resonances of the reaction $Mg^{25}(p,\gamma)Al^{26}$ and the E_p = 860 keV resonance of $Cl^{35}(p,\gamma)Ar^{36}$ were used. It should be noted that for the present discussion the precise proton energies of these resonances are irrelevant. A relatively large number of resonances were used to reduce the risk of a misinterpretation of a decay scheme (especially close doublets may be pitfalls). The preliminary result for the iron doublet is: E_γ = 7 646.63 ± 0.16 keV and 7 632.22 ± 0.16 keV. The 14.41 keV difference of these energies was used as an input datum. It should be noted that these values are still based on the Marion table of low energy calibration lines. A recalculation on the basis of the forthcoming, and partly already published,[21] Idaho energies is underway.

One may question the relevance of these high-precision measurements, if nuclear theory reproduces the nuclidic masses and the excitation energies of nuclear levels with errors that are at least a few orders of magnitude larger than the errors discussed above.

At present such precision is indeed not required for general theoretical reasons, but mainly for specialized applications.

A few of the following applications will be discussed:

1. Reaction Q-values and Coulomb displacement energies.
2. Resolution of doublets.
3. Calibration of resonance energies.
4. Coordination of results from different reactions.
5. The vector-coupling constant in beta decay.
6. Lifetime measurements in the attosecond and femtosecond range.
7. Calibration of high energy γ rays independent from the present common standard for all γ ray energies.

V. CONCLUSION

A synthesis of the results obtained with a large variety of experimental techniques and through the combined efforts of many groups of experimentalists, enables us to give a quantitative estimate of the progress in the spectroscopy of light nuclei. The progress, which has been impressive especially during the last five years, was discussed here mainly from an experimental viewpoint. It is unthinkable, however, without the continuous impetus from different types of theoretical calculations that will get more attention in several of the following lectures.

Since there are no signs of fatigue, an extrapolation of the trends indicated above, may lead us to expect, means and time permitting, a number of exciting years in nuclear spectroscopy.

REFERENCES

1. D. E. Alburger, E. M. Hafner, Revs. Mod. Phys. 22, 373 (1950).
2. P. M. Endt, J. C. Kluyver, Revs. Mod. Phys. 26, 95 (1954).
3. P. M. Endt, C. M. Braams, Revs. Mod. Phys. 29, 683 (1957).
4. P. M. Endt, C. van der Leun, Nucl. Phys. 34, 1 (1962).
5. P. M. Endt, C. van der Leun, Nucl. Phys. A105, 1 (1967).
6. P. M. Endt, C. van der Leun, Nucl. Phys. in press.
7. R. Klapisch *et al.*, Phys. Rev. Letters 23, 652 (1969).
8. A. G. Artukh *et al.*, Nucl. Phys. A176, 284 (1971).
9. E. G. Morpurgo, Phys. Rev. 110, 721 (1958), E. K. Warburton, J. Weneser, in *Isospin in Nuclear Physics*, ed. D. H. Wilkinson, North-Holland, Amsterdam (1969), p. 173.
10. F. W. Prosser, G. I. Harris, Phys. Rev. C4, 1611 (1971).
11. C. van der Leun, *Proc. Symp. Structure of Low-Medium Mass Nuclei*, Kansas (1964) p. 109.
12. C. A. Alderliesten *et al.*, Nucl. Phys. (to be published).
13. R. G. Allas *et al.*, Nucl. Phys. 61, 289 (1965).
14. G. A. P. Engelbertink (Utrecht), private communication.
15. P. J. M. Smulders, Thesis Univ. of Utrecht (1965).
16. M. J. A. de Voigt *et al.*, Nucl. Phys. A170, 449 (1971).
17. M. J. A. de Voigt *et al.*, Nucl. Phys. A170, 467 (1971).
18. M. J. A. de Voigt *et al.*, Nucl. Phys. A186, 365 (1972).
19. C. van der Leun and P. de Wit. Phys. Letters 30B, 406 (1969), and *Proc. 4th Int. Conf. on Atomic Masses*, Teddington (1971), p. 131.
20. J. B. Marion, Nucl. Data A4, 301 (1968).
21. R. G. Helmer, R. C. Greenwood and R. J. Gehrke, *Proc. 4th Int. Conf. on Atomic Masses*, Teddington (1971), p. 112.

DISCUSSION

MALIK: I am very happy to see your comments on the situation about measurements of transition rates. I want to point out that one should probably emphasize the situation regarding quadrupole moments. Many of the old measurements, but not all, probably need some revision because many of then didn't consider or even today don't consider the Sternheimer correction. Primarily, measurements that are done by using atomic hyperfine structure spectroscopy are affected by this correction. The type of effective charge in any kind of calculation should also be noted. That's one parameter which we would like to understand very much and which probably tells us something about the structure.

The first beautiful curve shows how much new work has come out. Clearly in that case an outlay for the manpower to tabulate the results and make them available quickly to the scientific community and to analyze them theoretically should also increase in that proportion.

VAN DER LEUN: Let me answer the last comment first. A compilation and evaluation of the data will probably be available within half a year (Reference 6). The results presented in Figure 1 are due to the continuous efforts of many groups all over the world. It is my impression that on a worldwide scale the manpower in nuclear physics is not decreasing. It may be the case in this country, but in others, *e.g.*, in Germany, it is still increasing. We are experiencing an exciting period in nuclear physics. We are harvesting the fruits of the work done over the past decades. The data are coming in faster than ever, and consequently the quality of the theoretical interpretations is increasing. Intriguing new questions are popping up. I do not see a valid reason to pull out at this exciting moment. Therefore I trust that manpower will be available also in years to come.

Since I don't have any experience in measuring quadrupole moments I would like to refer your first question to an expert. Indeed one finds in the literature discrepancies that are larger than the quoted experimental errors which may be due to these corrections. On the other hand I think that the precision of the present theoretical calculations of the quadrupole moments is not that high that we have to worry very much about the experimental precision.

SESSION CHAIRMAN (GOLDHAMMER): Well, maybe if we had better values we would have better interpretations?

VAN DER LEUN: You are saying that we have a sufficiently precise interpretation of the present experimental values?

GOLDHAMMER: No, what I mean is, it is a question of which comes first, the chicken or the egg. Poor experimental numbers give a theorist little encouragement to make a sophisticated calculation. In many critical examples E2 moments are not even known to an accuracy of 20%.

VAN DER LEUN: OK, as long as we are talking about errors I agree, but in general the experimental errors are smaller in the sd-shell.

JACKSON: When you limit the selection of strengths to only those for which the uncertainty is less than 50% of the measured value you probably introduce a bias in favor of the stronger transitions because you ignore a lot of very weak transitions which are now known only as limits. The mean value of these distributions may have relatively little significance. As you point out, probably it's the top end of the distributions that really are useful.

VAN DER LEUN: Right, it is one of the reasons for my suggestion to concentrate our attention on the right hand wing of the histograms instead of on average values.

JACKSON: For instance in the El distribution there is one that is something less than about 10^{-8} W.u. in Cl^{35} that is presumably not included in that distribution.

VAN DER LEUN: Yes, I did not include upper limits, but many transitions are extremely weak. I didn't have time to compile all the upper limits but in view of the many upper limits given in published decay schemes I'm sure that many of them will eventually strengthen the left hand wing of the histograms. But, once more, I would prefer to concentrate on the more interesting other side of the histograms. Non-observation of a possible transition is at most a weak argument that should not be used.

JACKSON: One other simple question: What was the shell model calculation for Al^{27}?

VAN DER LEUN: It is a many particle shell model calculation by de Voigt *et al.*

JACKSON: How big?

VAN DER LEUN: It's a calculation with up to four holes in the $1d_{5/2}$ shell.

DUNNAM: With regard to the gamma ray calibration sources, did I hear you say that there were troubles with the Pu-C^{13} sealed source which gives about a 6 MeV gamma ray from O^{16}, or did you mention that?

VAN DER LEUN: No, I was talking about the 6.13 MeV gamma rays that are due to surface contamination on the target.

DUNNAM: One other question: with respect to the neutron capture gamma rays from iron, what size source of neutrons were you using and what sort of times were required to take adequate calibration spectra?

VAN DER LEUN: We were using an Am^{241} source which gives 2.6×10^6 neutrons per second, the only source we had available in the lab. It turns out to be a very adequate source. With a 40 cm^3 Ge(Li) detector we observe a counting rate of 500 per hour in the 7.6 MeV peak of Fe(n,γ). That is a typical counting rate for a peak in the spectrum of a strange (p,γ) resonance. In (α,γ) experiments the source was placed at a distance of about one meter from the detector. All calibration peaks are measured simultaneously with the studied spectra.

I.B. NEW APPLICATIONS OF HARTREE-FOCK CALCULATIONS TO NUCLEI WITH 20 < A < 60

H. C. Lee
Chalk River Nuclear Laboratories
Atomic Energy of Canada Limited
Chalk River, Ontario, Canada

I. INTRODUCTION

In the past decade or so the Hartree-Fock (HF) method has become increasingly popular as a tool for the study of nuclear physics. First it was used on calculations of the binding energy of nuclear matter and of finite nuclei as well. In time, these calculations have become more sophisticated in technique and more encompassing in aim. In recent years, one has become familiar with terminologies such as Brueckner-Hartree-Fock[1], deformed HF[2], HF-Bogoliubov[3], density-dependent HF[4], HF with three-body interactions[5], and multi-configuration HF.[6] All these are methods prescribed for specific but complementary improvements over the basic HF method, and it is a fact that significant progress has been made towards a better understanding of nuclear physics through these endeavors. Neither time nor my knowledge would allow me to do justice to all work related to these methods if I were to review them here. Rather I will report only on some work that I have been personally associated with, or have a special interest in.

The type of HF calculations I will discuss today is somewhat old-fashioned, and because of this, it is perhaps also one of the most advanced. It is what is generally called the deformed projected HF (PHF) which employs spherical harmonic oscillator wavefunctions as a basis, and uses a density independent two-body nucleon-nucleon (NN) interaction. Some of the elements that make our calculations new are the following:

(a) A large, five major shell basis is used in order to allow the nucleus to assume its optimum self-consistent shape;

(b) The symmetries imposed on the single-particle (sp) wavefunctions have been reduced to a minimum. The only conserved quantum numbers of the sp wavefunction are parity and isospin. This enables us to study nuclei other than the $n\alpha$ type. Separate treatment for the protons and neutrons also allow us to study effects induced by the isospin symmetry breaking, electromagnetic interactions;

(c) An efficient angular moment projection technique has been developed, thus making PHF a viable tool for investigating the structure of axially asymmetric nuclei, as well as the effects of configuration mixing, in large bases.

In short, the comparison of the present day PHF calculation with the earliest deformed HF calcualtions is similar in character to the comparison of the shell model (SM) calculation of Zuker, Buck and McGrory[7] with that of Elliot and Flowers.[8] In both cases the basic concepts have not been changed. Rather, they have been enlarged and implemented to a higher degree.

In Section II, after a brief description of the specifics of the Hamiltonian, sp basis, *etc.* and a short discourse stating the reason for choosing the nuclei that we shall study, we present results of conventional, structural PHF calculations for four representative nuclei, namely Mg^{24} (even-even, N = Z), Na^{22} (odd-odd, N = Z), Mg^{25} (odd-even) and Fe^{56} (even-even, N > Z). In many cases, these results agree well with measured data. In other cases, especially when concerned with the energy spectra for nuclei other than Mg^{24}, the results are unsatisfactory. In these latter cases we believe the NN interaction is at fault, since there are no other adjustable parameters. In fact, we are led through the PHF method to believe that rotational nuclei not of the $N\alpha$ type are good testing grounds for the effective interaction.

In Section III we present results for two kinds of nonstructural HF calculations. The first is concerned with the "neutron-halo" in neutron rich nuclei, and the second with the mirror asymmetry in allowed beta decays. In these topics we study effects due to the charge asymmetry in the electromagnetic interaction. Because of the interplay between the strong, attractive NN interaction and the repulsive, weak symmetry breaking Coulomb interaction, the self-consistent HF method is one of the most suitable tools for the study of these effects.

In Section IV we summarize the lessons learnt and conclusions arrived upon in the studies reported.

II. STRUCTURAL CALCULATIONS

1. Preliminaries

The Hamiltonian used in our calculation is

$$H = T - T_{CM} + V_{NN} + V_{EM} \tag{1}$$

where T is the total kinetic energy, T_{CM} is the kinetic energy of the center of mass (CM), V_{NN} is the two-body interaction represented by the semi-realistic interaction No. 2 of Saunier and Pearson,[9] and V_{EM} is the two-body coulomb interaction plus the one-body reduced electromagnetic spin orbit interaction.[10] The basic HF method is well documented in the literature, and will not be repeated here. Specifics concerning our work, such as phase conventions, are to be found in Ref. 11. A five major-shell basis[11] is used for the sp wavefunctions. For the latter, only the parity and the τ_z quantum numbers are conserved. The method used for angular momentum projection are reported in Ref. 12 and will not be discussed here.

In the following subsections, we shall study the structure of the four nuclei, Na^{22}, Mg^{24}, Mg^{25} and Fe^{56}. The low energy properties of these nuclei all have rotational characteristics and are therefore suitable for HF calculations. Their structures are not so simple as to be dominated by one axially symmetric rotational band, but are simple enough so that they can be described either by one asymmetric band, or by two bands.

2. Mg^{24} (even-even, Z = N)

This nucleus is a classic example of an asymmetric rotor. Its ground state structure has two prominent K-bands, the $K \simeq 0$ ground band and the $K \simeq 2$ γ vibration band. In our calculation the HF intrinsic state has four K-bands, with K = 0,2,4 and 6, with probabilities 0.389, 0.492, 0.115 and 0.004, respectively. Figure 1 shows the calculated and observed spectra for Mg^{24}. We can see that the observed moments of inertia for the ground and $K \simeq 2$ bands are well reproduced by theory. However, the predicted $K \simeq 2$ band head is relatively too low by $\sim$2 MeV. The cause of this is not well understood at the moment. In Strottman's SU(3) based SM calculations[16] the depression of the $K \simeq 2$ band was also noticed. Presumably this is related to a defect in the effective NN interaction. Although this defect does not appear to be very serious for the present case, later we shall see that it seems to persist and indeed grow in the other nuclei studied.

The γ transition strengths are shown in Table 1. These are E2 transitions, and the agreement between theory and data is good. Equally good are the SM results of McGrory and Wildenthal,[17] and that of Strottman.[16] In the SM calculations a charge enhancement of Δe = 0.5e was assigned to the valence protons and neutrons. In our calculations no charge enhancement is needed, but *all* protons participate in the transitions. The success in E2 strengths can be considered as a small triumph of the self-consistent HF theory. The calculated static quadrupole moment of the first 2^+ state is -17.1 eF^2, the measured[18] moment is -21 ± 4 eF^2.

How does the theory fare when the photons emitted are off-the-energy-shell? Such transitions occur in electron scattering when the three-momentum transfer, q, is non-zero. First we shall look at the elastic scattering form factors shown in Figure 2. All form factors are calculated in the Born approximations, with corrections for effects due to the CM motion and the finite size of the proton.[21] As we all know the elastic form factor before the first diffraction minimum is essentially a measure of the rms radius of the charge distribution in the target. We notice that starting from about $q^2 \simeq 1.2\ F^{-2}$ to the first dip at $q^2 \simeq 2\ F^{-2}$ the

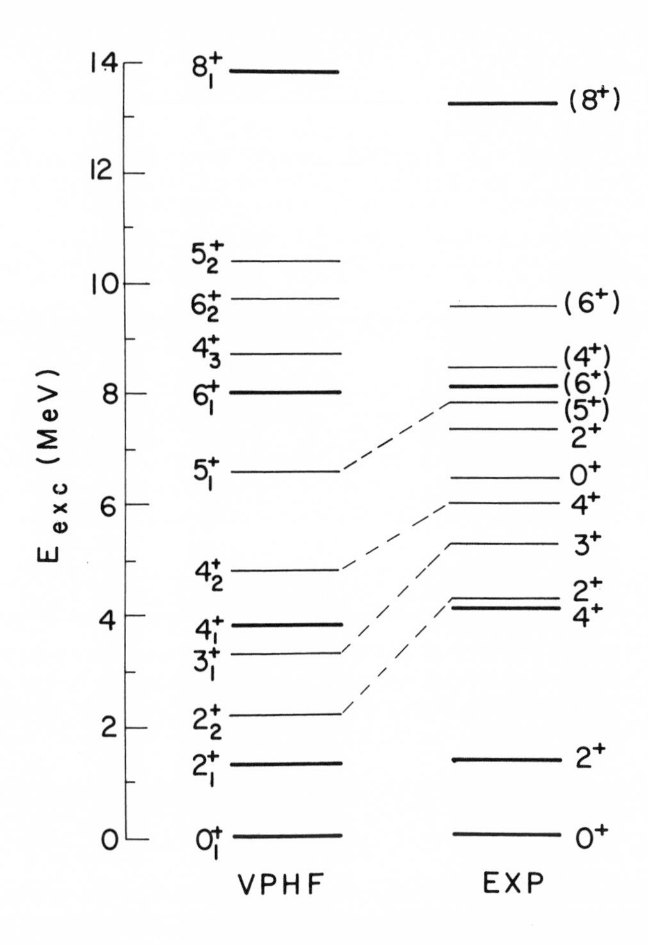

Fig. 1. Spectrum for Mg^{24}.

TABLE I

EM Properties of Low-Lying Levels in Mg^{24}

	Branching Ratio (%)		B(E2)(WU)[a]				Lifetime	
			Shell Model					
$J_i^\pi \to J_f^\pi$	exp[b]	th[c]	SU(3)[d]	J-J Coupling[e]	PHF	exp[b]	th[c]	exp[b]
$2_1^+ \to 0_1^+$	100	100	17.4	15.3	19.1	20.8 ± 0.8	2.18 ps	2.0 ± 0.08 ps
$4_1^+ \to 2_1^+$	100	100	28.9	20.0	26.4	23 ± 4.5	50 fs	59 ± 12 fs
$2_2^+ \to 0_1^+$	73	32	1.0	2.0	0.41	1.0 ± 0.15	113 fs	128 ± 15 fs
$\to 2_1^+$	23	68	8.2	1.7	6.21	2.2 ± 0.4		
$3_1^+ \to 2_1^+$	∿100	92	3.4	2.9	0.74	2.05 ± 0.4		
$\to 4_1^+$	<2	3			10.6	<20	289 fs	119 ± 25 fs
$\to 2_2^+$	<2	5		2.67	34.0	<33		

TABLE I (Continued)

$4_2^+ \to 2_1^+$	88	85	0.2	1.9	1.02	$1.0 {+0.3 \atop -0.2}$		
$\to 2_2^+$	11	11	9.5	9.7	9.0	$16 {+5 \atop -3}$	77 fs	77 ± 18 fs
$\to 4_1^+$	1	4			4.4	≤ 7		
$5_1^+ \to 4_1^+$	24	6			0.86	≤ 2.7		
$\to 3_1^+$	68	79			17.1	$33 {+6 \atop -5}$	80 fs	35 ± 5 fs
$\to 4_2^+$	8	15			19.5	≤ 17		
$6_1^+ \to 4_1^+$		100	29.6	17.5	31.0	$34 {+47 \atop -12}$	6.3 fs	6 ± 3
$6_2^+ \to 4_1^+$	≤ 20	32	0.5	0.95	0.98	$0.7 {+0.4 \atop -0.7}$	14 fs	12 ± 5 fs
$\to 4_2^+$	≤ 80	68	14.4	12.2	17.7	$25 {+18 \atop -9}$		
$8_1^+ \to 6_1^+$			23.2	2.4	32	$18 {+36 \atop -7}$	1.8 fs	3 ± 2 fs

a) 1 WU = 4.11 e^2F^4; b) experimental values from References 14-16; c) empirical energies are used to compute branching ratios and lifetimes; d) Reference 16; e) Reference 17.

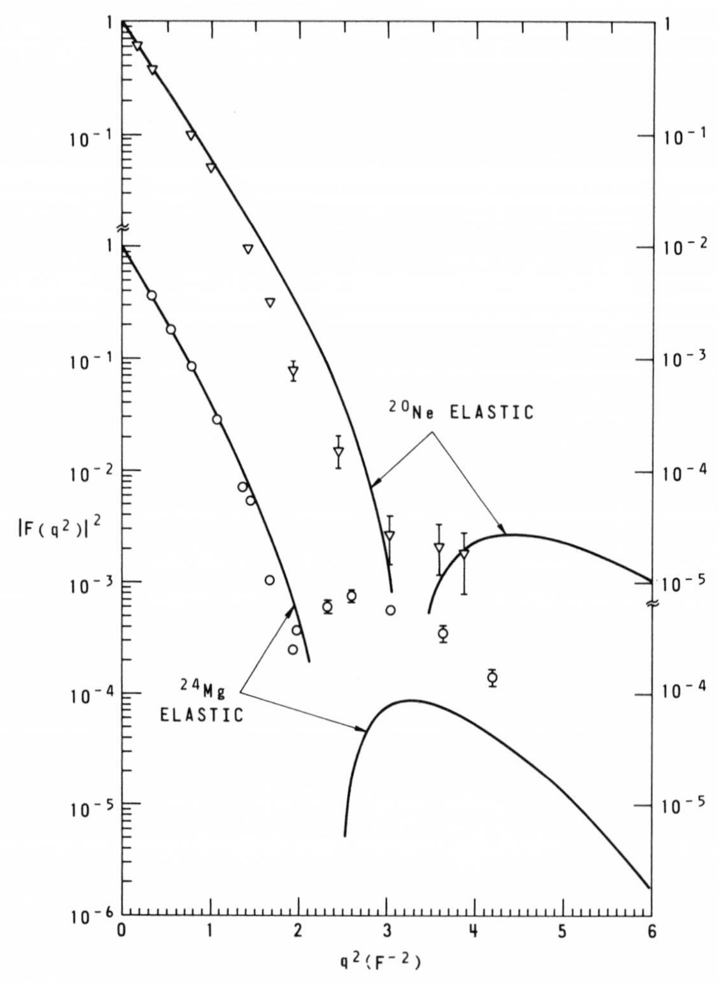

Fig. 2. Elastic electron scattering form factors for Ne^{20} and Mg^{24}. The Ne^{20} results are shown for comparison only. Data are from References 19 and 20.

data points lie inside the theoretical curve. This is an indication that the predicted rms radius is too small. It is well known that in HF calculations where a density independent two-body interaction is used, the radius predicted tends to be slightly too small, if the binding energy is approximately accounted for. This is believed to be one of the symptoms of a density distribution which is overly concentrated at the center of the nucleus, resulting in a central density much higher than that in nuclear matter. In some recent calculations where either a local density-dependent (required by Brueckner's K matrix theory), two-body interaction,[4b] or a density independent two-body plus three-body interaction[5] (the latter simulates a density dependent two-body interaction in HF) are used, the central peaking of the density in the finite nucleus have been eliminated. Coming back to our result for Mg^{24}, the predicted rms charge radius calcualted with respect to CM is 2.98 F. The radius extracted from muonic X-ray[22] data is 3.02 ± .04 F. The HF form factors do not fit the data beyond the first dip. This is partly due to the central peaking of the density discussed above, as processes with higher momentum transfer are more sensitive to smaller variations in the density distribution, and partly due to the fact that we have calculated the form factors only in the Born approximation.

Figure 3 shows the $0^+ \rightarrow 2_1^+$ form factor. Other than being slightly out of phase, which is another indication that the predicted size is too small, the HF form factors agree with data very well. We might say that the nuclear deformation, or in the SM language, that the effective charge[23]

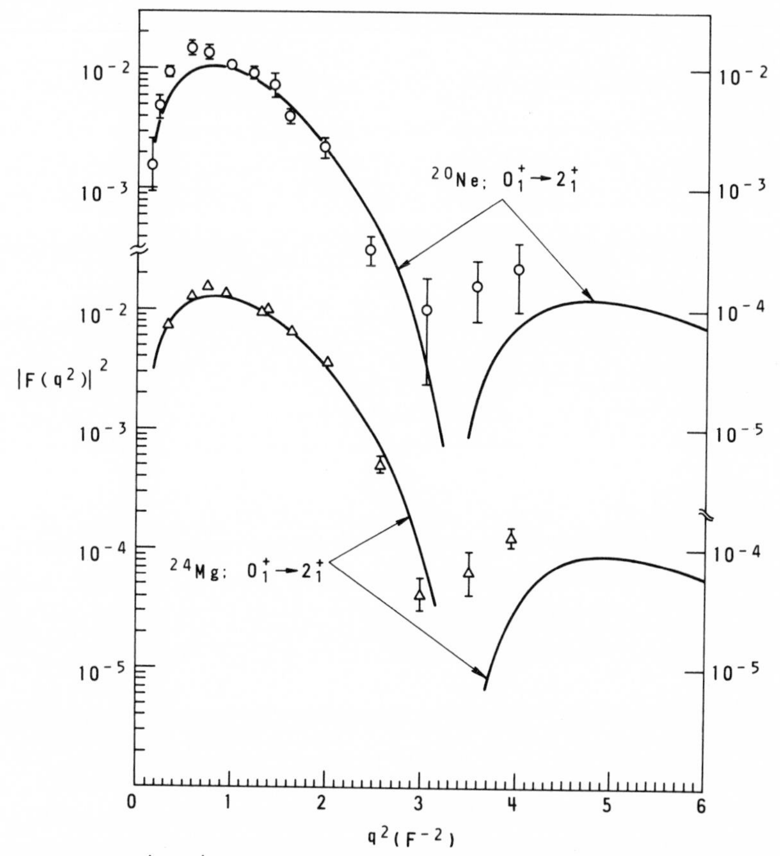

Fig. 3. $0_1^+ \rightarrow 2_1^+$ inelastic electron scattering form factors. Data are from References 19 and 20.

is q-independent, at least up to the first minimum.

In Figure 4 we show the $0_1^+ \rightarrow 2_2^+$ and the $0_1^+ \rightarrow 4_1^+$ form factors, compared with the experimentally unresolved $0 \rightarrow 4.2$ MeV data, which represents the sum of the two transitions. We see that the $0_1^+ \rightarrow 2_2^+$ calculated form factor alone already accounts for the data, suggesting that there is really no E4 strength from the ground state to the first 4^+ state.[20] This is contrary to our prediction, and here we see a distinct failure of the theory. This failure would not have been detected had we studied only the static transitions, since there the E4 strengths are too weak to be seen. Experimentally the E4 strength is mostly concentrated in the 4_2^+ (6.00 MeV) state, as shown in Figure 5. The sum

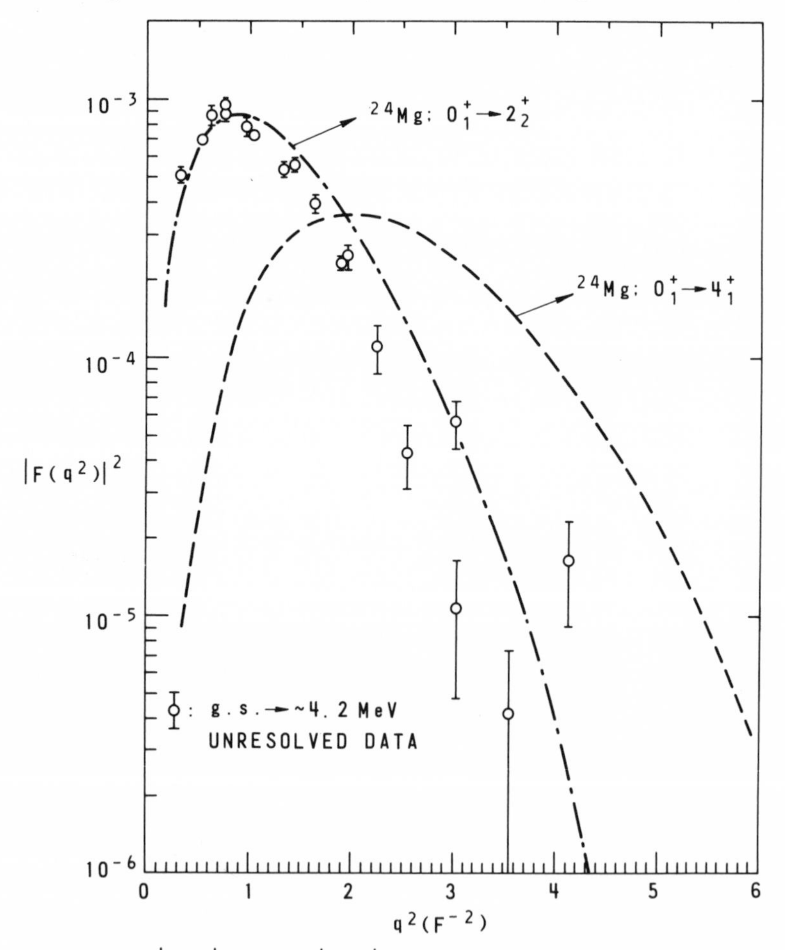

Fig. 4. $0_1^+ \rightarrow 2_2^+$ and $0_1^+ \rightarrow 4_1^+$ inelastic electron scattering form factors for Mg^{24}. Unresolved $0 \rightarrow 4.2$ MeV data are from Reference 20.

of the theoretical E4 strengths to the 4_1^+ and 4_2^+ states, indicated by the dashed line in Figure 5, accounts for approximately 70% of the empirical E4 strength. This percentage is very similar to that obtained in PHF for Ne^{20}, as shown in the upper half of Figure 5. We mention here that in the large basis SM calculation[23] the distribution of the E4 strength is correctly predicted.

3. Na^{22} (odd-odd, N = Z)

This is one of the odd-odd, rotational light nuclei that has been well studied experimentally.[24,25] The energy spectrum is shown in Figure 6. For classification purposes we have grouped the levels into four bands. The $K \simeq 3$ and $K \simeq 1$, $T = 0$ bands are projected from an axially asymmetric HF state. The $K = 0$, $T = 1$ and $T = 0$ bands are projected from an axially symmetric state. The moment of inertia for the $K \simeq 3$ band is significantly underestimated, causing the level spacings in this band to be too large. The moments of inertia for the other bands are correctly predicted. However, the $K = 0$, $T = 0$ band head is too high. This indicates a

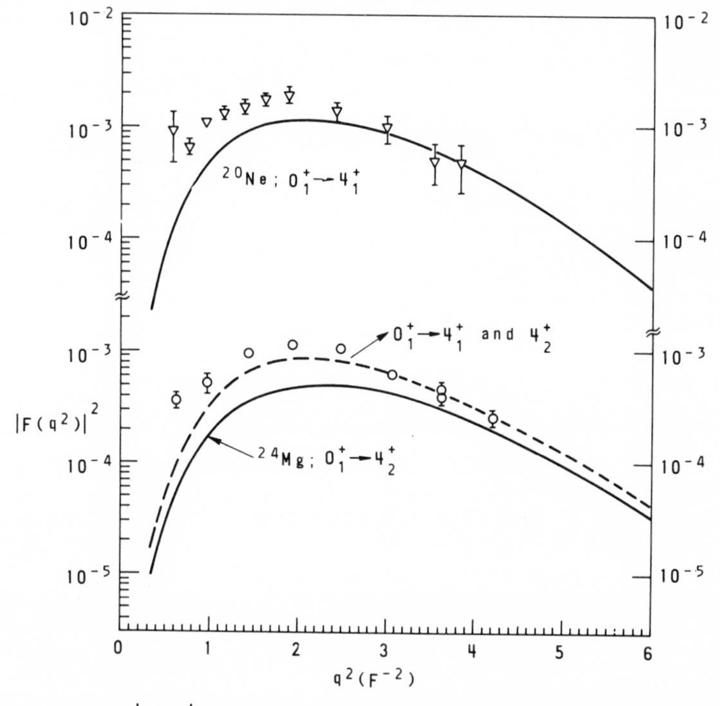

Fig. 5. $0_1^+ \to 4_1^+$ inelastic electron scattering form factors for Ne^{20} and Mg^{24}. The dashed line uses the sum of the $0_1^+ \to 4_1^+$ and $0_1^+ \to 4_2^+$ strengths in Mg^{24}. Data are from References 19 and 20.

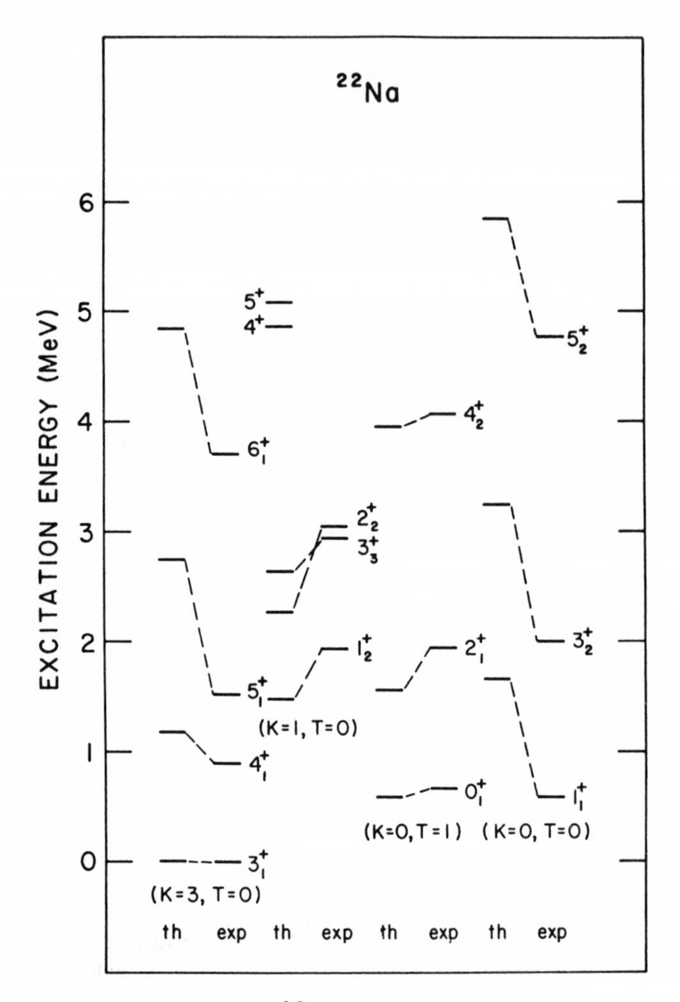

Fig. 6. Spectrum for Na^{22}. Levels are separated into four bands for convenience only.

possible defect in the relative strengths of the T = 0 and T = 1 components of the NN-interaction.

In Table 2 we show some E2 and M1 decay parameters. Again all observed E2 strengths are well accounted for. For the inhibited $\Delta T = 0$, M1 transitions ($4_1^+ \rightarrow 3_1^+$ and $5_1^+ \rightarrow 4_1^+$) the theory predicts a hindrance factor of $\sim 10^3$, compared to the observed hindrance factor of $\sim 10^4$. No measured data are available for strengths of "allowed" $\Delta T = 1$, M1 decays. However, if we assume that the E2 strengths in the $5_2^+ \rightarrow 3_2^+$ has been more or less correctly predicted, then the branching ratio for the $5_2^+ \rightarrow 4_2^+$ indicates that the M1 strength for the latter decay is only slightly overestimated by theory. This could be verified by a measurement of the lifetime for the 5_2^+ state, which we predict to be 30 fs. Another lifetime measurement that should be experimentally feasible is that of the $2_1^+ \rightarrow 1_1^+$ decay. This lifetime is predicted to be 7.1 fs. The magnetic moments of the 3_1^+ and 1_1^+ states are calculated to be 1.805 and 0.526 nm, respectively. The corresponding measured values[24] are 1.75 ± .01 and 0.54 ± .01 nm. The predicted static quadrupole moment of the 3_1^+ state is +22.81 eF^2.

4. Mg^{25} (even-odd, Z = N-1)

The low energy structure of Mg^{25} has long been recogniced to be mainly composed of two positive parity rotational bands;[26-28] with K = 5/2 being the ground band and K = 1/2 the excited band starting at 584 keV. Our lowest HF solution is triaxial, containing several K-bands. In the projected spectrum, the ground band has K = 1/2 and the first excited band at ∿900 keV has K = 5/2. However, contrary to experiment the cross band transitions are predicted to be quite strong. This rules out the triaxial solution. We then try to describe the nucleus by two axial intrinsic states allowing, however, these to mix. Again the lowest band has K = 1/2, but the lowest K = 5/2 band has an excitation of almost 10 MeV! This is in grave disagreement with experiment and is a clear indication of some very serious deficiency in the effective interaction we use. Without going into details we believe this is because either the spin-orbit or the tensor component in the interaction is too weak, causing the $d_{3/2}$ orbit to come too low in energy. However, leaving the band head energies aside, the two bands seem to account for some of the properties of Mg^{25}, as shown in Figure 7. In this figure the theoretical band heads have been shifted to match the corresponding observed levels. The level spacings for the first four levels in the K = 1/2 band are quite well reproduced. Based on this we indicate in the figure that the $9/2^+$ state in this band should correspond to the observed $9/2^+$ state at 4.7 MeV. The $11/2^+$ state should be at ∿5.9 MeV. The level spacings in the K = 5/2 bands are slightly underestimated. The interaction matrix elements, between states of the same spin belonging to the two band respectively, are of the order of 1∿2 MeV. Since the relative position of the band heads has been incorrectly predicted the mixing between the two bands cannot be reliably computed. If the two projected states are separated by ∿2 MeV, then the smaller amplitude of the mixed state will be ∿0.1. The beta decay strengths are listed in Table 3. The measured B(E2)'s are reasonably well reproduced. The weak inband E2 transitions, with $\Delta J = 1$, $J \geq 5/2$ in the K = 1/2

TABLE II

Electromagnetic Decay Properties of Low Lying States in Na^{22}

$J_i \to J_f$	Branching ratio (%)		$B(E2)(e^2f^4)$		$B(M1)(nm^2)$	
	exp[a]	th[b]	exp	th	exp	th
$4_1^+ \to 3_1^+$	100	100	98±8	111	$5.4\pm1.2\ 10^{-4}$	$4.07\text{X}10^{-3}$
$5_1^+ \to 4_1^+$	5	13	78±42	103	$5.5\pm3.0\ 10^{-4}$	$5.99\text{X}10^{-3}$
$\to 3_1^+$	95	87	25±9	27.1		
$6_1^+ \to 5_1^+$	35±10	37	135±61	82.4		$6.62\text{X}10^{-3}$
$\to 4_1^+$	65±10	63	57±21	48.1		
$0_1^+ \to 1_1^+$	100	100				7.91
$2_1^+ \to 1_1^+$	100	100		.0017		3.12
$\to 0_1^+$	0	0		64.7		
$3_2^+ \to 2_1^+$	0	0.4		.0035		3.33
$\to 1_1^+$	100	99.6	90±18	82.9		
$4_2^+ \to 3_2^+$	100	100		.010		3.36
$\to 2_1^+$	0	0		92.0		
$5_2^+ \to 4_2^+$	40	46		.010		3.37
$\to 3_2^+$	60	54		97.1		

a) Experimental data are from Reference 24, and references quoted therein. Weighted means are taken if several peices of recent data are available for one transition.

b) Empirical energies[24,25] are used to compute branching ratios, mixing ratios and lifetimes.

TABLE II (Continued)

Electromagnetic Decay Properties of Low Lying States in Na^{22}

$J_i \to J_f$	mixing ratio exp	mixing ratio th[b]	lifetime exp	lifetime th[b]
$4_1^+ \to 3_1^+$	-3.19±.26	-1.23	13.6±1.0 ps	7.87 ps
$5_1^+ \to 4_1^+$	±2.00±.15	-0.70	80± 27 ps	32.2 ps
$\to 3_1^+$			4.3±1.4 ps	3.62 ps
$6_1^+ \to 5_1^+$		-2.03	52± 17 fs	60.2 fs
$\to 4_1^+$				
$0_1^+ \to 1_1^+$				17.7 ps
$2_1^+ \to 1_1^+$		-2.7×10^{-4}		7.1 fs
$\to 0_1^+$				
$3_2^+ \to 2_1^+$		-1.5×10^{-4}	1.74±.34 ps	1.82 ps
$\to 1_1^+$				
$4_2^+ \to 3_2^+$		-9.5×10^{-4}		1.9 fs
$\to 2_1^+$				
$5_2^+ \to 4_2^+$		-8.9×10^{-4}		30.0 fs
$\to 3_2^+$				

band and with $\Delta J = 2$ in the $K = 5/2$ band, are also well accounted for. These are actually simple consequences of the rotational model.[27] Since the E2 strength between the two pure K-bands is calculated to be much smaller than one sp unit, the small amount of mixing between the two bands mentioned earlier is mainly responsible for the cross band transitions. Their strengths therefore should not follow those prescribed by the rotational model, as indeed they do not.[27] The more inhibited M1 transitions in the K = 1/2 band are overestimated by about a factor of five. The less inhibited M1 transitions in the K = 5/2 band are better reproduced.

5. Fe^{56} (even-even, Z = N-4)

Recent studies[29,30] have shown, among other things, that the stable even-even nuclei in the pf-shell all go through some degree of shape transition. For a given proton number, some isotopes show vibrational characteristics while others behave approximately like a rotor. The nuclei $Ti^{46,48}$, Cr^{50} and Fe^{56} fall into the second group. However, only Fe^{56} seems to be rather stiff against collective vibrational effects.[30] It would be nice if shape transitions could be

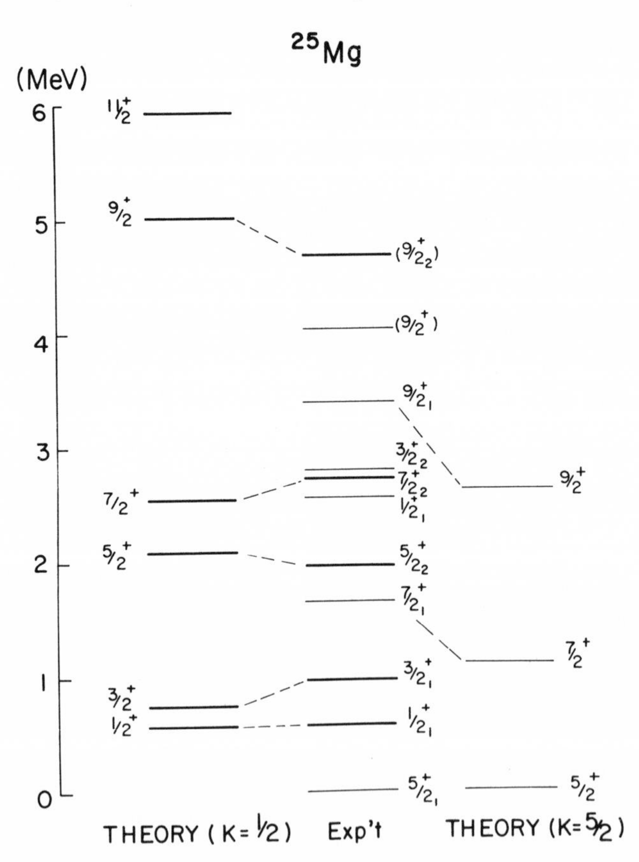

Fig. 7. Spectrum for Mg^{25}. Theoretical band heads have been shifted to match their respective observed counterparts.

TABLE III

E2 and M1 Strengths in Mg^{25}

	B(E2)(WU)[a]		B(M1)(WU)[b]	
	PHF	exp't[c]	PHF	exp't[c]
$1/2_1^+ \rightarrow 5/2_1^+$	$\sim.2$[d]	$.8\pm.1$		
$3/2_1^+ \rightarrow 5/2_1^+$	$\sim.2$	$1.2\pm.3$	$\sim10^{-3}$[d]	$(1.5\ ^{+1.3}_{-\ .9}) \times 10^{-3}$
$\rightarrow 1/2_1^+$	18.2	22^{+17}_{-12}	.10	$(2.5\ ^{+2.0}_{-1.4}) \times 10^{-2}$
$7/2_1^+ \rightarrow 5/2_1^+$	28.5	25 ± 6	.35	$.30 \pm .06$
$\rightarrow 3/2_1^+$				
$5/2_2^+ \rightarrow 5/2_1^+$	$\sim.2$	$.65\pm.20$	$\sim10^{-3}$	$(2.05\pm.5) \times 10^{-3}$
$\rightarrow 1/2_1^+$	18.2	35^{+6}_{-5}		
$\rightarrow 3/2_1^+$	5.2	5 ± 2	.16	$(1.2\pm.3) \times 10^{-2}$
$7/2_2^+ \rightarrow 5/2_1^+$	$\sim.2$	.3	$\sim10^{-3}$	$<4 \times 10^{-4}$
$\rightarrow 3/2_1^+$	23.5	32.6 ± 3.3		
$\rightarrow 5/2_2^+$	2.6	<160	.12	$<1.7 \times 10^{-2}$
$9/2_1^+ \rightarrow 5/2_1^+$	8.2	$2.1^{+5.8}_{-1.4}$		
$\rightarrow 7/2_2^+$	23.6	$5.6^{+11.8}_{-2.2}$	.50	$.17\ ^{+.20}_{-.10}$
$9/2_2^+ \rightarrow 7/2_1^+$	1.6		.18	
$\rightarrow 5/2_2^+$	26.1			

a) $1WU = 4.34e^2F^4$; b) $1WU = 1.8e^2F^2$; c) data from References 27 and 28; d) assuming $\sim.10$ admixture.

accurately accounted for in microscopic calculations. This would however necessitate the inclusion of at least pairing correlations in the calculation. HF-Bogoliubov calculations[3] are suitable for such purposes. But as stated earlier, these are beyond the scope of the present discussion. Here we only study the structure of Fe^{56}, assuming that it is rotational. Having noticed the erratic performances of the NN-interaction in previous cases, we should not now be surprised to learn that the lowest HF state for Fe^{56} does not exhibit the observed behavior. The orbit corresponding to the Nilsson orbit $1/2^-[310]$ instead of the $7/2^-[303]$ orbit was occupied by neutrons, and the $1/2^-[310]$ orbit instead of the $5/2^-[312]$ orbit was occupied by the protons. As a result the HF state was much too deformed, and the projected spectrum was much too compressed. To remedy this we force the occupied HF orbits to be those approximately corresponding to the lowest Nilsson orbits at $\beta \simeq 0.20$ and then iterate until self-consistency is achieved. The results are shown in the first column of Figure 8. All calculated quantities are now in reasonable agreement with experiment,[30] shown in the second column. The weak transitions from the 2_2^+ state to the 2_1^+ and 4_1^+ states suggest that Fe^{56} may be an asymmetric rotor. For technical reasons we can produce a "correct" triaxial state for Fe^{56} only when we use sp wavefunctions which are more compact than they should be. This has the effect of artificially increasing the kinetic energy, which in turn results in a less deformed self-consistent field, thus causing the desirable sp orbits to be populated. The results shown in column 3 of Figure 8 are calculated with $\hbar\omega$ = 13.5 MeV, but the E2 matrix elements are calculated with a dilated length parameter determined by equating the $\langle r_c^2 \rangle^{1/2}$ with the value given in the first column (calculated with $\hbar\omega$ = 10.5 MeV). The K = 2 band head is now below the 0^+ state. This is a recurrence of the trouble already noticeable in Mg^{24}. Compared to the data the B(E2) values are now approximately a factor of two too small. Very interestingly these values are quite similar to the SM results[30] calculated with the configuration $(f_{7/2})^6(f_{5/2},p_{3/2},p_{1/2})^2$, and with effective charges e_p = 1.5e and e_n = 1.0e. The HF prediction of the $B(M1,\ 2_1^+ \rightarrow 2_2^+)$ strength failed badly. This is because for even-even nuclei, as usual, the HF produces almost pure L-S coupling states, with S = 0.

III. NONSTRUCTURAL CALCULATIONS

We shall now mention some nonstructural calculations to which the HF method has been applied. Here we are not so worried that the nuclear structure be correct in detail as that the average nuclear field be self-consistent, in which case the effects we are interested in are guaranteed to be correct to first order.

1. Neutron Halo

Most stable nuclei heavier than Ca^{40} have neutron excesses. For these nuclei we expect the rms radius of the neutron distribution to be larger than that of the proton distribution. In other words, the neutron halo,

$$h = \langle r_n^{\ 2}\rangle^{\frac{1}{2}} - \langle r_p^{\ 2}\rangle^{\frac{1}{2}} \tag{2}$$

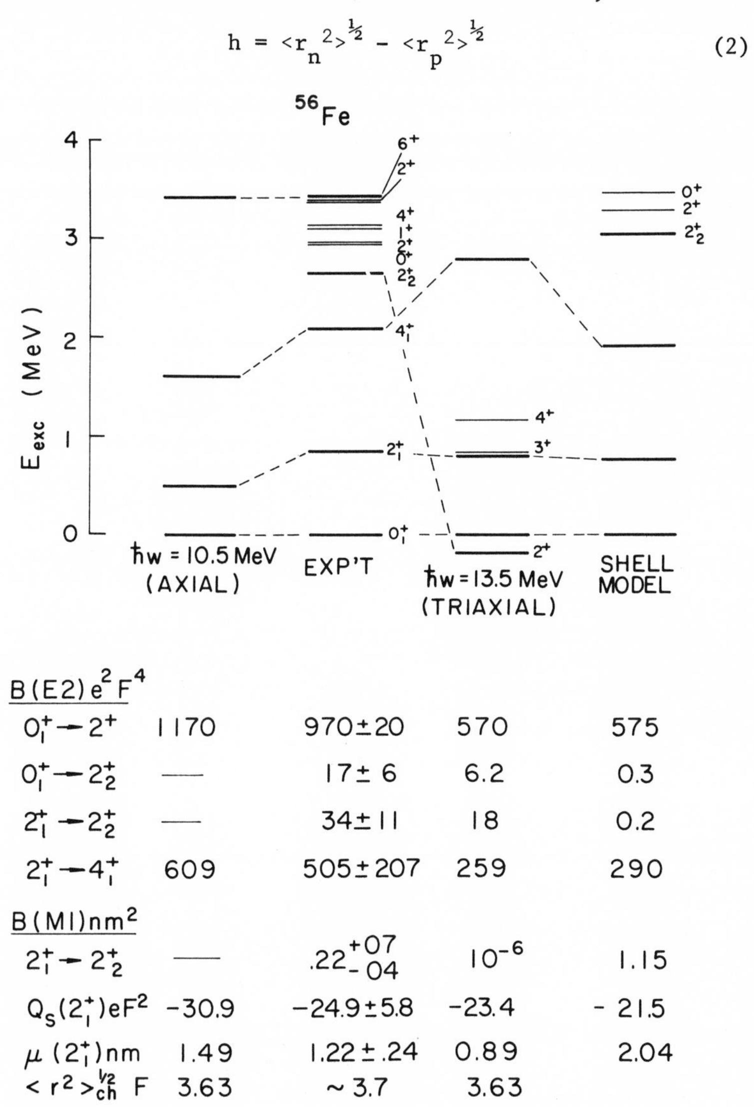

	ħw = 10.5 MeV (AXIAL)	EXP'T	ħw = 13.5 MeV (TRIAXIAL)	SHELL MODEL
$B(E2)\,e^2F^4$				
$0_1^+ \to 2^+$	1170	970±20	570	575
$0_1^+ \to 2_2^+$	—	17± 6	6.2	0.3
$2_1^+ \to 2_2^+$	—	34± 11	18	0.2
$2_1^+ \to 4_1^+$	609	505± 207	259	290
$B(M1)\,nm^2$				
$2_1^+ \to 2_2^+$	—	$.22^{+07}_{-04}$	10^{-6}	1.15
$Q_s(2_1^+)\,eF^2$	-30.9	-24.9±5.8	-23.4	- 21.5
$\mu\,(2_1^+)\,nm$	1.49	1.22±.24	0.89	2.04
$\langle r^2\rangle_{ch}^{1/2}$ F	3.63	~ 3.7	3.63	

Fig. 8. Spectrum, electromagnetic decay parameters and moments for Fe^{56}. Experimental data and shell model results are from Reference 30.

is expected to be positive. Because of the repulsive Coulomb interaction, which acts amongst the protons alone, h is actually significantly smaller than would be expected from a simple single particle model, where the effect of the Coulomb field is not properly accounted for.

In Table 4 we show neutron halos calculated for some light and medium nuclei. The predictions of HF,[31] and the droplet model of Meyer and Swiatecki,[32] and those deduced from Coulomb displacement energies by Janeck[33] agree extremely well. These are all substantially smaller than those calculated from single particle potentials[34] determined from the separation energies of the neutrons. Neutron halos deduced from elastic proton scattering data[35] agree with the HF results but have large error bars. A precise measurement of the neutron halo is very difficult. Unlike the proton distribution, which can be probed by electrons (or muons), the neutron distribution can be probed only through strong interactions, which are not so well understood as the electromagnetic interaction. Besides, multiple scattering processes must be considered, thus introducing more uncertainties in the analysis of experimental data. Nevertheless the excellent agreement between results from the different theoretical methods lends support to our belief that the predictions cannot be far from being correct.

2. Mirror Asymmetry in Allowed Beta Decays

In the absence of electromagnetic interactions, and ignoring any small charge asymmetry that may exist in the NN interaction[36], nuclei with $N \neq Z$ should appear in charge symmetric mirror pairs. In nature this symmetry is slightly broken, due mainly to the Coulomb interaction, and a small mirror asymmetry is expected. The mirror asymmetry in beta decay is characterized by the parameter

$$\delta \equiv [(ft)^{+}/(ft)^{-}] - 1 = |M_{-}|^2/|M_{+}|^2 - 1 \qquad (3)$$

where M_{+} is the positron decay matrix element, and M_{-} the mirror negatron decay matrix element. Recent experiments have shown that δ for some Gamow-Teller decays is significantly non-zero, and the question was raised whether this can be explained by the breaking of charge symmetry in nuclear structure alone.[37] An alternative is that departure of δ from zero is caused by fundamental second-class induced tensor currents[38] in the weak interactions.

Here we shall report PHF calculations of δ for some Gamow-Teller decays in $A = 4n$ systems and some Fermi decays in $A = 4n + 2$ systems. For the $A = 4n$ system the $T_z = 0$ nucleus is even-even, and the ground state has $J^\pi = 0^+$, $T = 0$. The $T = 1$, $T_z = \pm 1$ nuclei are odd-odd with $|N - Z| = 2$, and the ground state spins are irregular. In the HF description

TABLE IV

Neutron Halos (in Fermis) for Light and Medium Nuclei

Nucleus	HF[a]	Droplet model[b]	Coul. disp. energy[c]	single particle potential[d]	elastic p-scattering[e]
Ca^{48}	0.137	0.160	0.13 ± 0.03	0.45	
Fe^{56}	0.074	0.062	0.07 ± 0.01	0.24	-0.04 ± 0.15
Ni^{58}	0.020	0.023	0.03 ± 0.01	0.14	0.01 ± 0.18
Co^{59}	0.088	0.077	0.08 ± 0.02		-0.03 ± 0.16
Ni^{60}	0.065	0.057	0.06 ± 0.01	0.23	0.03 ± 0.16
Ni^{62}	0.114	0.090	0.09 ± 0.02	0.12	

a) Reference 31; b) Reference 32; c) Reference 33; d) Reference 34; e) Reference 35.

the simplest possible structure for, say the $T_z = -1$ nucleus, relative to that of the $T_z = 0$ nucleus, is shown in Figure 9. We see that two simple HF intrinsic states are possible for the $T_z = -1$ nucleus; one will form a $K = \Omega_L + \Omega_F$ band and the other a $K' = |\Omega_L - \Omega_F|$ band. Ω_L and Ω_F are the magnetic quantum numbers of the last occupied and first unoccupied orbits, respectively, in the $T_z = 0$ nucleus. In our calculation we have found that in most cases a normalized equal mixture of the two states projected respectively from the two bands usually has the appropriate beta decay strength and the approximately correct magnetic moment, whereas neither of the projected states alone has these properties. We thus use one of the equally mixed (symmetric or antisymmetric) states to represent the $T_z = \pm 1$ nucleus in all cases. The moments and log(ft) values (for Gamow-Teller decays), calculated from

$$(ft)^{\pm} = \frac{6146}{(1.226)^2 |M_{\pm}|^2} \tag{4}$$

for some nuclei investigated are shown in Table V. The predicted quantities are in reasonable agreement with experiment. The exceptions are: i) the predicted magnetic moment for N^{12} has the wrong sign; ii) for F^{20}, the magnetic moment and log(ft) value cannot be correctly predicted simultaneously. It should be emphasized that here we are not overly concerned with the details of the nuclear structure, as these are cancelled out to first order in calculating δ, but

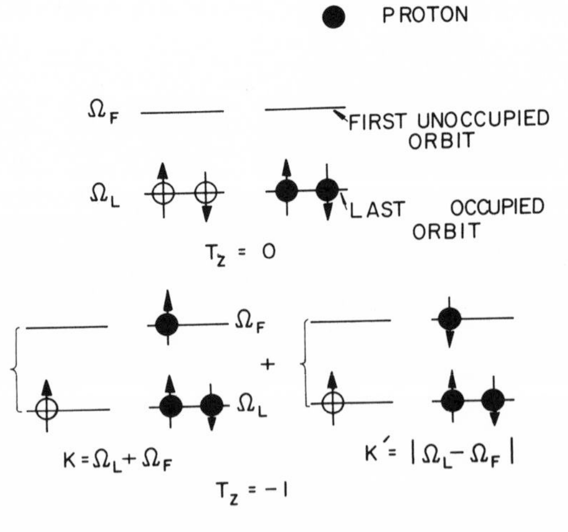

Fig. 9. Single particle population scheme for the ground state of the $T_z = -1$ nucleus, relative to that of the $T_z = 0$ nucleus. $\Omega_F(\Omega_L)$ is the magnetic quantum number of the first unoccupied (last occupied) orbit in the $T_z = 0$ nucleus.

are only seeking some assurance that we are not looking at the wrong state completely. In Table VI the predicted asymmetries are listed, together with the experimentally[42] deduced asymmetries and those calculated by Wilkinson *et al.*[42] using coefficients of fractional parentage calculated in the shell model but substituting Woods-Saxon wave functions in the single particle beta decay matrix elements[43]. It should be pointed out that the HF and SM methods emphasize different but complementary effects. On the one hand because of the finite basis the sp wave functions of the least bound particles in HF do not have the correct asymptotic behaviors as determined by the Woods-Saxon potentials used in the SM approach. On the other hand the HF method takes into account the Coulomb effect on all particles, resulting in different sets of coefficients of fractional parentage, which are assumed to be identical in the SM, for the $T_z = +1$ and $T_z = -1$ nuclei, respectively. Our results do not agree as well as the shell model results with the experimental δ. In particular we predict a very large δ for the A = 24 system, which is measured to be much smaller and possibly negative. Clearly much is left to be improved in our rather crude calculations. All we can say now is that electromagnetic effects are capable of producing large asymmetries, in some cases as large as or even larger than 10%, in Gamow-Teller decays. Also the asymmetry does not have a simple mass dependence.

We now turn to the super-allowed Fermi-decays. The $T_z = 0$ nucleus considered is odd-odd, and the $T_z = \pm 1$ nuclei are even-even. The initial and final states all have zero spin and form T = 1 triplets. Every state is now projected from one HF determinant. The results are shown in Table VII. All δ's are predicted to be less than 0.01. The only existing data are for[45] A = 26, which is essentially zero, and for[45] A = 34, which is somewhat larger and possibly negative. In general, without configuration mixing, the theory does not predict a negative value for δ. This is because the Coulomb effect is larger in the proton-rich system. It then follows that the reduction in M_+ (from its charge symmetric value of $\sqrt{2}$) is larger than that in M_-.

IV. SUMMARY

We have discussed some applications of HF calculations, both structural and non-structural, for light and medium nuclei. In some cases the results obtained agree with experimental data very well, and in other cases, not so well. The calculation almost invariably succeeds in accounting for The E2 strengths. It also accounts for about 70% of the E4 strengths, and we believe that this could be improved upon had the basis been increased beyond the five major shell basis used here. These are interesting results in view of the difficulties met in calculating effective charges in perturbation theory. As for the M1 strengths, the HF predictions

TABLE V

Moments and log(ft) Values for Nuclei Investigated for Mirror Asymmetry in Gamow-Teller Decays

Nucleus	$J_i^\pi \to J_f^\pi$	$\mu(J_i)$(nm)		$Q_s(J_i)(eF^2)$		log(ft)	
		PHF	exp't[a]	PHF	exp't	PHF	exp't[c]
Li^8		1.40	1.63	0.75			
	$2^+ \to 2^+$					5.8	5.6
B^8		4.06		0.52			
B^{12}		1.25	1.00	0.48	1.71[a]		
	$1^+ \to 0^+$					4.4	4.1
N^{12}		-0.46	+0.46	0.63			
	$\to 2^+$					5.0	5.1
F^{20}		-2.20 (1.58)[d]	2.09	1.88 (1.88)	1.2±.2[b]		
	$2^+ \to 2^+$					5.1 (5.8)	5.0
Na^{20}		4.24 (0.94)		3.07 (3.76)			

TABLE V (Continued)

Moments and log(ft) Values for Nuclei Investigated for Mirror Asymmetry in Gamow-Teller Decays

Nucleus	$J_i^\pi \to J_f^\pi$	$\mu(J_i)$ (nm)		$Q_s(J_i)$ (eF^2)		log(ft)	
		PHF	exp't[a]	PHF	exp't	PHF	exp't[c]
Na^{24}	$4^+ \to 4^+$	1.23	1.69	5.24		5.9	6.1
Al^{24}		3.19		5.85			
Al^{28}	$3^+ \to 2^+$	0.10		-2.64		4.9	4.9
P^{28}		2.94		-2.46			

a) Reference 39.
b) Reference 40.
c) Reference 41.
d) Alternative solution.

TABLE VI

Mirror Asymmetries in Gamow-Teller Decays for A = 4n Triads

	δ		
A	PHF	exp't[a]	Wilkinson *et al.*[a]
8	.073	.107 ± .011	.048
12(0$^+$)	.042	.115 ± .009	.098
12(2$^+$)	.008	.06 ± .04	.048
20	.086 (.104)[b]	.04 ± .018	.013
24	.156	−.03 ± .06	.011
28	.011	$-.04 \pm {}^{.04}_{.03}$	.014

a) Reference 42; b) see footnote d) of Table V.

are not so successful. In general, the HF calculation does not produce a large enough hindrance factor. However, the difference between theory and experiment diminishes with the hindrance factor. Thus M1 transitions with about unit sp strengths are predicted to be only slightly stronger. The failures of HF with regards to some M1 strengths are not unexpected as the M1 operator is sensitive to small components in the wave function but not to long range correlations, which are the only correlations emphasized by the HF theory.

Regarding level sequences, we have shown that the lowest HF solution quite successfully accounts for the level scheme of Mg^{24}, but did not do quite as well for Na^{22}, and failed rather badly for Mg^{25} and Fe^{56}. Since the low energy structure of these nuclei show, through one or two underlying intrinsic states, distinct rotational characteristics, and since we can be quite confident that we are not suffering from difficulties caused by the basis being too small, we can only conclude that the effective interactions we use are failing us. We noticed however, that for the K = 5/2 band in Mg^{25} and for Fe^{56}, satisfactory results are obtained if sp orbits which correspond approximately to the lowest Nils-

TABLE VII

Mirror Asymmetry in Fermi-Decays for
A = 4n + 2 Iso-triplets

	δ	
A	PHF	experiment
22	.002	(not feasible)
26	.001	-.009 ± .016[a]
30	.007	(not feasible)
34	.007	-.025 ± .017[b]
38	.002	
42	.000	

a) Reference 44; b) Reference 45.

son orbitals of appropriate deformation are populated instead of the lowest HF orbits. This is another tribute to the Nilsson prescription, but it also points out that the correct average sp fields exist for the representative nuclei studied, if only an effective interaction can be found to produce them. Indeed the creators of the interaction used here had noticed its defects and efforts[45] have already been made to correct them. For practical reasons we have not used the revised interaction in this work. For the same reasons, we have not used any other of the many interactions currently available, although such should be done, as we believe that through HF calculations, many light and meduim non-nα type nuclei provide excellent testing grounds for effective interactions. I would like to take the opportunity here to say that our PHF computer program will be available for general use, as soon as it has been properly documented.

Turning now to nonstructural calculations we have used the HF method to study the neutron halos for some N > Z, light and medium nuclei. The HF results agree extremely well with results from other methods where the Coulomb effect has been properly treated, and are factors of two to five smaller than those computed in a method where the Coulomb effect has not been adequately treated. We have also used the present method to calculate the mirror asymmetry in allowed beta decays. For the Fermi decays the theoretical asymmetry is always positive and a fraction of one percent. The two measured asymmetries are also small but one of them may be negative (see Table VII). For the Gamow-Teller decays both theory and experiment show asymmetries with fluctuating magnitudes, but the HF predictions do not agree with data in detail (see Table VI). However we feel that eventually the major part of the measured asymmetry will be explained in terms of nuclear electromagnetic effects and probably will not provide strong evidence for the existence of second class currents in the weak interaction.

ACKNOWLEDGMENTS

Most of the work reported here was done in collaboration with R. Y. Cusson. The work on mirror asymmetry was carried out in collaboration with I. S. Towner. I wish to thank them for their contributions. I would also like to acknowledge fruitful conversations with T. K. Alexander, M. Harvey, O. Häusser, F. C. Khanna and J. F. Sharpey-Schafer.

REFERENCES

1. K. A. Brueckner and C. A. Levinson, Phys. Rev. 97, 1344 (1955); K. A. Brueckner, Phys. Rev. 97, 1353 (1955); K. T. R. Davies, M. Baranger, R. M. Tarbutton, T. T. S. Kuo, Phys. Rev. 177, 1519 (1969); K. T. R. Davies and M. Baranger, Phys. Rev. C1, 1640 (1970); R. J. McCarthy and K. T. R. Davies, Phys. Rev. C1, 1644 (1970).
2. J. Bar-Touv and I. Kelson, Phys. Rev. 138, B1035 (1965); G. Ripka in *Advances in Nuclear Physics*, M. Baranger and E. Vogt, Eds., Vol. 1 (Plenum Press, New York, 1968) and references therein; D. R. Tuerpe, W. H. Bassichis and A. K. Kerman, Nucl. Phys. A142, 49 (1970).
3. A. L. Goodman, G. L. Struble, J. Bar-Touv and A. Goswami, Phys. Rev. C2, 380 (1970); M. R. Gunye and S. B. Khadkikar, Nucl. Phys. A165, 508 (1971).
4a. K. A. Brueckner, A. M. Lockett and M. Rotenberg, Phys. Rev. 121, 255 (1961).
4b. J. W. Negele, Phys. Rev. C1, 1260 (1970); X. Campi and D. W. L. Sprung, Nucl. Phys. to be published.
5. D. G. Vautherin and D. M. Brink, Phys. Rev. C5, 1990 (1972); S. J. Krieger and S. A. Moszkowski, Phys. Rev. C5, 1990 (1972).
6. L. Satbathy and Q. Ho-Kim, Phys. Rev. Lett. 25, 123 (1970); L. Satpathy, K. W. Schmid and A. Faessler, Phys. Rev. Lett. 28, 832 (1972).
7. A. Zuker, B. Buck and J. McGrory, Phys. Rev. Lett. 21, 39 (1968).
8. J. P. Elliott and B. H. Flowers, Proc. Roy. Soc. (London) A229, 536 (1955).

9. G. Saunier and J. M. Pearson, Phys. Rev. C1, 1353 (1970).
10. R. J. Blin-Stoyle in *Isospins in Nuclear Physics*, D. H. Wilkinson, Ed. (North-Holland, Amsterdam, 1969).
11. H. C. Lee and R. Y. Cusson, Ann. Phys. 72, 353 (1972).
12. H. C. Lee and R. Y. Cusson, Phys. Lett. 39B, 453 (1972) and R. Y. Cusson and H. C. Lee, to be published.
13. J. A. Haskett and R. D. Bent, Phys. Rev. C4, 461 (1971).
14. C. P. Swann, Phys. Rev. C4, 1489 (1971).
15. B. Branford, A. C. McGough and I. F. Wright (Nucl. Phys. in press), and I. F. Wright, private communication.
16. D. Strottmann, Phys. Lett. 39B, 457 (1972).
17. J. B. McGrory and B. H. Wildenthal, Phys. Lett. 34B, 373 (1971), and B. H. Wildenthal, private communication.
18. A. Christy and O. Häusser, preprint (submitted to Nuclear Data).
19. Y. Horikawa *et al.*, Phys. Lett. 36B, 9 (1971).
20. A. Nakada and Y. Torizuka, Jour. Phys. Soc. (Japan) 32, 1 (1972), and A. Nakada, private communication.
21. H. C. Lee, Phys. Lett. 41B, 421 (1972).
22. C. S. Wu and L. Wilets, Ann. Rev. Nucl. Sc. 19, 527 (1969).
23. G. R. Hammerstein, D. Larson and B. H. Wildenthal, Phys. Lett. 39B, 176 (1972).
24. J. W. Olness *et al.*, Phys. Rev. C1, 958 (1970), and references therein.
25. J. D. Garrett *et al.*, Phys. Rev. C4, 1138 (1971), and references therein.
26. H. E. Gove *et al.*, Nucl. Phys. 2, 132 (1956).
27. J. F. Sharpey-Schafer *et al.*, Can. J. Phys. 46, 2039 (1968) and references therein.
28. D. C. Kean and R. W. Ollerhead, Can. J. Phys. 50, 1539 (1972).
29. D. Cline, *Proc. Conf. on Intermediate Nuclei*, (Orsay, 1971).
30. P. M. S. Lesser *et al.*, Nuc. Phys. A190, 597 (1972).
31. H. C. Lee and R. Y. Cusson, Nucl. Phys. A170, 439 (1971).
32. W. D. Myers and W. J. Swiatecki, Ann. Phys. 55, 395 (1969); W. D. Myers, Phys. Lett. 30B, 451 (1969).
33. J. Jänecke (private communication).
34. C. J. Batty and G. W. Greenlees, Nucl. Phys. A133, 673 (1969).
35. G. W. Greenlees *et al.*, Phys. Rev. C1, 1145 (1970).
36. E. M. Henley in *Isospin in Nuclear Physics*, D. H. Wilkinson, Ed. (North-Holland, Amsterdam, 1969); E. M. Henley and T. E. Keliher, Nucl. Phys. A189, 632 (1969).
37. D. H. Wilkinson, Phys. Lett. 31B, 190 (1970).
38. S. Weinberg, Phys. REv. 112, 1375 (1958).
39. V. S. Shirley, in *Hyperfine Interactions in Excited Nuclei*, G. Goldring and R. Kalish, Eds. (Gordon and Breach, New York, 1971).

40. H. Ackermann *et al.*, Phys. Lett. 41B, 143 (1972).
41. C. M. Lederer, J. M. Hollander and I. Perlman, *Tables of Isotopes*, (John Wiley & Sons, Inc., New York, 1968).
42. D. H. Wilkinson, D. R. Goosman, D. E. Alburger and R. E. Marrs, preprint (submitted to Phys. Rev. C).
43. J. Blomqvist, Phys. Lett. 35B, 375 (1971); D. H. Wilkinson, Phys. Rev. Lett. 27, 1018 (1971).
44. J. C. Hardy, H. Schmeing, J. G. Geiger, R. L. Graham, and I. S. Towner, Phys. Rev. Lett. 29, 1027 (1972).
45. J. C. Hardy and D. E. Alburger, to be published.
46. B. Rouben and G. Saunier, Phys. Rev. C5, 1223 (1972); B. Rouben and G. Saunier, Phys. Rev. C6, 591 (1972).

DISCUSSION

COZ: I have a general comment and a question. When somebody chooses in a Hartree-Fock calculation to disregard the realistic two body interaction in favor of any other kind of interaction, an effective interaction for instance, he introduces willingly or not a density dependence--now when a density dependent interaction is used he must explain how a variational method will still give him the ground state wave function and the ground state energy. My question is: can you specify your use of Projected Hartree-Fock?

LEE: You are not asking me a question on density dependence, is that right? Just making a comment. How do I start the Hartree-Fock prodedure? Well, as you know the first thing you have to do in a Hartree-Fock calculation is to provide the trial wave functions. Ron Cusson has a fantastic Nilsson code which calculates and catalogs for all nuclei the beta-gamma energy surfaces. First of all, one looks into the catalog and says O.K. for this nucleus, it should have this beta and this gamma. Then one takes these as input data and calculates the Nilsson wave functions and takes these as the trial wave functions.

DAVIDSON: In one of your earlier slides (Figure 1) on Mg^{24} I notice that there is no reproduction in the theory calculation of the 0^+ band, sometimes called a beta band. Is that an oversight in drawing the slide or is that a result of the calculation and could you explain why that is?

LEE: I think I did mention that there is a beta vibration band. So it cannot be included in one intrinsic triaxial state and we just did not calculate that band.

MARIPUU: I would like to make a comment about the mirror asymmetry and beta decay. The test being shown a few years ago by Blomqvist was that mirror asymmetry can be predicted pretty well as just being due to the different binding energy of the proton and the neutron. I would like to hear comments about that.

LEE: Yes, that is a good calculation. However I don't think that accounts for all. Blomqvist has two method, method 1 and method 2 which are almost identical to Wilkinson's method A and method B. Blomqvist was first, naturally, but as far as I know he calculated only one number. Wilkinson calculated all of them and that's the only reason I showed Wilkinson's. But if you compare the experimental δ, Wilkinson's numbers are not big enough. Now I say that in this approach one does not account for all the asymmetry in that one only looks at the binding energy effect of the last neutron or last proton, whereas this effect is not emphasized in HF. But Hartree-Fock looks at the difference between the core particles, and this difference is ignored in Blomqvist's or the shell model approach.

In this sense the Hartree-Fock results and the shell model results are complementary but there are some double countings. In fact, I think the shell model approach perhaps overestimates the binding effect a little bit because it really does not look at the Coulomb effect very carefully. For example the neutron halos calculated using Woods-Saxon wave functions are very, very much different from all other calculations and much larger.

WARBURTON: As you said, it is very difficult experimentally to compare the mirror decays in mass 42, but a related experiment can more easily be done and that is to look for the decay from the 0^+ S^{42} ground state to the first excited 0^+ state in Ca^{42}.

LEE: Do you have a number?

WARBURTON: No one as yet has a firm number, but an "interesting" limit does exist. Can you predict this Fermi decay in your calculations? I know that Ian Towner has worked on it in a simple gamma model.

LEE: Yes, one would do a shell model calculation first. At this point I believe the shell model calculation is easier. But you have to do it in the proton-neutron formalism which is very time consuming; for example if you do it in the isospin formalism, using the best code available, you can allow six orbits, but if you do it in the proton-neutron formalism, you can allow three orbits, and you also have to restrict the number of valence particles. Now, we know that the Coulomb interaction has a very long range and it affects all the particles, so if you restrict it to a small orbital space and to a few valence particles, you are only looking at a very small portion of the Coulomb effect. That's the only trouble.

MCCULLEN: I would just like to point out that the comments that Barry Malik made about Sternheimer corrections are of an entirely different category than the problem of extracting the sign of the magnetic moment from hyperfine structure. If the sign of the magnetic moment that was quoted was extracted from hyperfine structure, it is much better information than hyperfine quadrupole moments.

Note added in proof: Recently Feldmeier, Manakos & Wolff (*preprint, Inst. Theor. Phys. Tech. Hoch. Darmstadt, 1972*) showed that the predicted relative position of the K = 2 band in the energy spectrum of Mg^{24} is very sensitive only to the strength of the antisymmetric, spin-orbit components (with the spin part proportional to $(\underset{\sim}{\sigma}_1 \otimes \underset{\sim}{\sigma}_2)^1$ or to $(\underset{\sim}{\sigma}_1 - \underset{\sim}{\sigma}_2)$ in the residual interaction. Thus, in a shell model calculation using an SU(3) truncated space, the K = 2 band, usually predicted to be too low in energy, can be shifted upwards by $\sim$1.5 MeV if the antisymmetric spin-orbit components in the well known Kuo-Brown effective interaction are artifically strengthened by a factor of 5.

I.a. HIGH-SPIN STATES IN Al^{27} AND THE NUCLEAR STRUCTURE OF NUCLEI AROUND MASS 28

M. J. A. de Voigt*
Nuclear Structure Research Laboratory**
University of Rochester, Rochester, New York

Many attempts have been made to apply the collective model to nuclei in the mass region A = 27-29, where the nuclear shape changes from prolate to oblate. Several discussions of Al^{27} in terms of rotational bands (see Reference 1) have been published, but they fail to explain many different features simultaneously. More experimental information, especially on high-spin states, might remove some of the existing ambiguities and uncertainties.

Therefore levels in Al^{27} have been investigated by means of the $Na^{23}(\alpha,\gamma)Al^{27}$ reaction, where the strongly competing proton emission turns out to provide a very useful natural selection mechanism in the observation of high-spin (α,γ) resonances.[2,3] Angular correlation measurements (Figure 1) uniquely determine the spin of the E_α = 2.69 MeV resonance as J = 11/2. The strong alignment for high-spin resonances in this reaction makes the angular distribution quite insensitive for the relative occupation of the magnetic substates as shown in Figure 2. Investigation of four significant resonances combined with the results of additional Doppler shift lifetime measurements yield several unique spin and parity assignments.[3]

These results can be compared with recent shell-model calculations.[4] The configuration space includes a maximum of four holes in the $1d_{5/2}$ subshell. The parameters of the MSDI Hamiltonian, the two splittings of the single particle energies and the two strengths for the SDI interaction follow from least squares fits to excitation energies in Al^{27}, Si^{28} and Si^{29}. Transition rates have been calculated[5] using effective electromagnetic operators. Good candidates for the $J^\pi = 9/2^+$ levels calculated at E_x = 2.76, 4.91, 5.94 and 6.41 MeV are the E_x = 3.00, 5.43, 5.67 and 7.17 MeV levels, respectively. The lowest $J^\pi = 11/2^+$ levels calculated at E_x = 3.91 and 5.25 MeV might correspond to the E_x = 4.51 and 6.51 MeV levels, respectively. However, there are other candidates for the low-lying $J^\pi = 11/2^+$ levels, such as the E_x = 5.50 MeV, $J^\pi \geq 7/2^+$ excited state.

Besides the reasonable reproduction of these high-spin states, spherical shell model calculations in general account well for the level energies, spectroscopic factors and

---

*On leave from the University of Utrecht, the Netherlands.

**Work supported by the National Science Foundation.

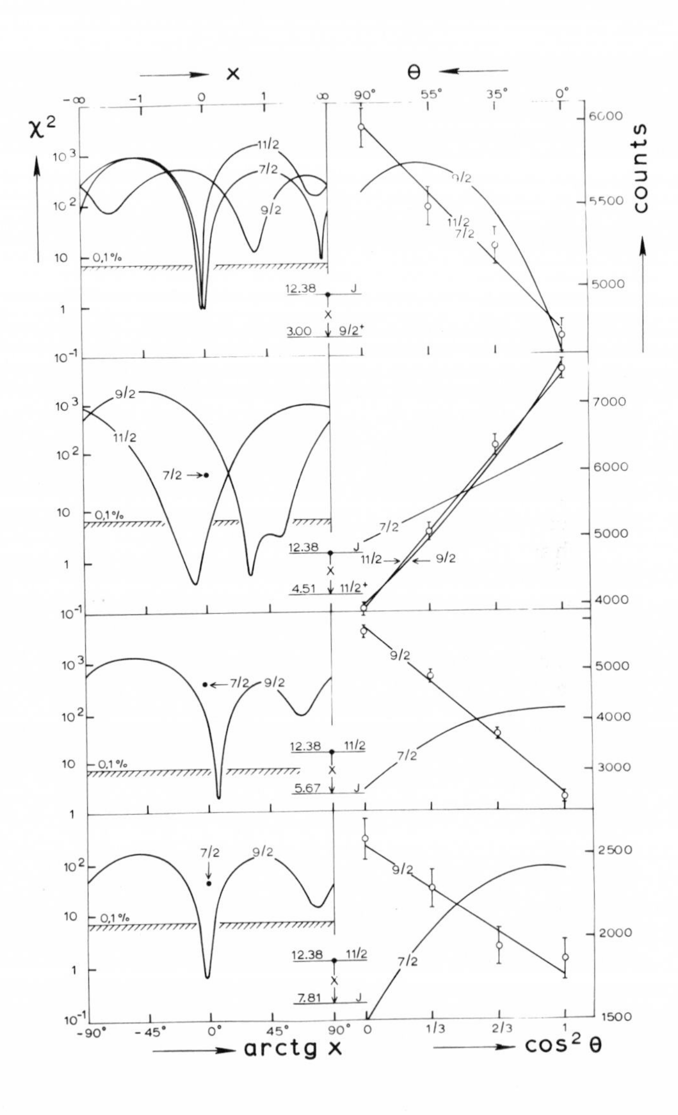

Fig. 1. The angular distribution data with theoretical fits and corresponding χ^2 curves at the E_α = 2.69 MeV $Na^{23}(\alpha,\gamma)Al^{27}$ resonance. The values of χ^2 are plotted against the amplitude mixing ratios with J as a parameter. The measurement uniquely determines the resonance spin as J = 11/2 and, in addition, the spins of the 5.67 and 7.81 MeV levels as J = 9/2.

electromagnetic decay properties of A = 27-29 nuclei simultaneously.[4-6] In contrast with these results the strong coupling Nilsson model fails to reproduce some important E2 transitions and excitation energies in Al^{27}, while the weak coupling model does not explain the spectroscopic factors.[1] Obviously the shell model offers a better representation for the complicated set of data of A = 27-29 nuclei than the collective model does.

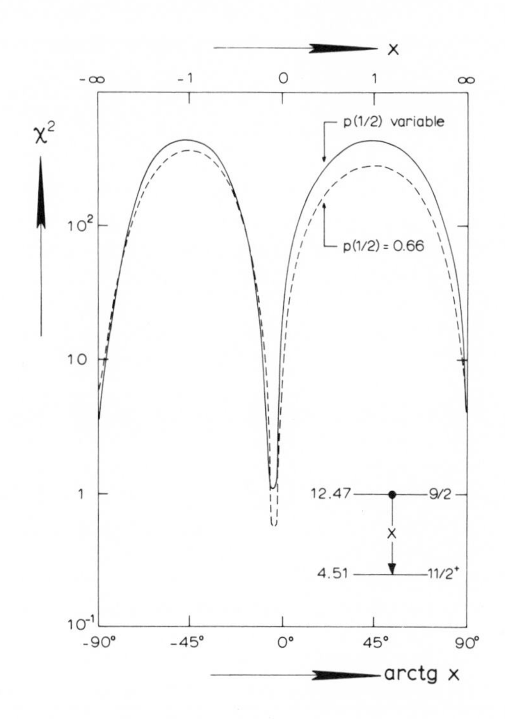

Fig. 2. The normalized χ^2 curve for the angular distribution of the 12.47 → 4.51 MeV transition at the E_α = 2.79 MeV $Na^{23}(\alpha,\gamma)Al^{27}$ resonance. The values of χ^2 are plotted against the quadrupole/dipole amplitude mixing ratio with fixed and variable population parameters. The differences in χ^2 are mainly due to the different number of degrees of freedom (2 and 1, respectively.)

REFERENCES

1. H. Röpke, V. Glattes and G. Hammel, Nucl. Phys. A156, 477 (1970).
2. M. J. A. de Voigt, J. W. Maas, D. Veenhof and C. van der Leun, Nucl. Phys. A170, 449 (1971).
3. M. J. A. de Voigt, J. Grootenhuis, J. B. van Meurs and C. van der Leun, Nucl. Phys. A170, 467 (1971).
4. B. H. Wildentahl and J. B. McGrory, Phys. Rev., to be published.
5. M. J. A. de Voigt, P. W. M. Glaudemans, J. de Boer and B. H. Wildenthal, Nucl. Phys. A186, 365 (1972).
6. M. J. A. de Voigt and B. H. Wildenthal, Nucl. Phys., to be published.

DISCUSSION

MALIK: I have several questions for the first speaker. What are the effective charges used? And what are the strong coupling calculations which were mentioned? And have you also used an orbital gyromagnetic ratio for the neutron because if an effective charge is given to a neutron, this should be included in calculating magnetic moments and M1 strengths? I wonder if you have seen the work of Dehnhard [Phys. Lett. 38B, 389 (1972)] published about 6 months ago, and did you calculate also the magnetic moment of Si^{29}? If so, what is the basis used and what is the result?

DE VOIGT: The simplest question is: what are the effective charges? Well, these are 1.6 e for the proton and 0.6 e for the neutron. Now, whether I have any idea of the work of Denhard? Indeed, I have seen this paper which concentrates on the description of some properties in Al^{27}. It is certainly possible to describe a limited set of features in A = 27-29 nuclei in terms of the collective model. But the band mixing in these nuclei seems to be so severe that the usual benefits of the Nilsson model are lost. Therefore many parameters are required to describe reasonably the observed properties. So far it seems to be impossible to find a consistent representation for the complicated set of data in A = 27-29 nuclei within the collective model framework. In our shell model calculations we only need four parameters to describe the excitation energies and spectroscopic factors, one isoscalar effective charge parameter for seventy-two E2 transitions and four parameters to reproduce thirty-five M1 transitions in the complete A = 27-29 mass region. As an answer to some of your other questions: concerning the collective models I refer to a paper of H. Röpke which contains also a review of different approaches published so far, such as the strong and weak coupling models for Al^{27}. Finally, the value of the presently calculated magnetic moment of the Si^{29} ground state is $\mu = -0.55$ n.m., to be compared with the experimental value of $\mu = -0.56$ n.m.

I.b. THE RESOLUTION OF ANGULAR CORRELATION AMBIGUITIES FOR HIGH SPIN STATES USING A 3 Ge(Li) DETECTOR COMPTON POLARIMETER

J. F. Sharpey-Schafer
University of Liverpool, England

Recent work[1] by P. J. Twin at Liverpool on the spin and mixing ratio ambiguities inherent in gamma ray angular correlation measurements shows that the most powerful method for resolving many of the ambiguities is to measure the linear polarization of the de-exciting gamma rays. The method is particularly successful for high spin states which decay by pure quadrupole radiation.

At Liverpool a Compton polarimeter has been constructed consisting of three large (7 to 9% relative efficiency) Ge(Li) detectors with resolutions for 1.33 MeV photons between 2.0 and 3.0 keV. The largest detector is positioned in the horizontal plane at 90° to the beam direction and is a scatterer of gamma rays into two Ge(Li) detectors, acting as absorbers, one in the reaction plane and the other normal to it. The absorber counters are shielded from direct gamma rays from the target by 70 mm of lead. Background is severely reduced[2] by requiring coincidence events to obey the Compton scattering formula. Polarization insensitive events associated with annihilation radiation are gated out by windows on the 511 keV peaks in the absorber spectra. The measured polarization sensitivity of this device is shown in Figure 1. Its advantages over a polarization sensitive planar (Ge(Li) detector have been recently discussed elsewhere.[3]

We have been using the (α,n) reaction on the spin zero targets Mg^{26}, Si^{30} and S^{34} to study the nuclei S^{29}, S^{33} and Ar^{37} respectively. Beam energies are selected to populate states of interest just a few hundred keV above threshold so that the neutrons emerge mainly as s-waves and the alignment of levels may be estimated using the MANDY program.[4] Measurement of the direct angular distributions of the de-exciting gamma rays sometimes allows a unique spin assignment for the level to be made, but often there is a choice of spins and/or mixing ratios which give equally good fits to the measured angular distributions. Measurement of the nuclear lifetime will often then determine the parity of the level, but still some ambiguities may remain. Table I gives the calculated gamma ray linear polarization predicted by level spins, parity and mixing ratio combinations that are compatible with the level lifetimes and gamma ray angular correlations for some levels in Si^{29} and S^{33}. Our experimental measurements of these polarizations (Table I) are able, in all cases shown, to select a unique spin and mixing ratio.

TABLE I

Comparison of the measured gamma ray linear polarization P_{ex} with the values P_p predicted by level spin J_i and mixing ratio combinations which give acceptable fits to the gamma ray angular distributions and are allowed by the level lifetimes. The spins selected by the polarization data are underlined.

Nucleus	Level No.	E_x (keV)	E_γ (keV)	J_i	J_f	P_p	P_{ex}	τ (fs)
Si^{29}	25	6781	3157	$\underline{11/2^-}$	$7/2^-$	0.70 ± 0.02	0.70 ± 0.19	< 10
				$9/2^-$		-0.55 ± 0.10		
				$7/2^-$		-0.32 ± 0.05		
	31	7139	3058	$\underline{11/2^+}$	$7/2^+$	0.56 ± 0.03	0.53 ± 0.18	< 40
				$7/2^+$		-0.19 ± 0.16		
	62	8761	3505	$\underline{13/2^-}$	$9/2^-$	0.70 ± 0.02	0.67 ± 0.39	< 15
				$9/2^-$		-0.03 ± 0.10		
S^{33}	4	2868	2868	$\underline{5/2^+}$	$3/2^+$	-0.63 ± 0.08	-0.65 ± 0.17	34 ± 14
				$3/2^+$ (δ_1)		0.62 ± 0.12		
				$3/2^+$ (δ_2)		0.09 ± 0.05		
	10	4050	2082	$\underline{9/2^+}$	$5/2^+$	0.82 ± 0.05	0.78 ± 0.14	304 ± 76
				$5/2^+$		-0.68 ± 0.10		
	12	4096	2128	$\underline{7/2^+}$	$5/2^+$	-0.68 ± 0.05	-0.73 ± 0.11	45 ± 12
				$5/2^+$		0.74 ± 0.05		
				$3/2^\pm$		∓0.25 ± 0.05		
	19	4868	1933	$\underline{11/2^-}$	$7/2^-$	0.75 ± 0.06	0.56 ± 0.15	350 ± 90
				$7/2^-$		-0.45 ± 0.13		

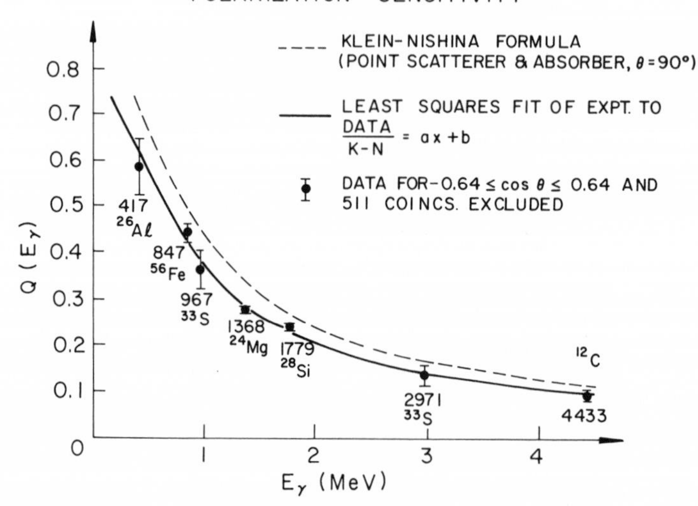

Fig. 1. The measured polarization sensitivity $Q(E\gamma)$ of the Liverpool 3-Ge(Li) Compton polarimeter plotted as a function of photon energy $E\gamma$.

By using a large (9% relative efficiency) Ge(Li) detector in a NaI(Tl) annulus[5] to detect the gamma rays and by careful measurements of the gamma ray thresholds, energies and Doppler broadening as a function of bombarding energy we have been able to work out the decay schemes for nuclei up to rather high excitation energies. Our record is the 62nd level in Si^{29} at 8761 keV which has $J^{\pi} = 13/2^-$ and decays by enhanced E2 transitions (Figure 2) to an $11/2^-$ at 6781 keV and the known[6] $9/2^-$ at 5255 keV. Our measurements suggest that these three levels belong to a $K = 7/2^-$ rotational band based on the $J = 7/2^-$ level at 3624 keV. The in band $\Delta J = 1$ transitions all have negative mixing ratios δ (using the sign convention of Rose and Brink[7]) which is indicative[6,8] of a band formed by a neutron outside a core with oblate deformation. We may therefore assume that this is a band associated[6] with the [303] Nilsson orbit. These negative parity levels lie on a straight line in the usual plot of excitation energy versus J(J+1) (Figure 3), and it can be seen that the $7/2^-$ [303] band has a very similar moment of inertia to the proposed[8,9] $3/2^+$ [202] band. Both these bands have moment of inertia parameters $\hbar^2/2\mathcal{J} \simeq 150$ keV which compares with 260 keV for the ground state band and 125 keV predicted for a rigid body. These data are in good qualitative agreement with the calculations of Ragnarsson and Nilsson[10] who have used the Strutinsky shell correction method to calculate the energies of different single particle orbits as a function of the core deformation. These authors

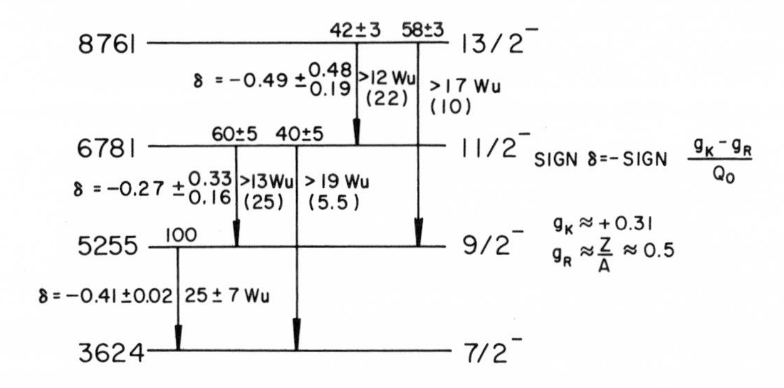

Fig. 2. Decay scheme of high spin negative parity levels in Si^{29} which are assumed to be members of the K = 7/2⁻ band based on the [303] Nilsson orbit. Mixing and branching ratios are given together with the measured E2 transition strengths in Weisskopf single particle units. Values expected for a pure rotational band are shown in brackets.

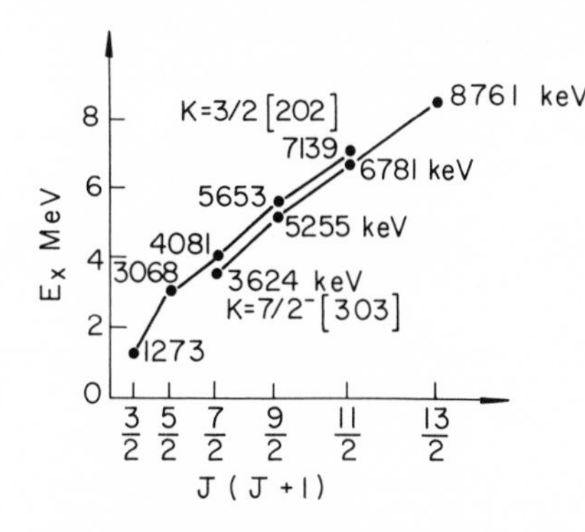

Fig. 3. Plots of excitation energy against J(J+1) for the proposed K = 7/2⁻ [303] and K = 3/2⁺ [202] rotational bands in Si^{29}.

predict a quadrupole moment of −62 fm^2 for a K = 7/2⁻ oblate band, which compares with $|Q_o|$ = 60 ± 6 fm^2 for the band shown in Figure 2 using the lifetime and mixing ratio of the 5255 keV level. A summary of the known band structure of Si^{29} is shown in Figure 4, a very tentative K = 1/2⁻ prolate band (predicted in Reference 10) being indicated by dotted lines.

The known[11] structure of the negative parity states in S^{33} do not open themselves at the moment to such a simple rotational interpretation in spite of calculations[11] similar to those of Reference 10 predicting bands identical to those of Si^{29}. The data on the positive parity levels in S^{33} give excellent agreement with the shell model calculations of Wildenthal *et al.*[12]

Preliminary measurements of lifetimes, angular correlations and gamma ray polarizations in Ar^{37} suggest that negative parity levels which decay to the 7/2⁻ level at 1611 keV (Figure 5) may well be represented by coupling an $f_{7/2}$

particle to a vibration of the Ar^{36} core. Figure 5 shows the measured E2 transition strengths from a J = 3/2 → 11/2 quintet of levels to the 1611 keV level and compares them with the $2^+ \rightarrow 0^+$ transition in Ar^{36}.

From these measurements it would appear that a good gamma ray polarimeter is a *sine qua non* for sd-shell spectroscopists wishing to investigate idiosyncrasies in their favorite nuclei. It is in any case a technique to which we are applying ourselves with great enthusiasm at Liverpool

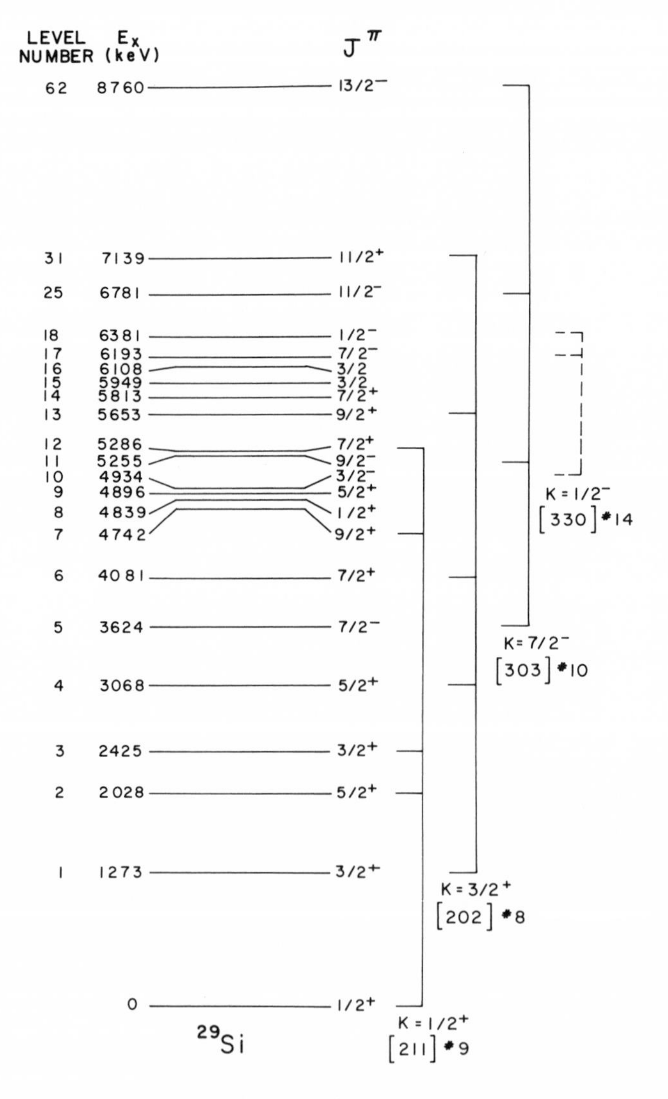

Fig. 4. Level scheme for Si^{29} showing the proposed rotational band structure.

and which we hope will help to pin down the structural changes and trends in these nuclei.

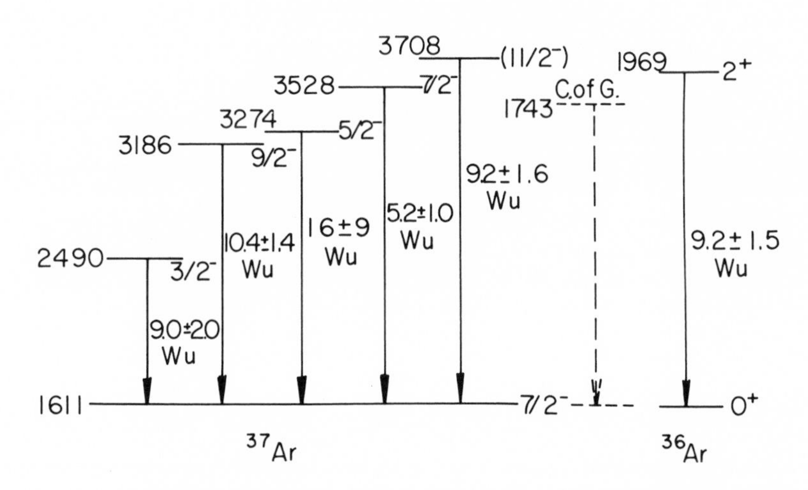

Fig. 5. Diagram showing the E2 strengths of negative parity levels decaying to the $7/2^-$ level at 1611 keV in Ar^{37}. The center-of-gravity of these levels is shown and is compared with the one phonon transition in Ar^{36}.

REFERENCES

1. P. J. Twin, submitted to Nucl. Instr. and Meth.
2. C. Broude *et al.*, Nucl. Instr. and Meth. 69, 29 (1969).
3. P. J. Twin, Nucl. Instr. and Meth. 103, 613 (1972).
4. E. Sheldon and D. van Patter, Rev. Mod. Phys. 38, 143 (1966).
5. J. F. Sharpey-Schafer *et al.*, Nucl. Phys. A167, 602 (1971).
6. T. T. Bardin *et al.*, Phys. Rev. Lett. 24, 772 (1970).
7. H. J. Rose and D. M. Brink, Rev. Mod. Phys. 39, 306 (1967).
8. I. G. Main *et al.*, Nucl. Phys. A158, 364 (1970).
9. D. C. Bailey *et al.*, J. Phys. A: Gen. Phys. 5, 596 (1972).
10. I. Ragnarsson and S. G. Nilsson, Nucl. Phys. 158, 155 (1970).
11. P. E. Carr *et al.*, submitted to J. Phys. A: Gen Phys.
12. B. H. Wildenthal *et al.*, Phys. Rev. C4, 1708 (1971).

DISCUSSION

DONAHUE: I would just like to ask the speaker if the interpretation of the polarization measurements depends on the fact that the neutrons have $\ell = 0$ going in?

SHARPEY-SCHAFER: Basically, no, because the errors on the calculated value that I showed in the table were due to things like the allowed limits on the population of the substates. It turns out that the Clebsch-Gordon coefficients in the angular correlations that also effect the polarizations here are very much the same even if you get quite a bit of population of the other substates. So it doesn't effect the angular distributions or the polarization much if you don't get complete alignment. It's very insensitive to the alignment.

II.A. NUCLEAR STRUCTURE CALCULATIONS WITH REALISTIC TWO-BODY FORCES

S. Maripuu*
Cyclotron Laboratory
Department of Physics
Michigan State University
East Lansing, Michigan

I. INTRODUCTION

The promising results of the early work of Kuo and Brown[1-4] have inspired many studies of the use of realistic two-nucleon interactions in nuclear structure calculations. Realistic is used here to mean that the two-nucleon interaction is derived from the phase shift data from free nucleon-nucleon scattering. The two-nucleon interaction so derived is then transformed into an effective interaction to be used in a small model space. The successful results obtained by employing this method are many. A large number of physical quantities have been explained accurately, usually at least as well as is managed with empirical interactions (see *e.g.* References 5 and 6). These latter interactions, deduced from energy spectra of simple nuclei or obtained as parameters from least squares fits to many energy levels often produce results which agree will with experimental data but their connections to the fundamental nucleon-nucleon problem are obscure. It has been shown that some matrix elements of empirically obtained two-body interactions are poorly determined by almost any feasible data set.[7]

An interesting alternative to the Kuo-Brown method for calculating realistic two-body interactions has been suggested by Elliott and his collaborators at Sussex.[8,9] They deduce relative harmonic oscillator matrix elements directly from the experimental scattering phase shifts without explicit construction of the potential. The method is a distorted wave Born approximation with reasonable (but not detailed) assumptions about the smoothness and range of the potential. In spite of a number of applications with the Sussex matrix elements (Reference 10), they have rarely been used in detailed large space nuclear structure calculations.

The present calculations employ the Sussex matrix elements to derive effective interactions in the model space of (a) the p-shell, (b) the sd-shell, and (c) the upper part of the sd-shell including some pf-excitations. The method of calculation is outlined in Section II and the different applications together with conclusions are presented in Sections III-V.

*Supported in part by the National Science Foundation.

II. METHOD OF CALCULATION

The relative harmonic oscillator matrix elements given in Ref. 9 are used to calculate the present two-body matrix elements. The necessary transformation from jj to LS coupling as well as the Brody-Moshinsky transformation between the laboratory frame and the center-of-mass and relative frame has been described by Kuo and Brown.[1] We have performed these transformations for different sets of b, the oscillator length parameter, which has been chosen to fit the nuclear size. The two-body matrix elements thus obtained are considered to correspond to the "bare" reaction matrix elements of Kuo and Brown. In order to be used in a small model space these matrix elements have to be corrected for space truncation effects. We have used perturbation theory (Ref. 11) up to second order to calculate 3p-1h, 2h and 2p corrections. Energy denominators up to $2\hbar\omega$ excitations have been included. The corrections are shown as diagrams in Figure 1. The first diagram represents the unperturbed interaction, V. The energy denominators (expressed in units of $\hbar\omega$) always have the same oscillator size parameter, as the one used in the calculation of the unperturbed matrix elements. In other words, once the calculation starts, there are no free parameters in the two-body matrix elements.

We have taken the single particle energies or one-body interactions from the spectra of closed shell plus one nucleon nuclei. This approach has been commonly used in earlier investigations. In the calculations of energy spectra in nuclei far from the closed shells we have also treated the

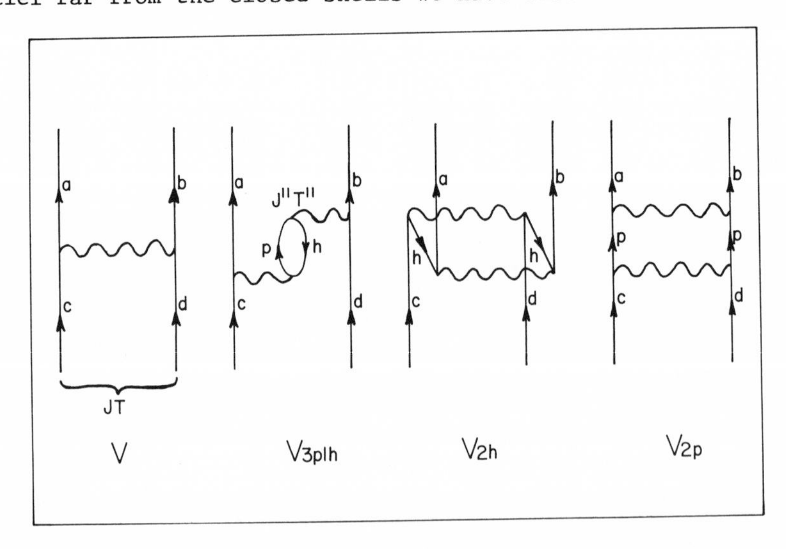

Fig. 1. The diagrams representing the perturbative corrections to the unperturbed interaction, V.

TABLE I

The Two-Body Matrix Elements in the form $\langle ab;JT||V||cd;JT\rangle$, are shown for the 1p shell along with their perturbative corrections. The numbers abcd represent $2J_a$, $2J_b$, $2J_c$ and $2J_d$ respectively. All matrix elements are calculated for the length parameter b = 1.7 fm and expressed in MeV.

a b c d - J T	$\langle V\rangle$	$\langle V_{3p-1h}\rangle$	$\langle V_{4p-2h}\rangle$	$\langle V_{2p}\rangle$	$\langle V_{eff}\rangle$
3 3 3 3 - 1 0	-1.541	-0.055	-0.112	-0.838	-2.546
- 3 0	-4.060	-0.158	-	-0.764	-4.982
- 0 1	-3.020	-0.276	-0.303	-0.414	-4.013
- 2 1	-1.453	+0.314	-	-0.170	-1.308
3 3 3 1 - 1 0	+3.528	+0.059	+0.316	+0.562	+4.465
- 2 1	-1.539	-0.276	-	-0.127	-1.942
3 3 1 1 - 1 0	+1.676	+0.026	-0.138	+0.437	+2.001
- 0 1	-3.606	-0.643	-0.214	-0.176	-4.639
3 1 3 1 - 1 0	-4.770	+0.196	-0.888	-0.923	-6.385
- 2 0	-5.350	+0.156	-	-1.255	-6.449
- 1 1	-0.728	+0.544	-	-0.051	-0.235
- 2 1	-2.542	+0.522	-	-0.261	-2.281
3 1 1 1 - 1 0	+0.942	+0.256	+0.389	-0.174	+1.413
1 1 1 1 - 1 0	-1.843	+0.065	-0.170	-0.722	-2.670
- 0 1	-0.470	+0.264	-0.152	-0.289	-0.647

single particle binding energy differences as free parameters (see Sections III and V).

The construction of Hamiltonians and diagonalization of the energy matrices has been performed with the codes described by French, Halbert, McGrory, and Wong.[12]

III. STRUCTURE OF A = 6-13 NUCLEI

We have calculated numerous properties of the p-shell nuclei with the method described in Section II. The two-body matrix elements obtained for the p-shell model shell are listed in Table I. As seen from Table I, the perturbative corrections change the two-body matrix elements significantly. Furthermore, they are definitely needed if one is to obtain satisfactory agreement between experiment and theory. The effective matrix elements shown in the last column of Table I are found to be similar to those determined from the least squares fittings of Cohen and Kurath[13] and of Goldhammer and coworkers.[14] A detailed comparison, however, is not quite in order since the one-body parameters used by the above mentioned authors are different from ours.

Recent experimental data indicate (somewhat surprisingly) that all p-shell nuclei have approximately the same *rms* radi-

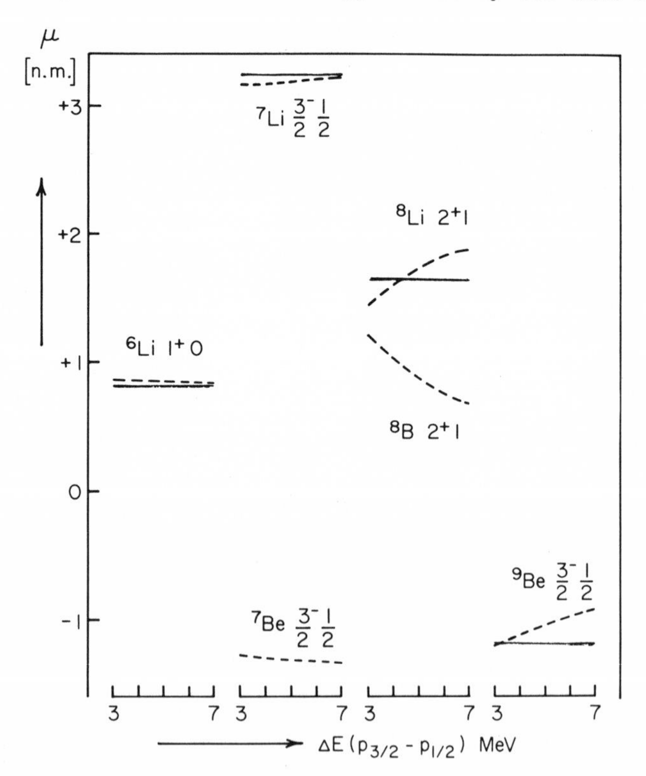

Fig. 2. Magnetic dipole moments of A = 6-9 nuclei. Solid lines indicate experimental values. The dashed lines show how the predicted values vary with the $p_{3/2}$ and $p_{1/2}$ single particle energy differences.

us.[15] Except for the absolute binding energies for different mass numbers (the accurate calculation of which we have not attempted), we find no significant changes in energy spectra for different values of the oscillator size parameter. We use therefore a fixed value, b = 1.7 fm, for all nuclei. On the other hand, we have found a definite need to increase the energy separation between the $p_{3/2}$ and $p_{1/2}$ single particle states (ΔE) with increasing mass number. For A = 6 we find ΔE = 3 MeV (the experimental spin-orbit splitting as observed in A = 5 is 2.6 ± 0.4 MeV), and for A = 13 we have adopted ΔE = 7 MeV.

Experimental and calculated energy spectra for A = 6-8 and A = 11-14 nuclei are shown in Figures 2 and 3, respectively. The results are encouraging, and perhaps much better than one might expect for a one parameter theory. We notice two minor discrepancies between the theoretical and experimental spectra.

(i) The three lowest levels in Be^8 are calculated about 2 MeV too high (in comparison with other levels of the Be^8 spectrum). Also, the 7.7 and

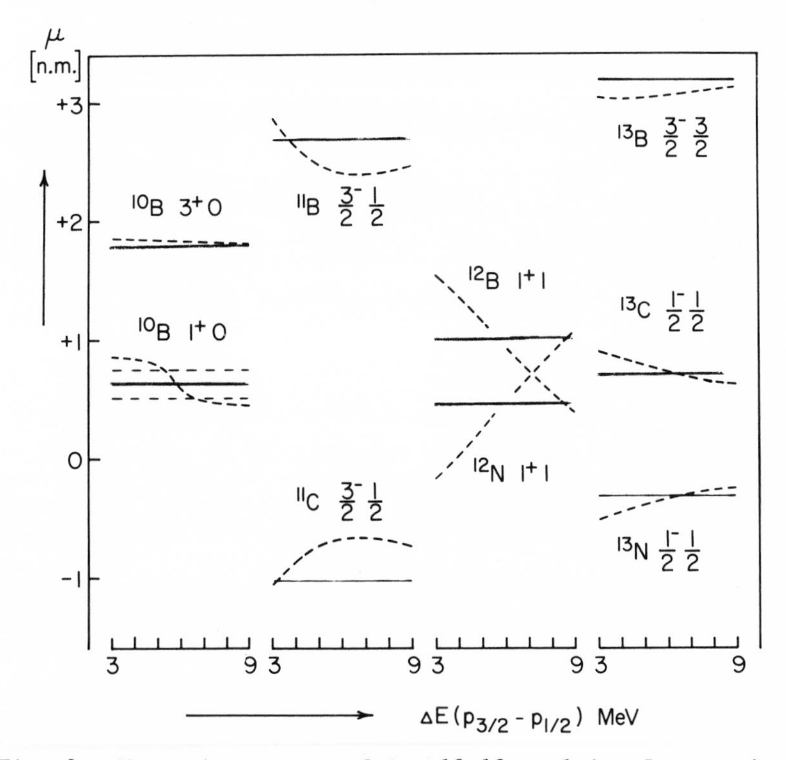

Fig. 3. Magnetic moments of A = 10-13 nuclei. See caption for Figure 2.

10.1 MeV levels in C^{12} are predicted to be higher than the experimental energies.

(ii) The lowest J^{π} T = $0^{+}1$ states in A = 6, 10 and 14 are predicted about 2 MeV too low.

The first discrepancy is easy to understand. The five experimental states involved have large reduced α particle widths, indicating deformed α particle type structures.[16] For an accurate description of these states in the shell model, we

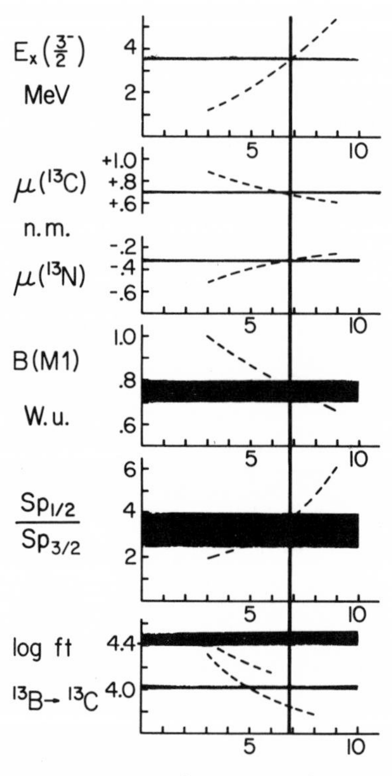

Fig. 4. Comparison between theory and experiment for several quantities in A = 13. Reading from top to bottom they are: The excitation energy of the first excited $3/2^-$ state in C^{13} (and N^{13}); the magnetic moments of the C^{13} and N^{13} ground states; the M1 transition strength from the 3.51 MeV, $J^{\pi} = 3/2^-$ state in N^{13} to the ground state; the ratio of the single nucleon spectroscopic factors between the ground state and first excited state of C^{13}; the log ft values for $B^{13} \rightarrow C^{13}$ $(1/2^-)$ and $B^{13} \rightarrow C^{13}$ $(3/2^-)$ beta decays. The horizontal solid lines show measured values, with the thickness of the line indicating experimental uncertainties. The dashed lines show how the predicted values vary with the spin-orbit splitting, $\Delta E(p_{3/2}-p_{1/2})$. See also text.

would need a larger configuration basis. The second discrepancy could possibly be rectified by including higher order corrections into the perturbation calculation or by including a monopole shift to increase the isospin splitting.[17]

With the wave functions obtained for a number of different ΔE values, we have calculated magnetic dipole moments, M1 transition strengths, log ft values for beta decay and single nucleon spectroscopic factors. In Figures 4 and 5 we show how the calculated magnetic dipole moments vary with ΔE. In most cases the experimental and theoretical values coincide for reasonable ΔE values. Furthermore, it appears that for T = 0 states the predicted magnetic moments are very insensitive to the variation of ΔE (or, in other words, to the amount of configuration mixing). This is due to the following reasons:

(i) The isoscalar reduced single nucleon matrix element $\langle p_{3/2}||\mu_{IS}||p_{3/2}\rangle$ is about five times larger than the $\langle p_{3/2}||\mu_{IS}||p_{1/2}\rangle$ and $\langle p_{1/2}||\mu_{IS}||p_{1/2}\rangle$ matrix elements.

(ii) In almost all many nucleon configurations of the 1p-shell there is at least one $p_{3/2}$ nucleon. All diagonal matrix elements will therefore be of about the same size (roughly the size of the matrix element $\langle p_{3/2}||\mu_{IS}||p_{3/2}\rangle$).

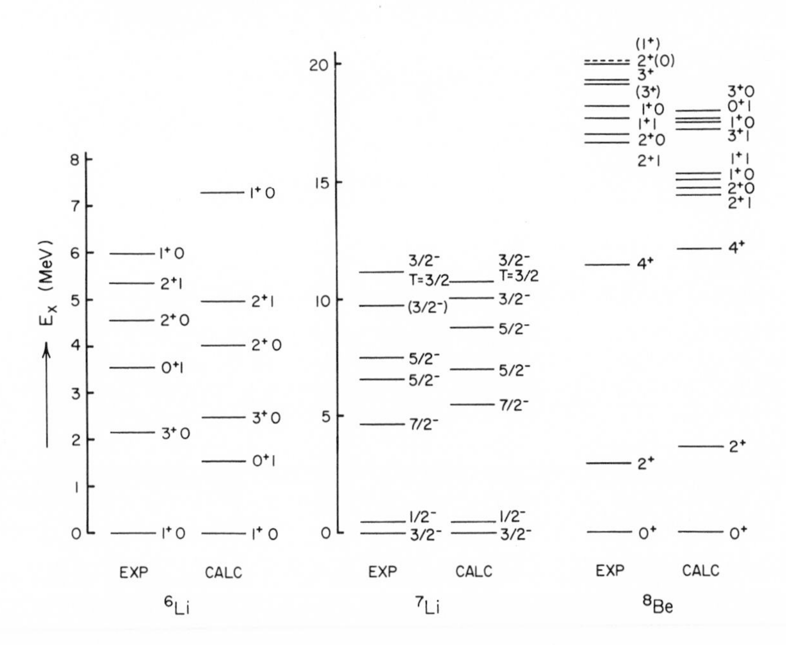

Fig. 5. Experimental and calculated energy spectra for A = 6-8 nuclei, all calculated with $\Delta E(p_{3/2}-p_{1/2})$ = 3.0 MeV.

We thus conclude that to a good approximation, configuration mixing does not affect the total isoscalar matrix elements. (This is not true for the isovector matrix elements because there the off-diagonal single-nucleon reduced matrix elements are comparatively large.)

Various quantities connecting the $J^{\pi} = 1/2^{-}$ and $3/2^{-}$ ground and first-excited states of A = 13 are shown in Figure 6. Again, the variation of the predicted values with ΔE is included. Most observables are predicted best at $\Delta E \simeq 6.8$ MeV, but the beta decay rates favor a value of $\Delta E \simeq 5.0$ MeV. This comparatively retarded beta decay might be explained with sd-shell admixtures in the B^{13} ground state.

From a detailed comparison of various calculated quantities we conclude that our calculations yield about the same results as those obtained in previous calculations with many parameters[13,14] (our energy level predictions are not quite as good). Therefore, we find it encouraging that a calculation, essentially without parameters and derived from first principles does as well as those with effective two-body matrix elements obtained from least squares fits with upwards of thirteen free parameters.

IV. ENERGY LEVELS OF A = 18-22 NUCLEI

Excitation energies in A = 18-22 nuclei have been calculated in full sd-model space. The calculation of the effective two-body interaction has, in principle, already been described in the previous sections. Again, perturba-

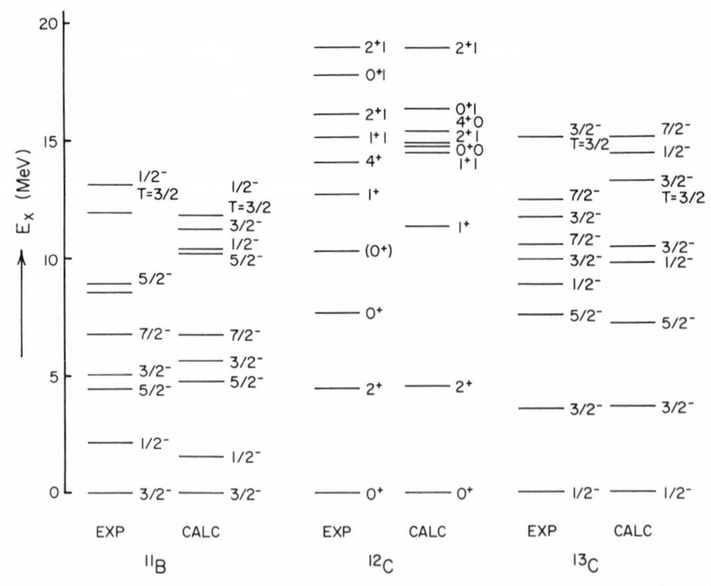

Fig. 6. Experimental and calculated energy spectra for A = 11-13 nuclei, all calculated with $\Delta E(p_{3/2}-p_{1/2}) = 7.0$ MeV.

tive corrections up to second order and to 2 $\hbar\omega$ excitations have been included. For A = 18-22 nuclei we have used the same harmonic oscillator size parameter, $\hbar\omega$ = 14.4 MeV, as in the A = 6-13 calculation. The single-particle energies have been taken from the O^{17} experimental spectrum. The absolute ground state binding energies calculated in this way turn out to be somewhat too large (interaction too attractive), indicating that better agreement would have been obtained with a somewhat smaller oscillator constant. (The absolute binding energies of the lowest T states in A = 19 and 22 are predicted about 0.5 and 4.0 MeV too low, respectively).

The spectra of F^{18} and O^{18} are shown in Figures 7 and

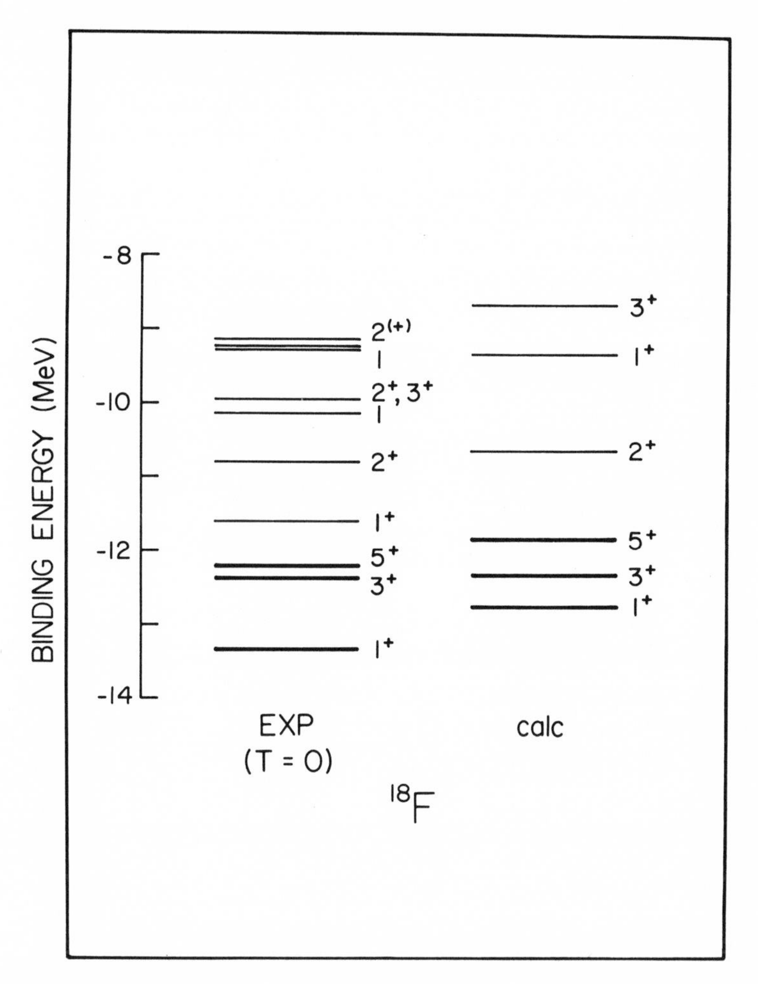

Fig. 7. The T = 0 spectrum of F^{18} as observed experimentally and as calculated with the present interaction.

8. The energy scales used in Figures 7 and 8 indicate binding energy with respect to O^{16}. The agreement between experiment and theory is good except for the $J^{\pi}T = 1^{+}0$ ground state of F^{18} which is predicted somewhat too high. The calculated and experimental spectra of F^{19}, Ne^{20}, F^{20} and O^{20}, all with ground state energies set equal are shown in Figures 9-12. The agreement between experimental and calculated excitation energies is again quite good. The calculated spectra are very similar to those obtained with the Kuo-Brown interaction.[5,6] A comparison is shown only for the O^{20} spectrum in Figure 12. (The Kuo-Brown two-body matrix elements contained some small errors,[18] therefore a detailed comparison has been left out.) In all spectra shown in Figures 7-12, the energy levels drawn with somewhat heavier lines indicate levels presumably belonging to the ground state bands. It is experimentally well established[19] that the *second* $7/2^{+}$ state in F^{19} (see Figure 9) belongs to the ground state band. Our calculation, however, predicts a fairly complicated $7/2^{+}$ state as the lowest state (and the second $7/2^{+}$ as predominantly of $d^{3}_{5/2}$ character).

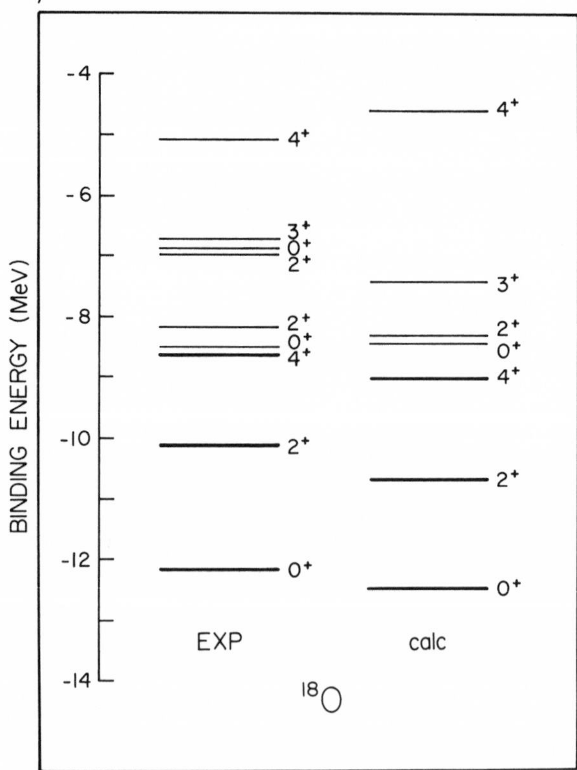

Fig. 8. The experimental and calculated spectrum of O^{18}.

It is well known that realistic interactions do not produce enough splitting between groups of energy levels with different isospin. In Figure 13 we have plotted the ground state binding energies for A = 19-22 nuclei. For each mass number we have normalized the spectrum to the binding energy of the ground state with lowest isospin. Included are also some levels close to the experimental (or predicted) ground states. For A = 22, we did not calculate the $J^{\pi}T = 3^{+}0$ level, the calculated energies are normalized to the lowest experimental $J^{\pi}T = 1^{+}0$ level. It can be concluded from Figure 13 that for most cases the predicted isospin splitting is only slightly smaller than the observed one. Larger discrepancies are found for the newly measured[20] binding energies of O^{21} and O^{22}.

We find that the inclusion of the 2p-corrections greatly improves the isospin splitting. Inclusion of higher order corrections might remove the remaining discrepancies provided the present model space is reasonable. As already mentioned, a smaller oscillator size would predict better absolute binding energies. However, it is not clear whether a physically reasonable size parameter will accomplish such an agreement.

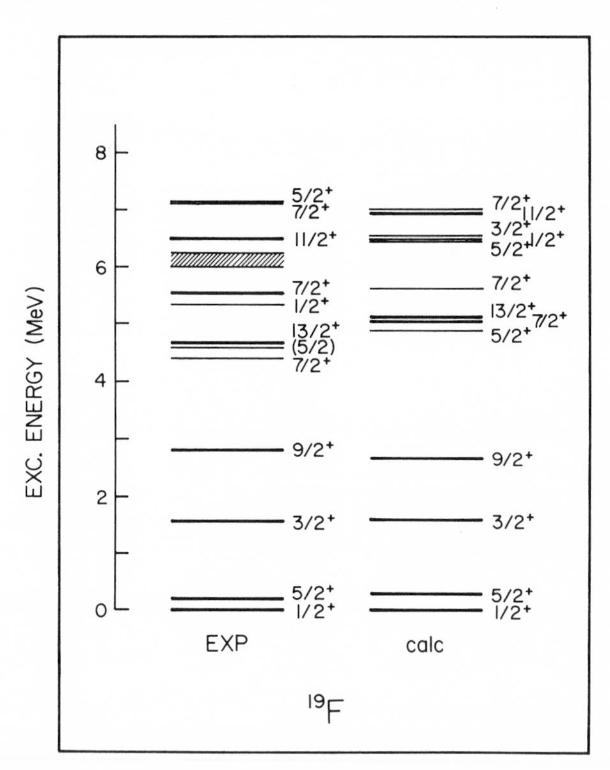

Fig. 9. Observed and predicted spectrum of F^{19}.

V. CONFIGURATION MIXING OF THE df AND dp MULTIPLETS OF Cl^{38}

The early description[21,22] of the low-lying negative-parity quartets of Cl^{38} and K^{40} has led to the consideration of these states as members of relatively pure $(\pi d_{3/2}, \nu f_{7/2})$ and $(\pi d^{-1}_{3/2}, \nu f_{7/2})$ configurations, respectively, with the spectra of the two nuclei almost perfectly related by the particle-hole transformation. More recent measurements[23,24] have revealed a higher lying multiplet with the $2p_{3/2}$ neutron replacing the $1f_{7/2}$ neutron, although the transformation of energies is not as accurate for this multiplet as for the lower set of states.

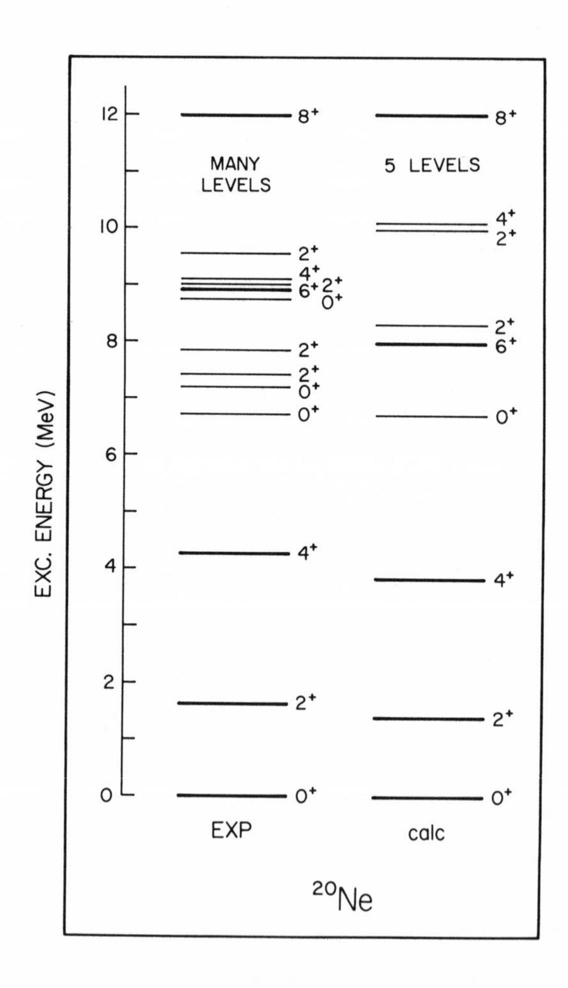

Fig. 10. Observed and predicted spectrum of Ne^{20}.

However, measurements[24,26] on M1 strengths in Cl^{38} cast serious doubt on this simple description of these states. A strong M1 transition is observed from the 3^- state of the dp-multiplet to the 4^- state of the df-multiplet. This transition is, of course, ℓ-forbidden to the extent that the wave functions of the states actually are made of such configurations. Other data that contradict the pure configuration picture are the spectroscopic factors for the $Cl^{37}(d,p)Cl^{38}$ reaction.[27] These experiments show some $\ell = 1$ mixing into the 3^- state of the lower multiplet.

We have calculated excitation energies, M1 transition strengths and single nucleon spectroscopic factors for A = 38-40 nuclei in a model space which includes active $d_{5/2}$, $s_{1/2}$, $d_{3/2}$, $f_{7/2}$ and $p_{3/2}$ particles.* The relative single-

---

* The method of calculation of the two-body interaction is the same as the one described in previous sections. The size parameter, $\hbar\omega$, has been chosen to be 12.8 MeV.

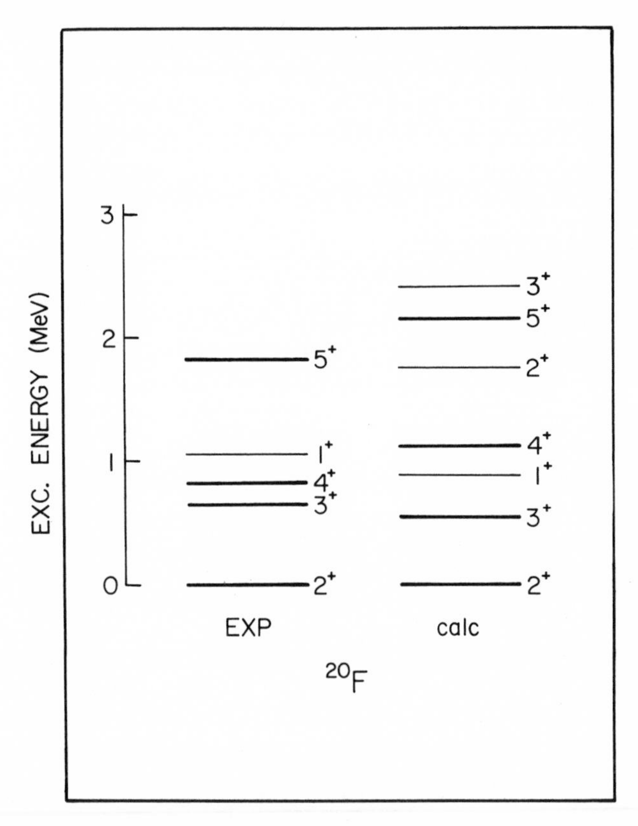

Fig. 11. Observed and predicted spectrum of F^{20}.

particle energies were chosen to yield calculated spectra for Cl^{38}, K^{39} and K^{40} in simultaneous best agreement with the experimental spectra. We found the values -9.40, -4.90, -2.77, -2.52 and 0.00 MeV for the $d_{5/2}$, $s_{3/2}$, $d_{1/2}$, $f_{7/2}$ and $p_{3/2}$, respectively. Our calculated Ml transition strengths for Cl^{38} are summarized in Figure 14 and Table II. The K^{40} Ml transitions are also included in Table II. The calculated spectroscopic factors are compared with the experimental ones in Figure 15. Our predictions for these observables (our Ml calculations use operators calculated from the bare nucleon g-factors) are in uniform good agreement with the observed values. In particular, for Cl^{38}, the mixing of the $\ell = 1$ and $\ell = 3$ strengths for the 3^- states and, most striking, the anomalously large $3^-_2 \rightarrow 4^-$ Ml transition are correctly predicted.

The difficulty of accounting for this Ml transition, enhanced when it should be severely retarded according to simple ideas, has been frequently noted of late.[28,29] No previous calculation has been able to reproduce the observed strength. It is therefore of interest to see how the correct strength emerges from the present wave functions. (In Table III we have listed the intensities of different con-

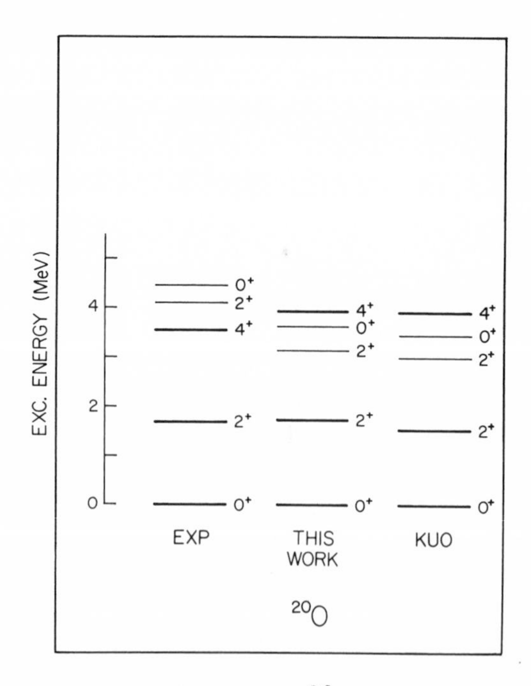

Fig. 12. Observed spectrum for O^{20} and shell model spectra calculated with the present method and as calculated by Kuo (Reference 6).

figurations in our model space for Cl^{38} and K^{40}.) The components in the wave functions of the 3_2^- and 4_1^- states which are important contributors to this strength are given in Table IV.* Also included in Table IV are the important con-

---

* Only isovector components have been included since no large isoscalar contributions have been predicted.

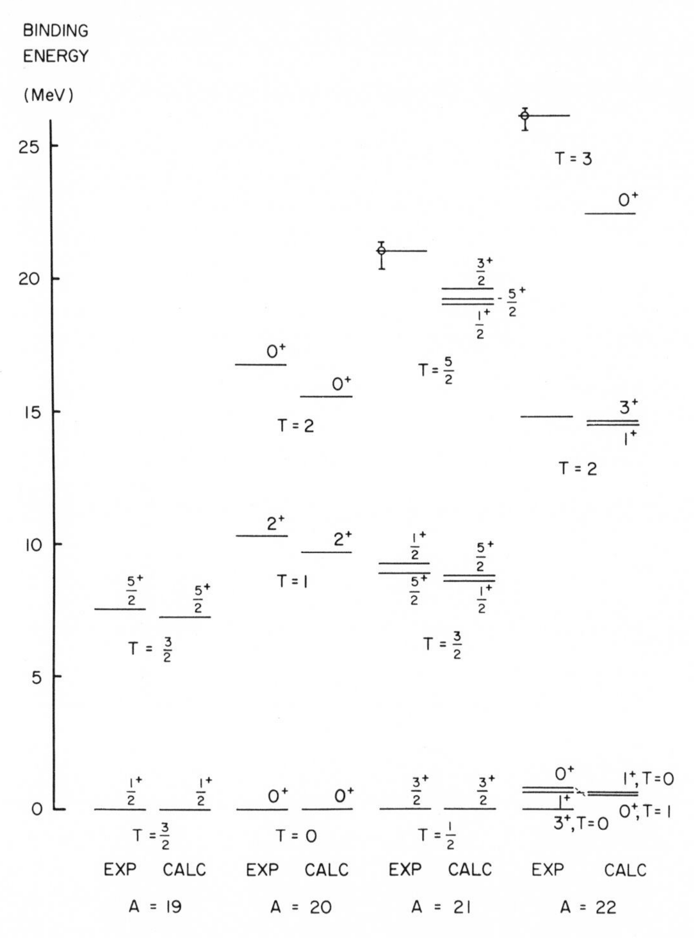

Fig. 13. Binding energies of ground states and a few other levels in A = 19-22 nuclei. For each mass number the spectra have been normalized to the level with the lowest binding energy. (The experimentally observed energy separations between different isobars have been adjusted by subtracting Coulomb energy differences estimated as explained in Ref. 6). For Ne^{22} the $J^{\pi}T = 3^{+}0$ level has not been calculated, the predicted spectrum has been normalized to the lowest $J^{\pi}T = 1^{+}0$ state.

tributions for the $3_2^- \to 3_1^-$ M1 transition. We see that the excitations from the $s_{1/2}$ to the $d_{3/2}$ orbit are of comparable importance to the mixing between $f_{7/2}$ and $p_{3/2}$ excitations in contributing to the M1 transition strength. (The $s_{1/2}$ particles, as well as $f_{7/2}$ and $p_{3/2}$, have their spins parallel to the orbital angular momentum, thus producing strong isovector contributions to the M1 strength[30].) Furthermore, it is important to notice that the strong contributions to the $3_2^- \to 4^-$ transition add up coherently. The individual contributions to the $3_2^- \to 3_1^-$ are even larger than those for the $3_2^- \to 4^-$ transition (as pointed out by Erné *et al.*[31]). However, in this case the signs of the individual components are such that the various contributions cancel each other.

We arrive at the following conclusions:

(i) the df- and dp-multiplets in Cl^{38} are about 80% pure, with considerable mixing only between the 3^- states.

(ii) The 5%-20% admixtures of the $s_{1/2}^{-1}$ configurations in these Cl^{38} states are important to account for the $3_2^- \to 4^-$ anomalous M1 strength.

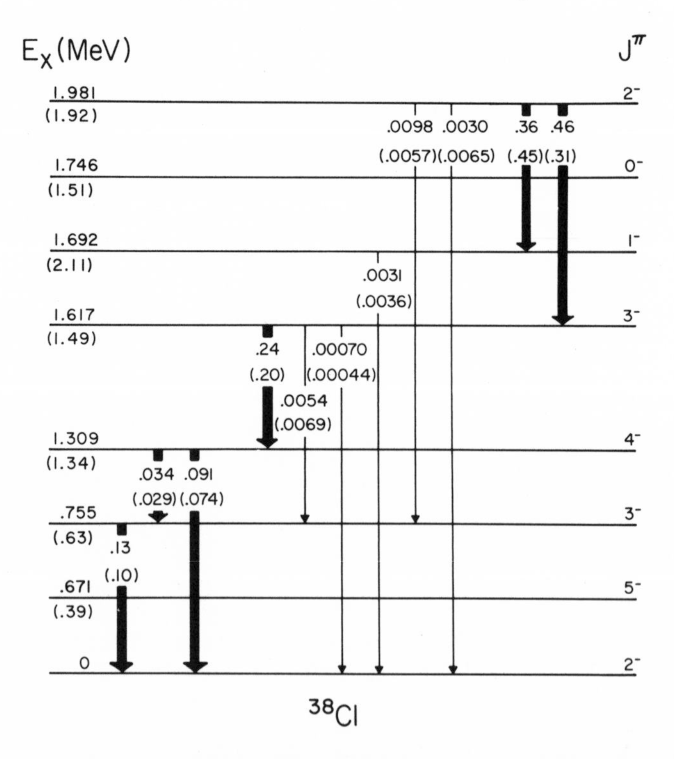

Fig. 14. The excitation energies and M1 transition strengths of Cl^{38}. Predicted quantities are shown within brackets.

TABLE II

M1 Transitions in Cl^{38} and K^{40}

Cl^{38}				K^{40}			
	M1 Strength (W.u.)				M1 Strength (W.u.)		
$J_i \to J_f$	Argonne[25]	Brookhaven[24]	Theory	$J_i \to J_f$	Argonne[25]	Frankfurt[26]	Theory
$3_1^- \to 2_1^-$	.23	.13	.10	$3_1^- \to 4^-$	.150	.170	.087
$4^- \to 3_1^-$	.062	.034	.029	$2_1^- \to 3_1^-$	.127	.140	.091
$4^- \to 5^-$	.170	.091	.074	$5^- \to 4^-$	.030	.030	.027
$3_2^- \to 4^-$	.38	.24	.20	$2_2^- \to 2_1^-$	.0130	.0100	.0045
$3_2^- \to 3_1^-$	.0084	.0054	.0069	$2_2^- \to 3_1^-$	.0026	.0023	.0016
$3_2^- \to 2_1^-$	.00105	.00070	.00044	$3_2^- \to 2_1^-$	.0033	.0020	.0048
$1^- \to 2_1^-$	.0050	.0031	.0036	$3_2^- \to 3_1^-$	.0046	.0025	.0053
$2_2^- \to 3_1^-$	.0174	.0098	.0057	$3_2^- \to 4^-$	.0030	.0016	.0012
$2_2^- \to 2_1^-$	.0053	.0030	.0065	$1^- \to 2_1^-$	.0076	.0038	.0222
$2_2^- \to 1^-$	.62	.36	.45	$0^- \to 1^-$	.47	.20	.91
$2_2^- \to 3_2^-$	.80	.46	.31				

TABLE III

Configuration Mixing in Cl^{38} and K^{40}
(intensities in per cent)

E_x(MeV) Exp	Calc	J^π	$d^5_{3/2}f_{7/2}$	$d^5_{3/2}p_{3/2}$	$s^{-1}_{1/2}d^6_{3/2}f_{7/2}$	$s^{-1}_{1/2}d^6_{3/2}p_{3/2}$
Cl^{38}						
0	0	2^-	82	3	7	2
.67	.39	5^-	86		5	
.76	.63	3^-	68	17	7	2
1.31	1.34	4^-	77		8	
1.62	1.49	3^-	14	45	28	1
1.69	2.11	1^-		81	12	3
1.75	1.51	0^-		97		1
1.98	1.92	2^-	1	77	16	1

E_x(MeV) Exp	Calc	J^π	$d^{-1}_{3/2}f_{7/2}$	$d^{-1}_{3/2}p_{3/2}$	$s^{-1}_{1/2}f_{7/2}$	$s^{-1}_{1/2}p_{3/2}$
K^{40}						
0	0	4^-	96		2	
.03	-.02	3^-	91	5	2	
.80	1.06	2^-	87	7		2
.89	.79	5^-	96			
2.05	1.99	2^-	6	89		
2.07	1.93	3^-	2	76	18	
2.10	2.54	1^-		96		
2.63	2.19	0^-		98		

(iii) The purity of the $d^{-1}f$- and $d^{-1}p$-multiplets in K^{40} is considerably higher (no anomalies in M1 strength are predicted) in agreement with observations. The differences between Cl^{38} and K^{40} are consequences of the freedom to excite particles within the sd-shell which exist for Cl^{38} but does not exist for K^{40}.

TABLE IV

Predominant contributions to the $J^\pi = 3^- \to 4^-$ and $3^- \to 3^-$ M1 transition strengths in Cl^{38}. Initial and final configurations are the same. The configuration amplitudes are denoted α_i and α_f, respectively. The isovector matrix elements are expressed in nuclear magnetons.

Initial and Final Configuration	α_i	α_f	Pure IV-Comp	Mixed IV-Comp	Predominant Contribution
$3_2^- \to 4^-$					
$(d_{3/2})_{3/2,3/2}f_{7/2}$	-.37	-.87	-3.94	-1.28	$f_{7/2}$
$s_{1/2}^{-1}(d_{3/2})_{21}f_{7/2}$	-.39	-.13	-13.2	- .67	$s_{1/2} + f_{7/2}$
$s_{1/2}^{-1}(d_{3/2})_{01}f_{7/2}$	+.33	+.25	-15.4	-1.29	$s_{1/2} + f_{7/2}$
			TOTAL	-4.43	
$3_2^- \to 3_1^-$					
$(d_{3/2})_{3/2,3/2}f_{7/2}$	-.37	+.83	+13.5	-4.13	$f_{7/2}$
$(d_{3/2})_{3/2,3/2}p_{3/2}$	+.67	+.42	+13.8	+3.86	$p_{3/2}$
$s_{1/2}^{-1}(d_{3/2})_{01}f_{7/2}$	+.33	+.21	+22.4	+1.52	$s_{1/2} + f_{7/2}$
			TOTAL	+ .69	

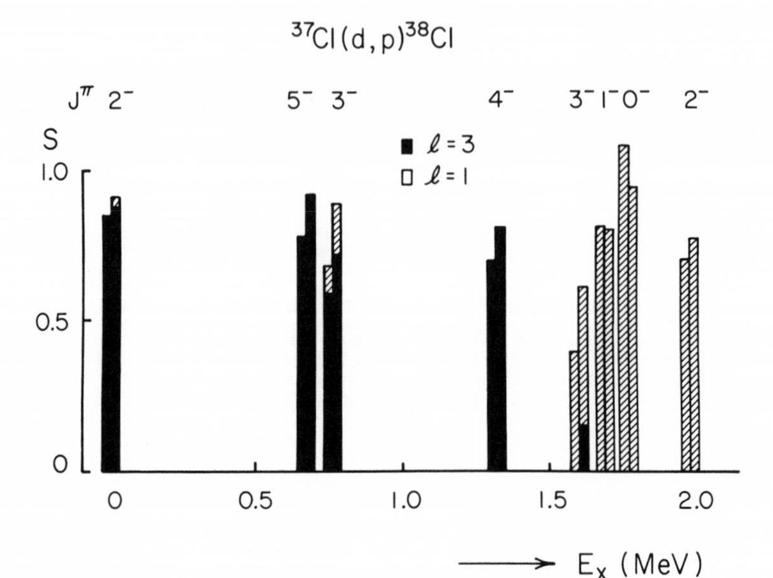

Fig. 15. Single-nucleon spectroscopic factors for the Cl^{37} (d,p)Cl^{38} reaction. For each level the experimental number is represented by the left hand bar; the bar immediately to its right is the calculated one.

ACKNOWLEDGMENTS

This talk describes work carried out by A. O. Evwaraye, Gale I. Harris, Paul S. Hauge, B. H. Wildenthal and myself.

Part of this investigation was initiated at the Aerospace Research Laboratories under a NRC-AFL contract.

REFERENCES

1. T. T. S. Kuo and G. E. Brown, Nucl. Phys. 85, 40 (1966).
2. T. T. S. Kuo, Nucl. Phys. A103, 71 (1967).
3. G. E. Brown and T. T. S. Kuo, Nucl. Phys. A92, 481 (1967).
4. T. T. S. Kuo and G. E. Brown, Nucl. Phys. A114, 241 (1968).
5. E. C. Halbert, in *The Structure of Low-Medium Mass Nuclei*, J. P. Davidson, ed. (University of Kansas Press, Lawrence, Kansas, 1968), p. 128.
6. E. C. Halbert, J. B. McGrory, B. H. Wildenthal and S. P. Pandya, *Advances in Nuclear Physics*, *Vol. 3*, M. Baranger and E. Vogt, eds. (Plenum Press, New York, 1969).

7. M. H. Macfarlane, in *The Two-Body Force in Nuclei*, S. M. Austin and G. M. Crawley, eds. (Plenum Press, New York-London, 1972), p. 1.
8. J. P. Elliot, in *The Structure of Low-Medium Mass Nuclei*, J. P. Davidson, ed. (University of Kansas Press, Lawrence, Kansas, 1968), p. 48.
9. J. P. Elliott, A. D. Jackson, H. A. Mavromatis, E. A. Sanderson and B. Singh, Nucl. Phys. A121, 241 (1968).
10. B. S. Cooper, J. B. Seaborn and S. A. Williams, Phys. Rev. C4, 1997 (1971).
11. T. T. S. Kuo, Nucl. Phys. A90, 199 (1967).
12. J. B. French, E. C. Halbert, J. B. McGrory and S. S. M. Wong, *Advances in Nuclear Physics, Vol. 2*, M. Baranger and E. Vogt, eds. (Plenum Press, New York, 1970) p. 193.
13. S. Cohen and D. Kurath, Nucl. Phys. 73, 1 (1965).
14. P. Goldhammer, J. R. Hill and J. Nachamkin, Nucl. Phys. A106, 62 (1968); J. L. Norton and P. Goldhammer, Nucl. Phys. A165, 33 (1971).
15. D. H. Wilkinson and M. E. Mafethe, Nucl. Phys. 85, 97 (1966).
16. T. Kanellopoulos and K. Wildermuth, Nucl. Phys. 14, 349 (1960); H. Morinaga, Phys. Lett. 21, 78 (1966).
17. M. Dworzecka and H. McManus, Bull. Amer. Phys. Soc. 17, 554 (1972).
18. T. T. S. Kuo, in *The Structure of Low-Medium Mass Nuclei*, J. P. Davidson, ed. (University Press of Kansas, Lawrence, Kansas, 1972), p. 75.
19. J. D. Garrett *et al.*, Bull. Amer. Phys. Soc. 16, 511 (1971).
20. A. G. Artukh *et al.*, Nucl. Phys. A192, 170 (1972).
21. S. Goldstein and I. Talmi, Phys. Rev. 102, 589 (1956).
22. S. P. Pandya, Phys. Rev. 103, 956 (1956).
23. R. M. Freeman and A. Gallman, Nucl. Phys. A156, 305 (1970).
24. G. A. P. Engelbertink and J. W. Olness, Phys. Rev. C5, 431 (1972).
25. R. E. Segel *et al.*, Phys. Rev. Lett. 25, 1352 (1970) and to be published.
26. R. Bass and R. Wechsung, Phys. Lett. 32B, 602 (1970).
27. J. Rapaport and W. W. Buechner, Nucl. Phys. 83, 80 (1966).
28. D. Kurath and R. D. Lawson, Phys. Rev. C6, 901 (1972).
29. P. Goodė, Bull. Am. Phys. Soc. 17, 34 (1972) and to be published.
30. S. Maripuu, Nucl. Phys. A123, 357 (1967).
31. F. Erne, W. A. M. Veltman and J. A. J. M. Wintermans, Nucl. Phys. 88, 1 (1966).

DISCUSSION

DE VOIGT: In a slide concerning the dependence of several observables on the spin-orbit splitting there was a nice consistency, except for the log ft value. It is our finding that in shell model calculations for mass 27 to 29 nuclei the log ft values turned out to be the worst while sometimes M1 transitions are reasonably reproduced. Do you or anyone in the audience have an idea how to explain this, taking into account that the operators for M1 and beta transitions are strongly related?

MARIPUU: I don't understand it. I know many cases where this discrepancy exists. One example is mass 32 where log ft is very much hindered. It might just be a matter of chance, as you know the operators for beta decay and M1 are not quite the same. Maybe there is an accidental cancellation or there are some coulomb energy effects you have to account for.

DE VOIGT: Maybe Lanford has some idea because he investigated a large set of log ft values in the whole sd-shell.

LANFORD: There is one very general comment that one can make. I think it was pointed out by Joe McGrory, and it is fairly obvious, that when calculating beta decay it is very important to include spin-orbit pairs in your shell model calculations. For example, in the sd-shell complete $d_{3/2}$ and $d_{5/2}$ basis states should be included in the original shell model calculation. If this is the case, then you do very well in predicting log ft values. I don't know in the particular case here what configurations may have been included in the calculation.

MARIPUU: In this particular case (A = 13), of course, we have introduced an extra level into the data, the ground state of mass 13, and maybe the wave function of that state is not well reproduced in our calculations.

LANFORD: In the case of other decays in the sd-shell it's with the calculations in the middle of the shell where one has to truncate and limit the number of particles or holes in the $d_{5/2}$ orbit which can give very poor results. For example, the mass 32 case is most interesting because the shell model does not predict the beta decay to be inhibited and yet both the transitions to and from the P^{32} ground state are two of the weakest decays in the sd-shell.

MARIPUU: I think it is incredible that some beta decays are hindered so much.

WARBURTON: I don't think that the leaving out of one of the spin orbit pairs is likely to cause the difference between the Gamow-Teller and Ml matrix elements because this will effect them both in the same way. There is another difference and that is that the same initial and final states are not usually involved in the two cases. If you do talk about the same initial and final states then the paper of Bardin and Becker which compares the beta and gamma rates shows that you do get pretty good consistency between the two.

MARIPUU: That's right. From our calculation you cannot draw any conclusions, since we do not compare the same initial and final states.

GOLDHAMMER: I'd like to bring up the rather thorny question of convergence when one works with the Sussex matrix elements or something similar. Now we use those matrix elements and suppose we are trying to derive a reaction matrix element between two nucleons. The Sussex matrix elements do have non-vanishing values going up to the next shell and the next so than one would really have a long chain calculation if one did it rigorously. All of the various versions of Brueckner-Goldstone theory indicate that this series is rather poorly convergent to say the least; in some versions you have alternating plus and minus infinities which are highly non-convergent. In others that are more similar to the Sussex approach, Moszkowski-Scott for instance, you can factor the matrix element into long and short range components. The first term gives a very good value and I think that is very close to the Sussex matrix element. Then the rest of it, which is smaller indeed but still substantial especially in the triplet S and triplet D states, is very poorly convergent and I think this explains actually why one gets the Li^6 spectrum that you showed first. It is those triplet S and triplet D matrix elements that are poorly convergent in this case.

MARIPUU: We have tried to enhance the triplet S as has been suggested by many authors. We haven't found any improvement in the spectra so we don't believe it's that. We would rather say that what we still have an open problem as to whether second order or third order corrections might improve it.

GOLDHAMMER: But, I'd even go further than asking if it is second order or third order to the really nasty question that I wouldn't want to answer at all myself. Namely, is it convergent?

MARIPUU: Yes, I think Philip Goode would be better suited to answer that question because I have not tried to calculate any third or fourth order corrections. We have just stopped the calculation at the second order corrections hoping that it is a good approximation. It is a tremendous task calculating these third and fourth order corrections. I hope that somebody will do that. Perhaps it would improve the situation.

LEE: Back to the M1 and beta decay case. The comment was that the two operators are very similar. But they are actually not the same. The M1 is $\vec{\sigma}$ and $\vec{\ell}$, and beta decay is just $\vec{\sigma}$, is that not right? So if you have a structure such that the ℓ doesn't come in then you would expect the two things to behave very much the same. But for odd-odd nuclei, for example, ℓ is not zero in general and the M1 matrix elements do not necessarily behave the same as the beta-decay matrix elements.

MARIPUU: I don't have any comment on that.

JACKSON: In the p-shell calculations I don't think you showed a spectrum for mass 10. Is the agreement approximately the same for that case?

MARIPUU: It's slightly worse. Just about the same discrepancy exists for the $J^{\pi} = 0^{+}$, T = 1 state which is about 2 MeV too low exactly as in the Li^6 calculation. The same goes for mass 14. We didn't want to include mass 14 as it is already pretty close to the sd-shell and might not be well represented in our model space.

JACKSON: In connection with that statement about the M1 matrix elements to the two $J^{\pi} = 1^{+}$, T = 0 states: they seem to have been a problem in the past even for the shell model calculations that have fitted two-body matrix elements.

MARIPUU: We have exactly the same problems as all previous calculations, so our results for all variables are similar to the calculations of Cohen and Kurath, and of Goldhammer and collaborators.

GOODE: I would like to make a comment about convergence. From calculations that I have done with Dan Koltun, it appears that the expansion for core polarization may be well behaved on the JT-average through third order, but fourth order does not seem to behave well and in fact may be larger than second order.

LEE: Recently Ellis and Osnes made a calculation of core polarization and instead of using harmonic oscillator wave functions they did a Hartree-Fock calculation and used Hartree-Fock wave functions to calculate the core polarization. What they found is that the convergence is very good. It does not diverge. The reason is quite simple. If you want to calculate core polarization to the highest possible order, the simplest thing to do is RPA. It is not the best thing to do but is the simplest thing to do and very often it will diverge. However if you do a Hartree-Fock calculation and it is stable it is guaranteed that RPA will not diverge. In fact the RPA calculation would be very much like the lowest order calculation. It converges very well.

MARIPUU: In which nuclei did they do this calculation?

LEE: I can't remember. Goode, do you remember?

GOODE: Well, I think they did it in oxygen or maybe in calcium. But I don't think it really matters too much. What you say is absolutely right but there are any number of classes of cancelling diagrams. For example, Kirson and Zamick have shown that there are other diagrams which cancel the RPA. The way to do the perturbation theory is to take everything in a particular order, and then if you do a JT-average calculation including the effects that Osnes and Ellis did the expansion would be well behaved through third order and probably would not behave in fourth order. Of course, you can do lots of calculations and take your own favorite class of diagrams. As I said there are other classes of diagrams that don't have Hartree-Fock insertions which cancel the RPA.

LEE: Right. My comment would be this: you can do perturbation calculations and depending on your starting point it may or may not converge. There may exist a set of single particle wave functions such that high order perturbation will converge. But if you don't hit on that set of wave functions it will not converge.

GOLDHAMMER: I guess the question of convergence is about the nastiest question you can ask a theoretical physicist but let us face it, when you try to do realistic calculations in this business now, I think it's come to that question. The result that Goode quoted a moment ago seems very logical to me for the following reasons: suppose you do the first order term and it's good; then as you do the next couple of orders (I don't know how many, I haven't done a calculation--he says fourth order--but say as you do the next couple of orders) the fact that the energy denominator is rapidly increasing helps you out and you get good convergence. Eventually you reach the point where those energy denominators just level out and they're not doing you much good any more. You've got to depend on convergence of the potential and I think at that point one is in extreme difficulty. Is that what does it? Or can you put your finger on what does it? Is it multiplication of diagrams? There's also the fact of course that as you go to higher order you get more possibilities.

GOODE: Well, that's right. It's a competition between the number of diagrams growing and $(1\hbar\omega)^n$ growing at the same time. I thought that maybe $(2\hbar\omega)^n$ would win but (afterwards, of course, there's no reason to think that) it appears that it doesn't even though it does through third order.

MALIK: Well there's another nasty question that can be posed. This one is: can the G-matrix of Kuo-Brown honestly be connected to the free G-matrix because the G-matrix equation defined for a bound system need not be the same one as it is in the case of free scattering? The theoretical connection between the two is by no means clear cut. In the Kuo-Brown calculation the G-matrix has again a projection operator. Its definition in case of a bound system also differs from the one used for a free non-interacting gas or the finite nucleus. When we are talking about corrections of the order of 200 keV, these questions are legitimate. Regarding the question of core polarization, I would like to point out that a study of the quadrupole moment may provide some clue. I'm not quite sure, did you mention anything about the quadrupole moments of Li^6 and Li^7?

MARIPUU: No, we haven't calculated any quadrupole moments.

MALIK: This is a very interesting question because in Li^6 the quadrupole moment is much less than the shell model single particle estimates, whereas Li^7 has a much larger one. My last comment is that we calculated some time back Na^{22} using the Coriolis coupling model and found that the ground states can be obtained properly which was not true in your case. We used Kuo-Brown matrix elements between the last two odd nucleons but we had the same problem, that is, the $T = 0$ and $T = 1$ splitting did not come out very well.

MARIPUU: Well, to partly answer your question we don't claim that the unperturbed matrix elements we deduce are similar to the bare reaction matrix elements of Kuo-Brown. Our unperturbed matrix elements should be suitable for applying the perturbation theory. Indeed, we would avoid some of the double counting that exists in the Kuo-Brown calculation but again I don't want to go into the details of the perturbative corrections. It's quite complicated.

GOODE: The matrix elements of Barrett, Hewitt, and McCarthy don't have the Pauli principle problem but they behave almost exactly the same in convergence through third order as do Kuo-Brown's, irrespective of the starting energy if you consider the ratio of second to third order.

ENDT: As the present speaker is perhaps the most pure proponent of the many particle shell model calculations, I have a question in general about this sort of calculation. One way to take the truncation of the model into account is of course to do the sort of perturbation theory calculation you've been telling about. Another would be to enlarge the shell model space you are working in to put the perturbation as far off as you possibly can. If one for instance would like to take the three subshells in the sd-shell then the hardest part would of course be in the middle at Si^{28}; for instance, to do the 3^+ T = 1 states requires the diagonalization of a matrix of (6,706) x (6,706) which is a hard problem. Now very recently two Glasgow people, Whitehead and Watt, have claimed that they can do that in a slightly different way. They use a different sort of basis function; I think they take the z component of angular momentum as a parameter. Has anybody gone into this? What can we really do with this method? Would you be able to comment on that?

MARIPUU: I don't think I am quite able to answer that. It's certainly interesting, what they are doing, but except for a brief report I have not heard of further applications with their method.

ENDT: They even claim they do it on a rather small computer. Why can they do what nobody else can do?

II.B. ROLE OF NUCLEAR COEXISTENCE IN SHELL MODEL SPECTRAL DIFFERENCE CALCULATIONS*

Philip Goode
Department of Physics, Rutgers University
New Brunswick, New Jersey

I. INTRODUCTION

There are many examples of anomalous behavior of nuclei in the mass-40 region. It is the purpose of this paper to investigate some of these anomalies in terms of difference calculations. In the course of the investigation, it is also our purpose to illustrate the interplay of the valence effective interaction and space truncation as well as some convergence properties of the valence effective interaction and its effective particle rank.

One of the oldest known anomalies in the mass-40 region is that of the Cl^{38}/K^{40} pair. Within the framework of the pure shell model, Cl^{38} is a proton particle in the $1d_{3/2}$-shell and a neutron particle in the $1f_{7/2}$-shell, while K^{40} is a proton hole in the $1d_{3/2}$-shell and a neutron particle in the $1f_{7/2}$-shell. Pandya[1] showed that the application of this pure shell model point of view was sufficient to show that

$$E_J(Cl^{38}) = -\sum_I (2I + 1) \begin{Bmatrix} \frac{3}{2} & \frac{7}{2} & J \\ \frac{3}{2} & \frac{7}{2} & I \end{Bmatrix} E_I(K^{40}) \equiv T_J^I\left(E_I(K^{40})\right)$$

where the quantity in brackets is a Racah coefficient, and where $E_J(Cl^{38})$ is the interaction energy of the spin J state in Cl^{38} and $E_I(K^{40})$ is the interaction energy of the spin I state in the particle-hole nucleus, K^{40}. The interaction energies in Cl^{38} and K^{40} are defined pictorially in Figure 1. The interaction energy in these cases is the effective energy of interaction between an $1f_{7/2}$-shell nucleon and $1d_{3/2}$-shell nucleon, the $1d_{3/2}$-shell nucleon being a particle in the case of Cl^{38} and a hole in the case of K^{40}. Of course, when the theorem is applied to the experimental spectra, the Cl^{38} spectrum determined from that of K^{40} differs very slightly from that of the true Cl^{38}. This kind of comparison introduces us to the notion of differences. The experimental spectral discrepancies or differences are amazingly small. In fact,

---

*Supported in part by the National Science Foundation.

attempts to calculate them in perturbation theory fail because the calculated results are so much larger than those determined from experiment. This does not mean that the shell model holds very well (the condition of shell purity that Pandya used to prove his theorem is a sufficient one, but not necessary), in fact we will see that the contributions to spectral differences in perturbation type calculation are cancelled by those due to effects of nuclear coexistence.

To be precise, we define the energy difference, E_J^1 as

$$E_J(Cl^{38}) - T_J^I\left(E_I(K^{40})\right) = \Delta E_J^1$$

where the interaction energies involved are identified with $d_{3/2} - f_{7/2}$ configurations (in the Cl^{38} case), and with

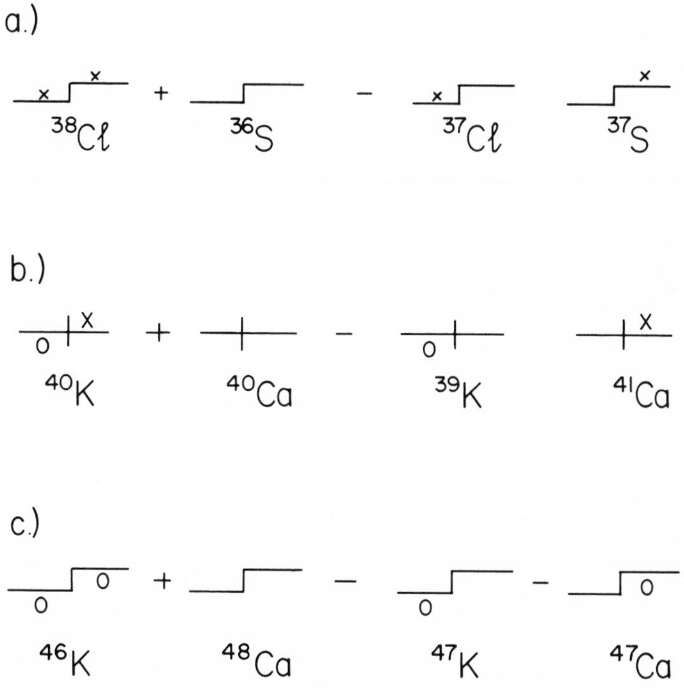

Fig.1. (a) represents the Cl^{38} $f_{7/2}$-$d_{3/2}$ interaction energy, (b) represents the K^{40} $f_{7/2}$-$d_{3/2}^{-1}$ enteraction energy, (c) represents the K^{46} $f_{7/2}^{-1}$-$d_{3/2}^{-1}$ interaction energy. All are in the pure model space.

$d_{3/2}^{-1}\ f_{7/2}$ configurations (in the K^{40} case). K^{46} is the third member of the $d_{3/2} - f_{7/2}$ triptych--being identified with the $d_{3/2}^{-1} - f_{7/2}^{-1}$ configuration. We therefore define ΔE_J^2 and ΔE_J^3 as

$$E_J(Cl^{38}) - E_J(K^{46}) = \Delta E_J^2$$

and

$$T_J^I\big(E_I(K^{40})\big) - E_J(K^{46}) = \Delta E_J^3$$

respectively. The results of the difference calculations will all be presented in terms of these quantities. In addition the results will be presented in multipole form. The Cl^{38}, K^{40}, and K^{46} interaction energies can be expanded in their multipole forms as follows:

$$\alpha_\lambda(Cl^{38}) = (-)^{\lambda+1}\sqrt{\frac{[\lambda]}{32}}\sum_J [J]\,(-)^J \begin{Bmatrix} \frac{7}{2} & \frac{3}{2} & J \\ \frac{3}{2} & \frac{7}{2} & \lambda \end{Bmatrix} E_J(Cl^{38}),$$

$$\alpha_\lambda(K^{40}) = \sqrt{\frac{[\lambda]}{32}}\sum_I [I]\,(-)^I \begin{Bmatrix} \frac{7}{2} & \frac{3}{2} & I \\ \frac{3}{2} & \frac{7}{2} & \lambda \end{Bmatrix} E_I(K^{40}),$$

and

$$\alpha_\lambda(K^{46}) = (-)^{\lambda+1}\sqrt{\frac{[\lambda]}{32}}\sum_J [J]\,(-)^J \begin{Bmatrix} \frac{7}{2} & \frac{3}{2} & J \\ \frac{3}{2} & \frac{7}{2} & \lambda \end{Bmatrix} E_J(K^{46}),$$

where λ is the multipole rank and the α_λ's are multipole coefficients. From these definitions we can define the multipole difference quantities $\Delta\alpha_\lambda^1$, $\Delta\alpha_\lambda^2$, and $\Delta\alpha_\lambda^3$ as follows:

$$\alpha_\lambda(Cl^{38}) - \alpha_\lambda(K^{40}) = \Delta\alpha_\lambda^1,$$

$$\alpha_\lambda(Cl^{38}) - \alpha_\lambda(K^{46}) = \Delta\alpha_\lambda^2,$$

$$\alpha_\lambda(K^{40}) - \alpha_\lambda(K^{46}) = \Delta\alpha_\lambda^3.$$

We analyze the results in terms of multipole discrepancies because the largest violations are in the monopole, $\Delta\alpha_0$, which is the discrepancy between the centroids of the spectra being compared. For example, in the Cl^{38}/K^{40} system, the experimental $\Delta\alpha_0$ is as large as all the $\Delta\alpha_{\lambda\neq 0}$ terms combined. To say this another way, if ΔE_J were J-independent, the entire spectral discrepancy would be in the monopole. The fact that the large experimental uncertainties come in the determination of the experimental monopole (*e. g.* the mass of S^{37}) is another reason for the analysis of the differences in terms of multipoles. The point is that all the ΔE's will carry the same uncertainty, but in terms of multipole differences all appreciable experimental uncertainty is in the monopole.

In the process of calculating the various ΔE_J's and $\Delta\alpha_\lambda$'s we will be forced to use standard approximations in nuclear physics and thereby investigate their usefulness and applicability in various cases. In particular, in the course of the difference calculations we will investigate the interplay of the valence effective interaction and space truncation, the relative roles of "low and high orders of perturbation theory" in nuclei. In addition, we will investigate the feasibility of substituting higher particle rank forces for configuration mixing and discuss additional applications of the difference calculations.

II. MODELS

In this paper we use three different models. In the first it is assumed that the strict shell model is nearly valid and therefore configuration mixing effects are due to simple excitations which have structure that is close to that of the relevant pure shell model state. The second model is at the opposite extreme. In this second model it is assumed that core excitations (multi-particle, multi-hole states) play a significant role in the difference calculations. Thus, we are assuming in this model that nuclear coexistence plays a crucial role. The third model is intermediate to the first two. In this last model we incorporated the effects of the first two models to try to obtain an intermediate model which is presumably more complete than either of the first two.

The examination of difference calculations in terms of these three fairly distinct models is necessitated by the fact that a complete shell model calculation which would contain all of the above mentioned aspects plus other effects, is simply too cumbersome to be dealt with. We are therefore forced, in difference calculations as well as other shell model calculations, to search for simpler models which we hope can be used to properly account for the effects of space truncation.

A. The Ordinary Shell Model

The first model we used was an ordinary shell model (OSM) by which we mean that corrections to the pure shell model state are fairly well treated in second order perturbation theory in which low-lying configurations admix. The meaning of this is quite simple and explicit, but perhaps is best illustrated by an example. In terms of the pure shell model, the first four states in K^{40} are said to belong to the $d_{3/2}^{-1}$ - $f_{7/2}$ multiplet with respect to a pure Ca^{40} core. The simplest corrections to the energies involved would be obtained in a second order perturbation theory calculation in which the $f_{7/2}$-neutron would instead be a neutron which is allowed to roam over the entire fp-shell and the $d_{3/2}$-proton hole would instead be a proton hole allowed to roam over the entire sd-shell. Two separate truncations are involved here, the first being that all active nucleons are confined to two major shells, and the second being that only single nucleon excitations are allowed. Both these truncations are consistent with the assumption that the shell model is valid. Nowadays, doing a second order perturbation theory calculation is outdated and instead we diagonalize shell model matrices in the OSM calculations.

Since we are interested in difference calculations, we need a Cl^{38} interaction energy and a K^{46} interaction energy as well as the aforementioned K^{40} interaction energy. Once we have calculated these three sets of interaction energies, we can proceed to the calculation of differences using the quantities defined in the previous section. In the OSM calculations, we need to specify single-particle energies, two-body matrix elements, and the configurations used. Since this is crucial and, at least in one case, a bit tricky, we will describe in detail how the calculation of the OSM interaction energies are set up. But before beginning this, we remark that our basic approach in these OSM calculations is to choose the single-particle energies from experiment, where possible, use a realistic residual two-body interaction, and choose the simplest configurations. In all cases for $Cl^{38}/K^{40}/K^{46}$, the realistic interaction is chosen to be the bare Kuo-Brown interaction as calculated by Kuo.[2]

Cl^{38}

In Cl^{38}, the $f_{7/2}$ - $d_{3/2}$ interaction energies in terms of the simplest configurations can be seen in Figure 1a. Thus, in the calculation of Cl^{38} itself, we start with a S^{36}

core and allow the valence neutron to roam over the $1f_{7/2}$ - $2p_{3/2}$-$1f_{5/2}$ single j shells. Similarly, we simultaneously allow one proton to excite from the lower sd-shell ($1d_{5/2}$, $2s_{1/2}$) into the $1d_{3/2}$-shell. The relevant proton single nucleon energies can easily be determined in the usual manner from the mass tables. This gives us all the relevant proton energies except that of $\varepsilon_{2s_{1/2}} - \varepsilon_{1d_{5/2}}$ for which we would need the experimental spectrum of P^{35}. Since this is not available we assume it to be 4 MeV and remark that our difference calculation results are quite insensitive to fairly wide variations of this number. This insensitivity is due to the fact that the $2s_{1/2}$ and $1d_{3/2}$ shells are split by 4.8 MeV which means that the perturbation calculation involving the $d_{5/2}$ orbit would have a large denominator.

The neutron excitations are quite another matter. That is, it is crucial to the difference calculations to know the spectrum of S^{37} so that neutron single-particle energies are known. Unfortunately the spectrum of S^{37} is unknown. At this point previous authors have taken the neutron single particle energies from the experimental spectrum of Ca^{41}. This approximation, it turns out, is quite poor. Within the last year the members of the pd- as well as fd-multiplets have been identified in Cl^{38} and K^{40}. If we knew the $\varepsilon_{p_{3/2}} - \varepsilon_{f_{7/2}}$ splitting in S^{37} we could determine the interaction centroids of Cl^{38} and K^{40} for both the pd- and fd-multiplets. As can be seen by examining Table I, the magnitude of the fd experimental centroids of Cl^{38} and K^{40} are the same to within experimental errors. If we *assume* this to be the case for the pd-multiplets, then we can deduce $\varepsilon_{p_{3/2}}$ - $\varepsilon_{f_{7/2}}$ in S^{37}. The validity of this assumption will be demonstrated in Section VI. At any rate, with this one assumption we deduce that $\varepsilon_{p_{3/2}} - \varepsilon_{f_{7/2}} = 0.36$ MeV, which is less than one-fifth the value of this splitting in Ca^{41}. Therefore the new single-particle energy will yield a profound effect on the difference calculations.

Using $\varepsilon_{p_{3/2}} - _{f_{7/2}} = 0.36$ MeV for S^{37} is consistent with using $\varepsilon_{p_{3/2}} - \varepsilon_{f_{7/2}} = 2.00$ MeV for Ca^{41}. In fact the largest source of error in the S^{37} single-particle energy is the uncertainty of the same single-particle energy in Ca^{41}. These uncertainties do not bother us too much for two reasons - one theoretical and one empirical. The theoretical reason is that within the framework of OSM the first excited $3/2^-$ state in Ca^{41} is the $p_{3/2}$ single-particle state. The empirical reason is that the difference calculations are

TABLE I

The experimental results in Cl^{38} and K^{40} expressed in terms of energy and multipole differences (in MeV).

(a) Energies

J	$E_J(Cl^{38})$	$E_J(K^{40})$	$T_J^I(E_I(K^{40}))$	ΔE_J^{exp}
2	-1.80 ± 0.06	+1.36	-1.77	-0.03 ± 0.06
3	-1.04 ± 0.06	+0.59	-1.03	-0.01 ± 0.06
4	-0.49 ± 0.06	+0.56	-0.45	-0.04 ± 0.06
5	-1.12 ± 0.06	+1.45	-1.081	-0.04 ± 0.06

(b) Multipoles

λ	$\alpha_\lambda(Cl^{38})$	$\alpha_\lambda(K^{40})$	$\Delta\alpha_\lambda^{exp}$
0	-1.03 ± 0.06	-1.00	-0.03 ± 0.06
1	-0.15	-0.17	0.02
2	-0.39	-0.39	0.00
3	0.04	+0.05	-0.01

fairly insensitive to about an 0.1 MeV variation in the single-particle energy which is more or less the size of the expected uncertainty. Now we need one further single-particle energy, $\varepsilon_{f_{5/2}} - \varepsilon_{f_{7/2}}$ in S^{37}. Since the $p_{3/2}$ single-particle state is closer to ground in S^{37} than Ca^{41} (and since the $df_{5/2}$-multiplet is unknown) we assume that the same is true for the $f_{5/2}$ shell. We thus assume that $\varepsilon_{f_{5/2}} - \varepsilon_{f_{7/2}} = 2.00$ MeV in S^{37}. We remark that the $p_{1/2}$ neutron excitations are deleted for technical reasons. Other calculations have shown that its effect is small in difference calculations. A complete list of the single-particle energies appears in Table II.

TABLE II

The single particle energies (in MeV) used in the OSM calculation.

	K^{46}	K^{40}	Cl^{38}
$\varepsilon_{1f_{5/2}}$	8.40	6.00	2.00
$\varepsilon_{1p_{3/2}}$	4.80	2.00	0.36
$\varepsilon_{1f_{7/2}}$	0.00	0.00	0.00
$\varepsilon_{1d_{3/2}}$	0.00	0.00	0.00
$\varepsilon_{2s_{1/2}}$	+0.38	-2.50	-4.80
$\varepsilon_{1d_{5/2}}$	-1.72	-5.40	-8.80

K^{40}

In K^{40}, the particle-hole $f_{7/2} - d_{3/2}^{-1}$ interaction energies are depicted in Figure 1b. The depiction is in terms of the pure shell model configurations. In the OSM calculation of K^{40} the proton hole is allowed to roam over the entire sd-shell with the single hole energies taken from the experimental spectrum of K^{39}. Similarly, the valence neutron in K^{40} is allowed to roam over the $1f_{7/2} - 2p_{3/2} - 1f_{5/2}$ single j-shells and the appropriate single-particle energies are taken from the experimental spectrum of Ca^{41}. We emphasize that the calculation of K^{40} interaction energies is performed with respect to a Ca^{40} core. The single-particle energies used in this calculation are depicted in Table II.

K^{46}

The pure shell model configurations for K^{46} $f_{7/2}^{-1} - d_{3/2}^{-1}$ interaction energies are depicted in Figure 1c. In the K^{46} calculations we are dealing with a Ca^{48} core. We again allow the proton single hole state to roam over the sd-shell and therefore take the single hole energies from the experimental spectrum of K^{47}. The ground state of K^{47} is $1/2^+$ with the first excited state being a $3/2^+$ at 0.38 MeV. Thus, within the framework of OSM it can be said that the $2s_{1/2}$ and $1d_{3/2}$ single hole states have crossed, that is, at K^{39} their difference is about 2.5 MeV, at K^{47} their difference is about -0.38 MeV. Therefore, we will expect a severe fragmentation of the J = 3 and J = 4 states in K^{46} (with $f_{7/2}^{-1}$ $s_{1/2}^{-1}$ basis states being strongly mixed with $f_{7/2}^{-1}$ $d_{3/2}^{-1}$ states). The location of a $1d_{5/2}$ single nucleon state in K^{47} is arbitrarily set by defining the separation of the $2s_{1/2}$- and $1d_{5/2}$-states to be 2.0 MeV. A decrease in this splitting in going from K^{39} to K^{47} is also expected.[5]

Simultaneously, with the proton hole excitation we allow a single neutron to excite from the $1f_{7/2}$-shell to the $2p_{3/2}$ and $1f_{5/2}$ single-j shells. The appropriate single particle energies can be determined from the mass tables and the spectrum of Ca^{49}. A complete list of the single-nucleon energies used in the K^{46} OSM calculations appears in Table II.

B. The fd-Model

The fact of coexistence of "spherical" and "deformed" states in light nuclei has been known for years. Perhaps the best known example of this is the case of O^{16} in which the ground state is, roughly speaking, a 0p-0h state whereas the first excited state, also roughly speaking, is a 4p-4h state. It is clear that the presence of coexistence would never be properly treated in an OSM-type calculation simply because the unperturbed energies of the configurations involved would be too large. Thus, we are led to approach the difference calculations from a completely distinct point of view. That is to say, a point of view in which the push of the core excited states on the spherical ones is treated as the origin of the spectral differences in the "spherical" states.

With the success of the Zuker, Buck and McGrory[6] sd-shell multi-particle multi-hole calculation in mind, we proceed to formulate an fd model in which the valence space is the $1f_{7/2}$ and $1d_{3/2}$ single-j shells. The push of the core excited (by which we mean multi-particle, multi-hole) states on the spherical (by which we mean pure shell model) states gives rise to whatever spectral discrepancies exist between the $Cl^{38}/K^{40}/K^{46}$ interaction energies of the so called spherical states.

All the fd-model calculations are performed in a particle basis in which all the valence particles are taken with respect to a S^{32} core. In this calculation we take the value of $\varepsilon_{f_{7/2}} - \varepsilon_{d_{3/2}}$ from the experimental spectrum of S^{33} ($\varepsilon_{f_{7/2}} - \varepsilon_{d_{3/2}}$ = 2.94 MeV). The most difficult problem is the selection of the residual interaction to be used. But first it is perhaps worthwhile to eliminate a possible source of confusion in these fd-model calculations. To wit, since the fd-model calculations are performed in a particle basis all of the effective interactions used are calculated in terms of a particle-particle interaction even though we may discuss the results in terms of a particle-hole interaction. For example, we discuss K^{40} as a particle-hole nucleus even though for convenience sake we calculated K^{40} in a particle basis. This is purely a technical matter of which the results of the calculations are independent.

Insofar as the actual calculation of the effective interaction is concerned, instead of choosing it in some *ad hoc* way, we directly calculate it in a series of renormalizations which are described in detail in Reference 7. Therefore rather than going into a detailed description of the calculation, we will instead briefly describe how the matrix elements $\langle f^2|v|f^2\rangle$ are calculated.

The starting point is the renormalized Kuo-Brown[2] two-body matrix elements for a Ca^{40} core--the tacit assumption here is that these matrix elements include all effects $2\hbar\omega$ or more in energy away from the valence space (this assumption is certainly open to question). The next renormalization is to account for the fact that we are approximating the pf-shell by the $f_{7/2}$-shell. To take care of this approximation we include $0\hbar\omega$ ladders by performing a Sc^{42} diagonalization $\left((pf)^2\right)$ and associating $\langle f^2_{JT}|v|f^2_{JT}\rangle$ with the state that has most of the f^2 strength ($\approx$90%) for the (JT)-state in question. Lastly, we must renormalize to account for the fact that we are using a S^{32} core. This renormalization is performed by allowing the f^2 matrix element deduced from the Sc^{42} diagonalization to mix with states which have a single nucleon excited out of the S^{32} core (plus the two additional valence nucleons). None of these renormalizations will overcount because they have different intermediate states, and will, in a low order perturbative sense, fold into the effective interaction the excitations which are outside the $d_{3/2}$-$f_{7/2}$ model space. Of course one can argue that many other renormalizations are possible or even that the renormalizations here included could be done in a different order--either way the calculated effective interaction will be different. We do not argue against this point, in fact we performed several different renormalizations and discovered that the calculated positions of the core excited states with respect to the spherical ones is fairly sensitive to wide variations in renormalizations. On the other hand, recall that we are not primarily interested in the position of core excited states, but rather their effect on the spherical states as seen in difference calculations. By calculating differences with several effective interactions, we determined that the position of the core excited states is of secondary importance in the calculation of spectral discrepancies. The reason being that an "error" would manifest itself in both sets of calculated spectral interaction energies being considered. Therefore, the "error" would tend to be cancelled in the difference calculation. A particular illustration of this was revealed by choosing the $<f^2|v|f^2>$matrix elements from the spectrum of Sc^{42} (clearly this results in pure double counting because the 4p-2h configurations appear both in the interaction and the matrix). In this example the $<f^2|v|f^2>$ two body matrix elements differed markedly from those we used (see Table III), yet the trend of the Cl^{38}/K^{40} spectral differences was the same as that when the calculated interaction was used. Parenthetically, we remark that the fd-model interaction we used positioned the core excited states fairly well in $A \simeq 40$ nuclei. We also remark that the fd-model difference calculations were truncated at the two $1d_{3/2}$-shell holes beyond the pure model state. The configurations with more than two holes are important in some cases, in positioning the core excited states, but these effects are of secondary consequence in the calculation of spectral differences. This result is consistent with the insensitivity of the difference calculation results to fairly wide variations in the effective interaction.

C. The Hybrid Model

It should not be surprising that the effects we are trying to describe within the framework of the OSM and fd-models both play a role in difference calculations. But it is clear that the two models are at opposite extremes, and therefore are, at the very least, collectively esoterically unsatisfactory. Therefore, we require a hybrid model which includes aspects of both models. As we have previously mentioned, the most straight forward approach would appear to be the diagonalization of huge matrices, but this approach is untenable because of the incredible size of the matrices involved. We are therefore led to a simpler model. Namely, we use the results of one calculation to modify the effective interaction in the second calculation.

Explicitly, we use the OSM results for $Cl^{38}/K^{40}/K^{46}$ interaction energies to modify the $<d_\pi f_\upsilon|v|d_\pi f_\upsilon>$ two-body matrix elements which are part of the input to the corresponding fd-model (now hybrid) calculations. Again recall that the modified two body matrix elements which appear in Table IV are presented, for the technical reasons previously discussed, in particle-particle form. This being the case even though, for example, it is entirely appropriate to interpret the K^{46} interaction energies deduced from hybrid calculation as being hole-hole interaction energies. The OSM results are used to renormalize the fd-model interaction because the OSM states which are interpreted as the members of the $f_{7/2}$-$d_{3/2}$ multiplet are fairly pure (more than 80% $f_{7/2}$-$d_{3/2}$ except for the J = 3,4 states in K^{46} for which the $f_{7/2}$-$d_{3/2}$ strength is fragmented into two states--for each of these two states the interaction energy is taken to be the centroid

TABLE III

The fd-model interaction in JT where $d \equiv 1d_{3/2}$ and $f \equiv 1f_{7/2}$. The energies are presented in MeV.

Matrix Element	Energy (MeV)
$\langle d^2 01 \| v \| d^2 01 \rangle$	-1.06
$\langle d^2 10 \| v \| d^2 10 \rangle$	-0.38
$\langle d^2 21 \| v \| d^2 21 \rangle$	0.39
$\langle d^2 30 \| v \| d^2 30 \rangle$	-1.73
$\langle f^2 01 \| v \| f^2 01 \rangle$	-2.79
$\langle f^2 10 \| v \| f^2 10 \rangle$	-2.08
$\langle f^2 21 \| v \| f^2 21 \rangle$	-1.00
$\langle f^2 30 \| v \| f^2 30 \rangle$	-1.22
$\langle f^2 41 \| v \| f^2 41 \rangle$	-0.16
$\langle f^2 50 \| v \| f^2 50 \rangle$	-1.24
$\langle f^2 61 \| v \| f^2 61 \rangle$	0.22
$\langle f^2 70 \| v \| f^2 70 \rangle$	-2.08
$\langle d^2 01 \| v \| f^2 01 \rangle$	2.62
$\langle d^2 10 \| v \| f^2 10 \rangle$	-0.45
$\langle d^2 21 \| v \| f^2 21 \rangle$	0.45
$\langle d^2 30 \| v \| f^2 30 \rangle$	-0.24
$\langle df\ 20 \| v \| df\ 20 \rangle$	-2.93
$\langle df\ 21 \| v \| df\ 21 \rangle$	0.19
$\langle df\ 30 \| v \| df\ 30 \rangle$	-1.40
$\langle df\ 31 \| v \| df\ 31 \rangle$	0.08
$\langle df\ 40 \| v \| df\ 40 \rangle$	-0.86
$\langle df\ 41 \| v \| df\ 41 \rangle$	0.49
$\langle df\ 50 \| v \| df\ 50 \rangle$	-1.37
$\langle df\ 51 \| v \| df\ 51 \rangle$	-0.56

TABLE IV

The $d_{3/2}$-proton $f_{7/2}$-neutron effective interactions (in MeV) used in the various hybrid model calculations.

	V_{eff} with no OSM effects	V_{eff} for mass-38	V_{eff} for mass-40	V_{eff} for mass-46
$\langle d^{\pi}f^{\nu}J=2\|v\|d^{\pi}f^{\nu}J=2\rangle$	-1.37	-1.37	-1.24	-1.74
$\langle d^{\pi}f^{\nu}J=3\|v\|d^{\pi}f^{\nu}J=3\rangle$	-0.66	-0.68	-0.66	-0.69
$\langle d^{\pi}f^{\nu}J=4\|v\|d^{\pi}f^{\nu}J=4\rangle$	-0.19	-0.24	-0.07	-0.21
$\langle d^{\pi}f^{\nu}J=5\|v\|d^{\pi}f^{\nu}J=5\rangle$	-0.95	-0.82	-0.62	-0.91

of the $f_{7/2}$-$d_{3/2}$ strength of the two states involved in each case). Therefore, the particle-particle effective interaction is different for the Cl^{38} set than it is for the K^{40} set, *etc*. This means that the effective interactions involved in the hybrid calculations do not obey the Racah transformation. To say it another way, in the fd-model there are effects outside the model space which do not obey the Pandya transform, thus, it is appropriate that the hybrid effective interactions include such effects. Therefore, in the limit of no core excitations the hybrid model will result in the prediction of OSM differences. When the core excitations are allowed to admix in the hybrid model the effects of both models are included. Perhaps it is worthwhile restating the obvious--because of space truncation we are forced to employ different effective interactions for nuclei having a very similar number of nucleons.

III. RESULTS

A. OSM Results

In Table V we present the results of the $Cl^{38}/K^{40}/K^{46}$ difference calculations. In the difference calculations in which one interaction energy corresponds to K^{46}, we cannot make a comparison between experiment and theory. Even though Sherr and Daehnick[8] have determined the spectrum of K^{46}, this is not sufficient to determine $d_{3/2}^{-1}$ $f_{7/2}^{-1}$ interaction energies because the lowest lying J = 3 and J = 4 states in K^{46} have considerable $2s_{1/2}$-hole impurities. Thus to determine experi-

TABLE V

The OSM results for the Cl^{38}/K^{40}, Cl^{38}/K^{46}, and K^{40}/K^{46} differences presented in MeV. The superscripts one, two and three, on the difference quantities correspond to Cl^{38}/K^{40}, Cl^{38}/K^{46}, and K^{40}/K^{46} differences, respectively.

(a) Energies

J	$E_J(Cl^{38})$	$E_J(K^{40})$	$T_J^I(E_I(K^{40}))$	$E_J(K^{46})$	ΔE_J^1	ΔE_J^2	ΔE_J^3
2	-1.98	1.73	-1.86	-2.36	-0.12	0.38	0.50
3	-1.12	0.64	-1.11	-1.14	-0.01	0.02	0.03
4	-0.59	0.61	-0.42	-0.56	-0.17	-0.03	0.14
5	-1.52	1.53	-1.33	-1.62	-0.19	0.10	0.29

(b) Multipoles

λ	$\alpha_\lambda(Cl^{38})$	$\alpha_\lambda(K^{40})$	$\alpha_\lambda(K^{46})$	$\Delta\alpha_\lambda^1$	$\Delta\alpha_\lambda^2$	$\Delta\alpha_\lambda^3$
0	-1.24	-1.11	-1.33	-0.13	0.09	0.23
1	-0.05	-0.10	-0.10	0.05	0.05	0.00
2	-0.48	-0.47	-0.60	-0.01	0.12	0.13
3	0.04	0.09	0.01	-0.05	0.03	0.08

mental interaction energies we need the relative $2s_{1/2}$- and $1d_{3/2}$-hole spectroscopic strengths. Lacking this evidence we can only present the OSM results for differences involving K^{46}. On the other hand, there is no such incompleteness in the Cl^{38}/K^{40} difference calculations as compared to experiment.

The OSM calculation for the Cl^{38}/K^{40} spectral differences yields unsatisfactory results for the difference quantities ΔE_J^1 and $\Delta\alpha_\lambda^1$ (see Table V). Explicitly, the ΔE_J's and $\Delta\alpha_\lambda$'s tend to be too large when compared to experiment. This poor agreement holds for other OSM type calculations. For example, when we deduce $\Delta\alpha_\lambda^1 > 0$ from the Cl^{38} and K^{40} calculation of Maripuu[9] we see (in Table VI) that the calculated multipole discrepancies are too large compared to experiment--the typical OSM result. To get a feeling for the residual discrepancies between experimental and OSM difference calculation results we concentrate on the monopole discrepancy. The fact that much of the experimental $\Delta\alpha_\lambda$ is concentrated in the $\lambda = 0$ term is seen in the fact that the experimental ΔE_J's are fairly constant. That is, the monopole is the [J]-weighted average of the energy levels; therefore, if the ΔE_J's are constant,

$$\Delta E_J = a$$

then there is only a monopole discrepancy,

$$\Delta\alpha_\lambda = \delta_{\lambda,0}a$$

The unsatisfactory results for the OSM are somewhat dependent on the choice of single-particle energies and two-body matrix elements. Still, the collective OSM results are reflective of the incomplete nature of a second order perturbation theory approach to the problem.

To further clarify the distinctions between the OSM type calculation and an fd-model calculation and to emphasize the need of the latter, we will consider the contribution of second order perturbation theory results to ΔE_J^1 and $\Delta\alpha_\lambda^1$. West and Koltun[10] have shown that the primary contribution, in second order perturbation theory, to the Cl^{38}/K^{40} monopole discrepancies can be obtained from a calculation of the graphs shown in Figures 2a and 2b. We will confine our discussion to the monopole because in the Cl^{38}/K^{40} system, it is the largest discrepancy. We remark that the role of the excited single-nucleon state shown in Figure 2 increases the closer it lies, energetically, to the model space, thus simplifying the calculations of West and Koltun.[10]

The expression for the energy of the term represented by Figure 2a is

$$E_J(jk) = \sum_{\alpha} \frac{<jkJ|v|\alpha kJ>^2}{-(\varepsilon_\alpha - \varepsilon_j)}$$

and that for Figure 2b is

$$E_I(jk^{-1}) = \sum_{\alpha} \frac{<jk^{-1}I|v|\alpha k^{-1}I>^2}{-(\varepsilon_\alpha - \varepsilon_j)}$$

where j and k are neutron and proton labels, respectively. We add that

$$\Delta E_J^1 = E_J(jk) - T_J^I(E(jk^{-1})).$$

For the sake of simplicity, discussion and illustration we will assume

$$<jkJ|v|\alpha kJ> = b$$

and

$$<jk^{-1}I|v|\alpha k^{-1}I> = c.$$

By this restriction, we fix ΔE_J to be a constant (J-independent) and therefore $\Delta\alpha_\lambda = 0$ if $\lambda \neq 0$. Therefore, we cannot discuss the discrepancies of higher multipolarity within

TABLE VI

The shell model discrepancies calculated by Maripuu,[7] compared to $f_{7/2}$-$d_{3/2}$ Cl^{38}/K^{40} experimental discrepancies (in MeV).

λ	$\Delta\alpha_\lambda^{Maripuu}$	$\Delta\alpha_\lambda^{exp}$
0	-	-0.03 ± 0.06
1	0.04	0.02
2	-0.05	0.00
3	-0.02	-0.01

this framework. However, we emphasize that it is quite adequate for a discussion of the monopole discrepancy.

If there are N single particle states each with the same single particle energy, ε_α, then

$$E_I(jk^{-1}) = \frac{Nc^2}{-(\varepsilon_\alpha - \varepsilon_j)}$$

and

$$T^I_J(E_I(jk^{-1})) = \frac{Nc^2}{(\varepsilon_\alpha - \varepsilon_j)}$$

Furthermore

$$\Delta\alpha_o = \Delta E_J = \frac{Nb^2}{-(\varepsilon_\alpha - \varepsilon_j)} + \frac{Nc^2}{-(\varepsilon_\alpha - \varepsilon_j)}$$

which is less than zero for all cases. It is clear that the

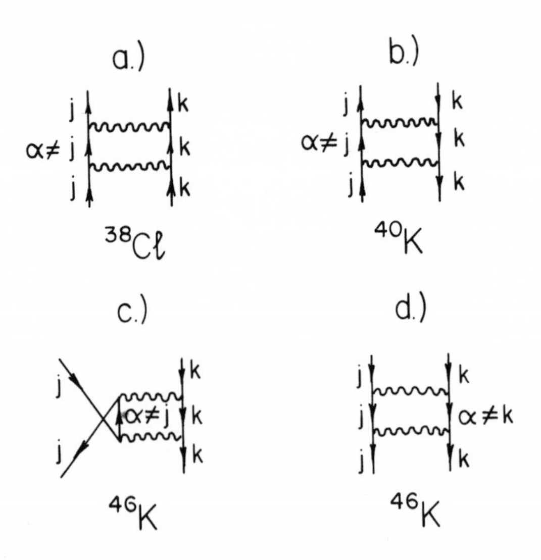

Fig. 2. In all cases, j = $f_{7/2}$-nucleon and k = $d_{3/2}$-nucleon. (a) and (b) represent the largest contribution to the Cl^{38}/K^{40} discrepancy in the $f_{7/2}$-$d_{3/2}$ case, and (c) and (d) represent incompletely cancelled contributions to the K^{46} discrepancies.

monopole is negative irrespective of our choice of single particle energies or two body matrix elements (which appear quadratically). We remark that this trend would not change even if we treated these type configurations in an OSM (matrix) calculation instead of second order perturbation theory. Since the monopole discrepancy has the same sign as the corresponding experimental discrepancy, it appears to be a matter of details as to whether or not the perturbation theory results are in agreement with experiment. However, there are some interesting aspects to the details. First of all, if we do the calculation using experimental single particle energies and realistic two-body matrix elements, it is not unexpected that the overwhelming contribution to $\Delta\alpha_o$ comes from the cases when α, the neutron excitation, is in the pf-shell. Perhaps more interesting is the fact that $\Delta\alpha_o$ would be -0.08 MeV if the single neutron energies for S^{37} were the same as those in Ca^{41}. This -0.08 MeV is barely within the experimental error bars. But as we have discussed previously, the one body neutron spectrum of S^{37} is more compressed than that of Ca^{41}. Clearly the $\Delta\alpha_o$ for the true S^{37} energies is larger, in fact it is -0.22 MeV, which is well outside the experimental limits. This result is somewhat moderated by the OSM calculation, in which some higher order terms are included. The OSM calculation yields $\Delta\alpha_o$ = -0.13 MeV. Still this is outside the experimental limits. This leads us to discuss the results of the fd-model calculation. But before we do that it is worthwhile discussing this same second order perturbation calculation for K^{46}.

A leading term in K^{46}, which does not transform back to Cl^{38}, is shown in Figure 2c. That is, when the valence hole lines (label corresponding to a $d_{3/2}$-proton hole) are transformed to particle lines, we see that there is not a corresponding (and hence) cancelling graph in Cl^{38} because the transformed K^{46} graph has particles and holes in the $d_{3/2}$-shell which does correspond to reality. Therefore this graph is uncancelled and will contribute directly to the difference. In this case the monopole discrepancy, as calculated in second order perturbation theory using Figures 2a and 2c with α in the pf-shell, is -0.06 MeV. This is in sharp contrast to the OSM value of 0.09 MeV. The primary reason goes back to another single particle energy argument. Namely, the $d_{5/2}$- and $2s_{1/2}$-proton holes. The largest second order contribution to the monopole of K^{46} comes from the graph represented by Figure 2d, where α is either a $d_{5/2}$- or $2s_{1/2}$-proton hole.

The expression for the contribution of the terms represented by Figure 2d to the energy of a state J is

$$\frac{\langle j^{-1}k^{-1}J|v|j^{-1}\alpha^{-1}J\rangle^2}{-(\varepsilon_\alpha-\varepsilon_k)}$$

For J = 2 and J = 5 only $\alpha = 1d_{5/2}$ contributes, and in these two cases it makes the monopole more negative than a similar term in Cl^{38} because $\varepsilon d_{5/2}-\varepsilon d_{3/2}$ = 1.72 MeV in K^{46} and 8.80 MeV in Cl^{38}. Even though the diagram represented in Figure 2d disobeys the Racah transformation between Cl^{38} and K^{46}, its contribution to the difference is partly cancelled by a similar Cl^{38} term represented by Figure 2c (with the roles of j and k interchanged and all particles changed to holes and *vice versa*), the cancellation is quite incomplete in the difference calculation ($\Delta\alpha_0$ due to the J = 2 and J = 5 terms is 0.13 MeV, which is considerably larger than the -0.02 MeV which is the entire contribution of the relabelled Figure 2c to the monopole discrepancy). The situation is a bit more complicated for the J = 3 and J = 4 states because $\alpha = 1d_{5/2}$ and $\alpha = 2s_{1/2}$ can both contribute but they tend to cancel because $\varepsilon_{d_{5/2}} - \varepsilon_{d_{3/2}} < 0$, whereas $\varepsilon_{2s_{1/2}} - \varepsilon_{d_{3/2}} > 0$. We remark that a perturbation argument is difficult to make because of the close proximity of the $2s_{1/2}$ and $1d_{3/2}$ shells in K^{46}. However, in the OSM calculation the cancellation is fairly complete. Therefore the large change in the monopole from Cl^{38} to K^{46} is primarily a single particle energy effect. Therefore we could make the monopole discrepancy quite small with a different (more strongly resembling $d_{5/2}$ proton energy used in Cl^{38}) choice of single particle energy. However, such an approach is of little interest to us and we do not use it. We further remark that the large values of $\Delta\alpha^2_{\lambda>0}$ and $\Delta\alpha^3_{\lambda>0}$ would not be significantly changed with a new $d_{5/2}$ proton energy for K because of the intrusion of the $2s_{1/2}$ orbit.

B. fd-Model Results

As we have seen, the spectral differences as calculated on the OSM do not agree well with those that are known experimentally. In Table VII appear the results of the fd model calculation for the spectral differences between Cl^{38}/K^{40}, Cl^{38}/K^{46}, and K^{40}/K^{46} interaction energies. It is clear that the discrepancies, as calculated within the fd model, tend to cancel those from OSM (compare the $\Delta\alpha_\lambda$'s of Tables V and VII). For the Cl^{38}/K^{40} case this is the kind of effect that is needed to improve the agreement with experiment. Since the size of the $\Delta\alpha_\lambda$'s is comparable in OSM and the fd-model, it is clear that nuclear coexistence plays a central role in spectral differences. The OSM effects come from low orders in perturbation theory (we have seen this in our second order perturbation calculations of the OSM discrepancies). The fd model effects come from higher orders in perturbation theory--this is the case because it takes several interactions to create the core excited state, then make it interact with the spherical state. Due to the comparable size of OSM and and fd-model effects, the perturbation expansion for the spectral differences is not well behaved. This slow convergence becomes more rapid as OSM effects become relatively more important than fd-model effects.

Similar large fd-model affects appear in the differences involving K^{46} interaction energies (see Table VII). Again, as in the Cl^{38}/K^{40} situation, the fd-model effects tend to go in the opposite direction of those of the OSM. To estimate crudely the combined effects of OSM and fd-model difference calculations we can merely add the discrepancies of each for the appropriate case. However, to really combine the two models we use one (OSM) to generate the alterations in the effective interaction for the other. The hybrid matrix theory thus includes both the "low orders in perturbation theory" of the OSM and the "high orders of perturbation theory" of the fd-model.

C. Hybrid Model Results

The results of the hybrid model difference calculations for the Cl^{38}/K^{40}, Cl^{38}/K^{46}, and K^{40}/K^{46} pairs appear in Table VIII. It is clear that the expected cancellation of the effects of each model calculation does occur although the cancellation does not exactly follow what a simple addition of the effects of each model yields.

Again in the case of Cl^{38}/K^{40} we make a comparison with experiment and see that the agreement between experimental and theoretical multipole discrepancies is better than any of the previous calculations. Since the final result is sensitive to the cancellations discussed, we are satisfied to have done this well and believe we have explained the mechanisms by which the Racah transformation works so well between Cl^{38} and K^{40}.

In the cases involving K^{46} the OSM discrepancies are larger than those previously discussed. These discrepancies also are calculated in the fd-model and the hybrid model. The primary reason for the larger OSM and hybrid calculational discrepancies is the $2s_{1/2}$ shell. As we have previously discussed, the $s_{1/2}^{-1}\ f_{7/2}^{-1}$ configuration is prominent in the lowest lying J = 3 and J = 4 states in K^{46}($\approx$50%). Therefore there are strong effects which, for example, will make Cl^{38} and K^{46} have different spectra. This will manifest

TABLE VII

The fd-model results (in MeV) for the Cl^{38}/K^{40}, Cl^{38}/K^{46}, and K^{40}/K^{46} systems.

(a) Energies

J	$E_J(Cl^{38})$	$E_J(K^{40})$	$T_J^I(E_I(K^{40}))$	$E_J(K^{46})$	ΔE_J^1	ΔE_J^2	ΔE_J^3
2	-1.21	1.11	-1.25	-1.14	0.04	-0.07	-0.11
3	-0.73	0.04	-0.79	-0.60	0.06	-0.13	-0.19
4	-0.44	0.55	-0.51	-0.28	0.07	-0.16	-0.23
5	-0.80	1.06	-0.93	-0.71	0.13	-0.09	-0.21

(b) Multipoles

λ	$\alpha_\lambda(Cl^{38})$	$\alpha_\lambda(K^{40})$	$\alpha_\lambda(K^{46})$	$\Delta\alpha_\lambda^1$	$\Delta\alpha_\lambda^2$	$\Delta\alpha_\lambda^3$
0	-0.75	-0.83	-0.63	0.08	-0.12	-0.20
1	-0.09	-0.06	-0.09	-0.03	0.00	0.03
2	-0.23	-0.24	-0.26	0.01	0.03	0.02
3	0.01	0.02	0.01	-0.01	0.00	0.01

TABLE VIII

The hybrid model results (in MeV) for the Cl^{38}/K^{40}, Cl^{38}/K^{46}, and K^{40}/K^{46} systems.

(a) Energies

J	$E_J(Cl^{38})$	$E_J(K^{40})$	$T_J^I(E_I(K^{40}))$	$E_J(K^{46})$	ΔE_J^1	ΔE_J^2	ΔE_J^3
2	-1.28	0.80	-1.18	-1.41	-0.10	0.13	0.23
3	-0.73	0.45	-0.81	-0.52	0.08	-0.21	-0.29
4	-0.47	0.53	-0.45	-0.19	-0.02	-0.26	-0.24
5	-0.68	1.01	-0.68	-0.57	0.00	-0.11	-0.11

(b) Multipoles

λ	$\alpha_\lambda(Cl^{38})$	$\alpha_\lambda(K^{40})$	$\alpha_\lambda(K^{46})$	$\Delta\alpha_\lambda^1$	$\Delta\alpha_\lambda^2$	$\Delta\alpha_\lambda^3$
0	-0.73	-0.72	-0.58	-0.01	-0.15	-0.14
1	-0.14	-0.15	-0.20	0.01	0.06	0.05
2	-0.20	-0.18	-0.33	-0.02	0.13	0.15
3	0.00	0.03	0.04	-0.03	-0.04	-0.01

itself in large calculated OSM discrepancies. These discrepancies are propagated to the hybrid model because the OSM effects appear in the effective interaction for each K^{46} calculation. We remark that the fd-model interaction includes the $2s_{1/2}$ shell from the mass 40 region so that the differences in the fd-model are smoother and the observed variations are shell filling effects. We should also add that for the K^{46} interaction calculations the $\langle f^2|v|f^2\rangle$ interaction was, appropriately, taken from a calculation of the Sc^{48} interaction energies[4] rather than those of Sc^{42}. This choice for $\langle f^2|v|f^2\rangle$ is not crucial, that is, if we used the same $\langle f^2|v|f^2\rangle$ in K^{46} as in Cl^{38} and K^{40} fd-model calculations, the results would be substantially the same. The modification of the interaction is made from the point of view of consistency.

IV. PARTICLE RANK OF THE INTERACTION

In 1957 Pandya and French[11] saw the value of using a higher particle rank interaction in the calculation of spectral discrepancies between K^{40} and the Racah transform of Cl^{38}. In particular, they did a crude OSM type calculation in addition to which they allowed a three body contact force to admix in the pure model space (no configuration mixing space). Their results reveal that their residual three body interaction yielded a monopole discrepancy which tends to cancel the monopole discrepancy of their OSM calculation.

The point of this section is to show how our fd-model results can be presented in terms of an expansion in the particle rank of the interaction, thereby enabling us to understand the successes of Pandya and French type calculations. The point being that we will see how the effects of space truncation can be simulated in the particle rank of an effective interaction. Thus, the interaction energy of Cl^{38} as calculated in the fd-model will yield a monopole which we can view as the effective monopole in the pure shell model (no configuration mixing). Similarly, the fd-model calculation of the K^{40} (with respect to a S^{36} core) monopole can be viewed in the pure shell model space as resulting from a df- interaction which is 2- plus 3- plus 4-body in nature. Therefore we are simulating the effect of core excitation by including a linear combination higher particle rank components of the interaction in the pure shell model space. Due to the limited experimental information available and the size of the monopole discrepancies, we will restrict the discussion in this section to monopoles.

There is a new class of discrepancies of the difference type that are associated with the interaction energies $d_\pi f_\nu$, $d_\pi^2 f_\nu$, $d_\pi^3 f_\nu$ and $d_\pi^4 f_\nu$ which are represented in Figure 3. The monopole discrepancy is associated with the fact that the monopoles which are obtained from Figure 3 are not linear in the number of d-particles. This linearity would apply if the interaction were only two body in nature and there were only pure configurations.

Since the calculations are to be done in the fd-model and the results are to be reexpanded in terms of various particle ranks for the $d_\pi^n f_\nu$ (particle rank is ≥ 2 and $\leq n + 1$) interaction, we must explicitly account for OSM effects. That is, it appears that we should calculate a separate G-matrix for each set represented in Figures 3a-3d. Fortunately, we are saved at least part of this task because we already have calculated v_{eff} for the Cl^{38} and K^{40} sets. The calculation of the effective interaction which does not obey the Racah transformation has been described in detail in Section II. In addition, we still need a v_{eff} for the fd-model for the Ar^{39} and Ca^{41} cases, both including the relevant OSM effects. These two calculations of v_{eff} are done in the same manner as those previously described. The resultant interactions are nearly linear extrapolations of those for Cl^{38} and K^{40}. For example, if we average the particle-particle v_{eff} for Cl^{38} and K^{40} the interaction we would obtain is quite close to that which we calculated. Thus, for each set represented in Figure 3 a different effective interaction is used for $\langle d_\pi f_\nu J | v | d_\pi f_\nu J \rangle$.

Experimentally, the interaction monopole for Ar^{39} is unknown; however, for

$$Cl^{38}: \quad V_2^e = -1.03 \pm 0.06 \text{ MeV},$$

and for

$$K^{40}: \quad 1/3\ (3V_2^e + 3V_3^e + V_4^e) = -1.02 \pm 0.01 \text{ MeV}.$$

and for

$$Ca^{41}: \quad 1/4\ (4V_2^e + 6V_3^e + 4V_4^e + V_5^e) = -1.01 \pm 0.01 \text{ MeV},$$

where the subscript on V^e denotes the particle rank of the interaction. If we assume there is only a two body force and no configuration mixing, then V_2^e (Cl^{38}) should equal V_2^e (K^{40}) in the pure shell model limit. However, V^e (Cl^{38}/K^{40}) = -1.03 + 1.02 = -0.01 MeV which has about the same magnitude as the monopole violation in the pp- to ph-transformation of Cl^{38} to K^{40}, *i.e.* very small. The addition of the error bars is a bit trickier here; we might think that it is simply the sum of the error bars (± 0.07); but the largest uncertainty is in the mass of S^{37} (0.04 MeV) and the next largest is in the mass of S^{36} (0.01 MeV). The masses of S^{36} and S^{37} are subtracted so that the errors cancel and we obtain

$$\Delta V^e(Cl^{38}/K^{40}) = -0.01 \pm 0.05 \text{ MeV}.$$

To see how the hybrid model fares, we calculate the terms represented in Figure 3 to obtain for

$$Cl^{38}: \quad V_2 = -0.73 \text{ MeV},$$

and for

$$Ar^{39}: \quad 1/2\ (2V_2 + V_3) = -0.70 \text{ MeV},$$

and for

$$K^{40}: \quad 1/3\ (3V_2 + 3V_3 + V_4) = -0.68 \text{ MeV},$$

and for

$$Ca^{41}: \quad 1/4\ (4V_2 + 6V_3 + 4V_4 + V_5) = -0.67 \text{ MeV}.$$

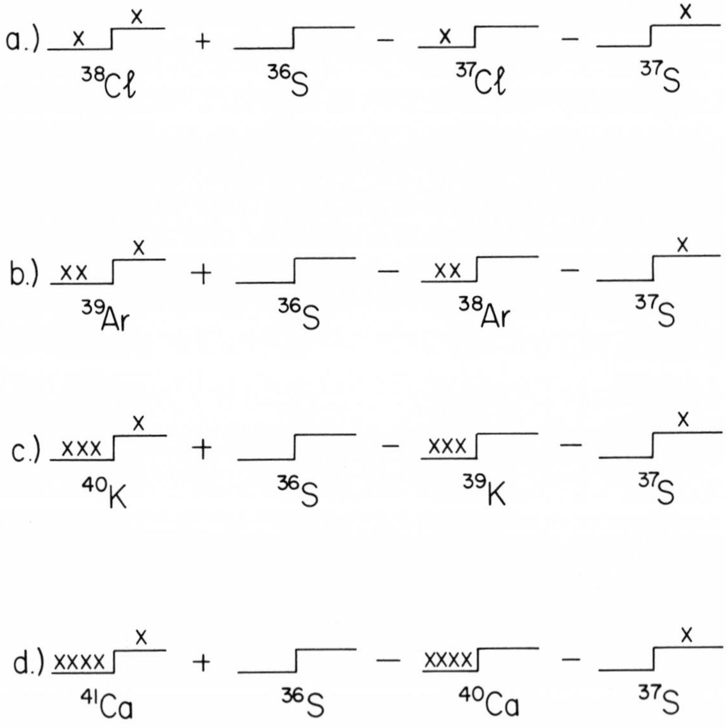

a.) ^{38}Cl + ^{36}S − ^{37}Cl − ^{37}S

b.) ^{39}Ar + ^{36}S − ^{38}Ar − ^{37}S

c.) ^{40}K + ^{36}S − ^{39}K − ^{37}S

d.) ^{41}Ca + ^{36}S − ^{40}Ca − ^{37}S

Fig.3. (a) represents the $f_{7/2}$-$d_{3/2}$ interaction, (b) represents the $f_{7/2}$-$d_{3/2}^2$ interaction, (c) represents the $f_{7/2}$-$d_{3/2}^3$ interaction, (d) represents the $f_{7/2}$-$d_{3/2}^4$ interaction. All are in the pure model space.

When we solve the equations we find that the monopoles are

$$V_2 = -0.73 \text{ MeV}$$
$$V_3 = +0.06 \text{ MeV}$$
$$V_4 = -0.03 \text{ MeV}$$
$$V_5 = 0.00 \text{ MeV}.$$

We remark that there is some accuracy (round off) problem associated with the calculation of the higher particle rank monopoles. Nonetheless, the effective three body monopole term is more than ten times smaller than the corresponding two body term. We further remark that the three body force has a repulsive monopole which is in agreement with the Pandya-French result and general expectations for such a force.

For the sake of comparison we present the results of the fd-model in which no OSM results are included (*i.e.* v_{eff} transforms),

$$V_2 = -0.75 \text{ MeV}$$
$$V_3 = -0.06 \text{ MeV}$$
$$V_4 = 0.03 \text{ MeV}$$
$$V_5 = 0.00 \text{ MeV}.$$

Notice that the three body monopole is attractive in contradistinction to our expectations. On the other hand for the OSM case, where

$$V_2 = -1.24 \text{ MeV}$$
$$V_3 = 0.06 \text{ MeV}$$
$$V_4 = -0.09 \text{ MeV}$$
$$V_5 = 0.12 \text{ MeV},$$

the three body term is repulsive. The large size of the V_4 and V_5 terms is not entirely meaningful because of accuracy (round off) problems. It is satisfying to note that the combined fd-model and OSM calculation results are more satisfactory than when compared with the results of either individual calculation. When comparing Cl^{38} and K^{40} in the pure model space we see the origin of the discrepancy between their monopoles can be attributed to an effective three plus four body interaction.

We can deduce V_3^e and V_4^e, if we assume that V_5^e is zero which is consistent with the hybrid model calculations. Doing this we obtain

$$V_2^e = -1.03 \pm 0.06 \text{ MeV}$$
$$V_3^e = +0.01 \pm 0.05 \text{ MeV}$$
$$V_4^e = +0.01 \pm 0.05 \text{ MeV}$$

We see that a comparison between calculated and "experimental" monopoles is hard to make because of the large error bars. We again emphasize that the origin of the error bars is primarily the uncertainty in the mass of S^{37}. Still we emphasize that we have shown in a particular example how it is that some effects of a two body interaction in a large space can be simulated by a higher particle rank interaction in a pure shell model valence space.

V. Sc^{42}/Sc^{48}

Sc^{42} and Sc^{48} are particle-particle and particle-hole partners in the $f_{7/2}$-shell. Like the Cl^{38}/K^{40} pair the experimental situation is well known. Unlike the Cl^{38}/K^{40} pair, the Sc^{42}/Sc^{48} pair has large experimental discrepancies which are about ten times those of Cl^{38}/K^{40} pair. A further contrast is found in the fact than an OSM calculation describes the Sc^{42}/Sc^{48} discrepancies quite well.[4,9] The fd-model effects are comparatively small, and hence the hybrid model results agree quite well with those of the OSM.[7] Still the hybrid model results yield a slight improvement over the OSM model results. Thus the Sc^{42}/Sc^{48} pair provide an example in which the "low orders of perturbation theory" dominate the "high orders of perturbation theory." Thus, there are situations in which a hybrid model represents more work than is necessary. Probably a good rule of thumb is the "smallness" of the discrepancies; *i.e.* if the discrepancies are small the core excited states probably play a role. The detailed results of the Sc^{42}/Sc^{48} difference calculation of Reference 7 appear in Table IX.

VI. THE $p_{3/2}$-$d_{3/2}$ MULTIPLET

As we have discussed earlier, the members of the $p_{3/2}$-$d_{3/2}$ multiplet have been experimentally determined (see Table X). We see that the $p_{3/2}$-$d_{3/2}$ experimental multipole discrepancies are "small" like those of Cl^{38}/K^{40} $d_{3/2}$-$f_{7/2}$ system rather than being "large" like those of Sc^{42}/Sc^{48}. It is for this reason that we assumed the monopole discrepancy was zero in the $p_{3/2}$-$d_{3/2}$ system (like it is in the $f_{7/2}$-$d_{3/2}$ to within error bars). This assumption enables one to deduce (see Reference 7) that $\varepsilon_{p_{3/2}} - \varepsilon_{f_{7/2}} = 0.36$ MeV in S^{37}.

Examination of Table X also reveals that an OSM calcu-

TABLE IX

A comparison of the experimental Sc^{42}/Sc^{48} $\Delta\alpha_\lambda$'s to those calculated (in MeV) in the OSM and hybrid models.

λ	$\Delta\alpha_\lambda^{hybrid}$	$\Delta\alpha_\lambda^{OSM}$	$\Delta\alpha_\lambda^{exp}$
0	-0.32	-0.37	-0.54
1	-0.14	-0.08	-0.16
2	0.06	0.05	0.03
3	0.06	0.06	0.02
4	-0.02	-0.10	-0.08
5	0.11	0.09	0.10
6	-0.03	-0.09	-0.07
7	-0.04	0.04	-0.06

lation does about as well for the $p_{3/2}$-$d_{3/2}$ discrepancies between Cl^{38}/K^{40} as the calculation does in the $f_{7/2}$-$d_{3/2}$ case. We add that we do not present $\Delta\alpha_\lambda$ for cases involving K^{46} because the $p_{3/2}$-$d_{3/2}$ strength in K^{46} is quite fragmented

VII. CONCLUSIONS

We have seen that the fact that the Racah transformed interaction energies in K^{40} strongly resemble the interaction spectrum of Cl^{38} is due to a cancellation of simple shell model effects by the push of coexisting core excited states on the pure shell model states. Thus we have seen a phenomenological illustration of poorly convergent behavior (in the perturbation sense), and the implied difficulty in a perturbative calculation of the effective interaction. This interplay of "low- and high-lying" configurations is further seen to result in a nucleus dependent valence effective interaction in a truncated space.

We have also shown that another way to simulate the effect of space truncation is to increase the particle rank of the interaction. That is, the hybrid model calculation of the differences in the interaction energies of Cl^{38} and K^{40} are calculated with a nucleus dependent two body effective interaction. The results of this calculation can be reexpanded in the pure model space in terms of two, three and four body interactions.

TABLE X

A comparison of the calculated experimental and multipole differences (in MeV) for the $1p_{3/2}$-$1d_{3/2}$ case in Cl^{38} and K^{40}. In part (a) the experimental quantities appear, in part (b) the OSM results appear, and in part (c) the experimental and OSM results are compared with the OSM-type results of Maripuu (Reference 9).

(a) Experimental Multipoles

λ	$\alpha_\lambda(Cl^{38})$	$\alpha_\lambda(K^{40})$	$\Delta\alpha_\lambda^{pd}$
0	-	-0.40	-
1	+0.07	+0.07	0.00
2	-0.13	-0.09	-0.04
3	+0.06	+0.08	-0.02

(b) OSM Calculated Multipoles

λ	$\alpha_\lambda(Cl^{38})$	$\alpha_\lambda(K^{40})$	$\Delta\alpha_\lambda$
0	-	-0.54	-
1	0.00	+0.12	-0.12
2	-0.18	-0.10	-0.08
3	0.00	+0.02	-0.02

(c)

λ	$\Delta\alpha_\lambda^{OSM}$	$\Delta\alpha_\lambda^{exp}$	$\Delta\alpha_\lambda^{Maripuu}$
0	-	-	-
1	-0.12	0.00	-0.10
2	-0.08	-0.04	0.00
3	-0.02	-0.02	+0.02

REFERENCES

1. S. P. Pandya, Phys. Rev. 103, 956 (1956).
2. T. T. S. Kuo and G. E. Brown, Nucl. Phys. A114, 241 (1968); T. T. S. Kuo, private communication.
3. G. A. P. Englebertink and J. W. Olness, Phys. Rev. C5, 431 (1972).
4. P. Goode, D. S. Koltun, and B. J. West, Phys. Rev. C4, 1527 (1971).
5. G. Sartoris and L. Zamick, Phys. Rev. 167, 1035 (1968).
6. A. Zuker, B. Buck, and J. McGrory, Phys. Rev. Lett. 21, 39 (1968).
7. P. Goode, to be published.
8. R. Sherr and W. W. Daehnick, preprint.
9. S. Maripuu, private communication.
10. B. J. West and D. S. Koltun, Phys. Rev. 187, 1319 (1969)
11. S. P. Pandya and J. B. French, Ann. Phys. 2, 166 (1957).

DISCUSSION

ENDT: I will start off with a small question on errors. You start off with the Cl^{38} errors which were all 60 keV. I didn't quite understand where they came from. I mean the energies are fairly well known and the mass of Cl^{38} is well known.

GOODE: Yes, well the mass of S^{37} is uncertain to about 50 kilovolts and the mass of S^{36} is uncertain to about 10 keV so that the two of them together give you an uncertainty of about 60 keV. Now those are in the centroids. That's why I talk about multipoles and then all the experimental error will be in the monopole.

LANFORD: The same $(f_{7/2} \times d_{3/2}{}^{-1})$ matrix elements should show up in the particle hole spectrum of Ca^{40}. There should be a complete set of T = 0 and T = 1 $(f_{7/2} \times p_{1/2}{}^{-1})$ matrix elements and so there should be another place to test your calculations. I don't know what the latest experimental situation is, but at one time it was thought that the complete set of all 8 states were known, *i.e.* four T = 0 and four T = 1 states.

GOODE: To compare Ca^{40} with Cl^{34}?

LANFORD: No, to compare the particle-hole matrix elements from Ca^{40} with the proton-neutron matrix elements in Cl^{38}.

GOODE: O.K., that's T = 0 against T = 1.

LANFORD: Well, take the T = 1 and the T = 0 matrix elements

and average them to get the proton-neutron matrix element which would then be compared with Cl^{38}. At one time I know it was believed that all of the T = 1 members of that multiplet were known.

GOODE: The Ca^{40} T = 1 states are the analogs of the K^{40} states that I've been talking about. Now to compare the T = 0 Ca^{40} states you find we have an $f_{7/2}$ particle and a $d_{3/2}$ hole with T = 0, so if we go to Cl^{34} instead of talking about dd we talk about fd. Then we have particle-particle T = 0 and we can compare that with particle-hole T = 0 in Ca^{40}.

LANFORD: I'm sorry, are you talking about the $(d_{3/2} \times f_{7/2})$ proton-neutron matrix elements?

GOODE: Right.

LANFORD: They should also be measureable in a Ca^{40} particle-hole spectrum.

GOODE: Isospin is good so that's why the Ca^{40} T = 1 states are the K^{40} ground states. So to compare T = 0 states you have to go Cl^{34} fd-states.

MARIPUU: You made a very interesting point in mentioning the $s_{1/2}$ and the $d_{3/2}$ crossing in K^{46} and I'm not sure I understood you there. What exactly did you mean?

GOODE: Well, that in K^{39} the ground state is $3/2^+$ and in K^{47} it's $1/2^+$.

MARIPUU: Just from the spectrum?

GOODE: That's right. In the calculation I try to be as conservative as possible so I take the single-particle energy from experiment. I don't want anyone to dispute my shell model calculation; I want to make it as pure and simple as possible to show that any kind of calculation of that sort will fail in calculated differences.

MCCULLEN: In the Sc^{42} and Sc^{48} cases is it clear that confining Sc^{48} to the fd shell still makes sense?

GOODE: No, and certainly you could say that now you should talk about an fp core excited model and that certainly might be valid. For example, the fd configurations start to get much fewer as you get up to Ca^{48} simply because of blocking. Supposing that these fp effects did appear in the discrepancies between experiment and simple shell model calculations. They are still small enough so that even if an fp model were used and the effects came out right compared to the experiment they would still mean that these higher orders in perturbation theory were less important that the lower orders of perturbation theory. But certainly what you say is absolutely true. And that's why the emphasis of the talk was on the Cl^{38}/K^{40} pair and not on mass 48.

II.a. ISOSCALAR MAGNETIC MOMENTS IN LIGHT NUCLEI

P. M. Endt
Fysisch Laboratorium, Ryksuniversiteit
Utrecht, Netherlands

The multicomponent wave functions provided by many particle shell model calculations can be used, amongst other things, to calculate magnetic dipole moments.

In the course of such calculations for the sd-shell, van Hienen and Glaudemans[1] observed that the g-factor for the isoscalar contribution to magnetic moments (in the following denoted by g_o) is remarkably constant. Almost all calculated g_o-values were found to lie in the 0.4-0.6 region, depending very little on the type of two body interaction employed or on the size of the shell model space. This is in contrast to the g-factor for the isovector contribution which was found to fluctuate widely from state to state and which shows an outspoken dependence on the size of the model space.

In trying to explain this constancy they found that g_o is given by a very simple expression if one makes the approximation that all active particles are in the same subshell (ℓ,j). This expression, which is independent of the number of active particles, reads:

$$g_o = \frac{1}{2} \pm \frac{0.38}{(2\ell+1)} \quad \text{for} \quad j = \ell \pm \frac{1}{2}$$

It can be derived without much difficulty from formulae given by de Shalit and Talami.[2] A comparison with experiment (shown in Figure 1) is possible for states in self conjugate nuclei, and for the average magnetic moment of two mirror states (the isovector contribution is zero in a self conjugated nucleus and cancels in the mirror average). The results of multishell calculations (for details, see Reference 1) are also indicated in Figure 1.

On the whole, it appears that the single shell approximation is doing very well, but there are three noteworthy exceptions. For Li^8 the experimental value, 0.82, is much higher than the $p_{3/2}$ single shell value, 0.63, which might be explained by an admixture of the $1s^2_{1/2}1p^4_{3/2}$ configuration. The experimental value for the F^{19}, $Ne^{19}(j^\pi = 1/2^+)$ ground-state average, 0.74, is also much higher than the $d_{5/2}$ single shell value, 0.58, which, again, is easily explained by $d^2_{5/2}$, $s_{1/2}$, and/or $s^3_{1/2}$ admixtures. The too low value for the Si^{29}, P^{29} average is explained by means of $d_{5/2}$ and/or $d_{3/2}$ admixtures. In general, the experimental points seem to "round off" the sharp edges of the single shell expression. Point (c) is the average for the $7/2^-$ excited states in Ar^{37} and

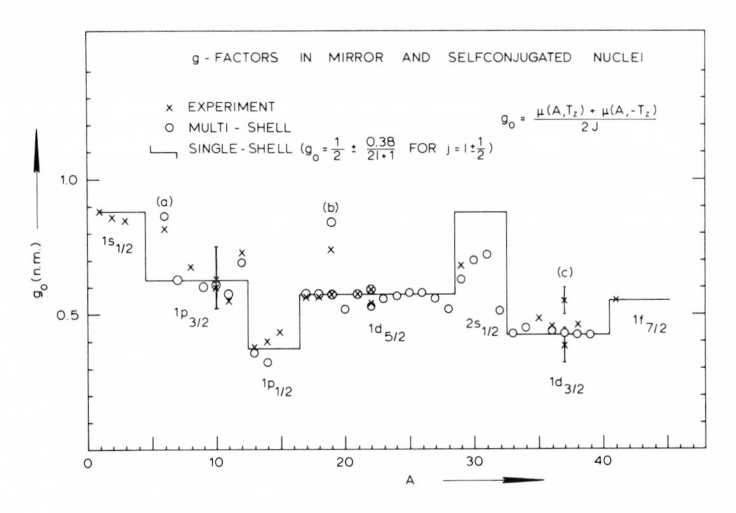

Fig.1. The g-factors in self conjugate and in mirror nuclei. The points (a)-(c) are discussed in the text.

K^{37} and agrees beautifully with the $f_{7/2}$ single shell value.

REFERENCES

1. J. R. A. van Hienen and G. W. M. Glaudemans, submitted for publication in Phys. Letts.
2. A. de Shalit and I. Talmi, *Nuclear Shell Theory*, Academic Press, New York, 1963.

DISCUSSION

GOLDHAMMER: You have a charming explanation there for the magnetic moment of Li^6 but I don't think you have to get that exotic. There's a lot of supporting evidence that LS coupling is good in Li^6 and I think if you take a very simple wave function in LS coupling you come quite close to the experimental magnetic moment.

III.A. A PEDESTRIAN APPROACH TO THE INTERMEDIATE STRUCTURE IN PHOTONUCLEAR REACTIONS IN LIGHT NUCLEI

F. B. Malik
Department of Physics, Indiana University
Bloomington, Indiana

and

M. G. Mustafa*
Physics Division, Oak Ridge National Laboratory
Oak Ridge, Tennessee

(Presented by F. B. Malik)

I. INTRODUCTION

In talking about the photonuclear or any other type reaction theory in nuclear physics, it has become customary to classify them into two distinct catagories, *viz* the direct reaction (referred to as DR) and the compound process (denoted further as CP). The basic underlying picture of a DR is outlined in Figures 1a and 1b. A photon of energy $h\nu$ being incident on a nucleus knocks out one of the nucleons, leaving the rest of the system essentially unperturbed. At best, the effect of the rest of the nucleus can be incorporated in terms of a smooth background of an optical potential distorting the outgoing nucleons. This is reminiscent of the dominating process in the photoemission in an atom and this approach was applied to the $C^{12}(p,\gamma)N^{13}$ by Breit and Yost[1] in attempting to explain the experiment of Reference 2. Since then, this model has been applied to the capture of nucleons in the MeV region[3] and recently[4] to the emission of high energy nucleons from C^{12} and O^{16}. The excitation functions characterized by direct emission exhibit broad resonances having widths of the order of a MeV in light nuclei.

The other extreme model widely used in understanding thermal neutrons (or neutrons in keV region) is the concept of the compound nucleus, originally introduced by Bohr and Kalcker[5] to understand the neutron capture cross section observed by Fermi and his collaborators.[6] According to the picture of a compound nucleus, photons entering a target nucleus excites the entire nucleus and all nucleons participate in this excitation. When a nucleon has accumulated enough momentum to escape, the compound nucleus emits this nucleon and in the process gets deexcited. The excitation functions, where this physical picture dominates, are

---

*On a leave of absence from Bangladesh Atomic Energy Center.

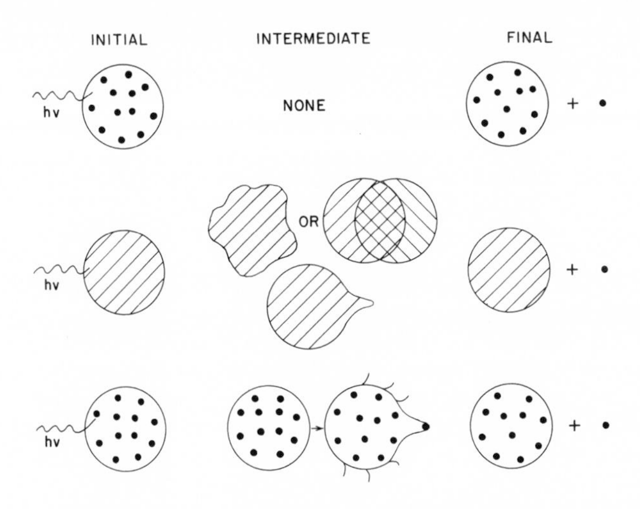

Fig. 1a. A schematic drawing of the physical situation underlying a direct process (top), and a compound nuclear process (center), and an intermediate process (bottom). In a direct reaction a photon essentially knocks out a single nucleon without affecting other nucleons appreciably. In a compound process, a photon interacts with a nucleus as a whole and this nucleus remains in an excited state until a nucleon has gathered enough momentum to escape. In an intermediate process a photon can interact with a number of nucleons. A part of the emitted neutron flux is by a direct process and another part is emitted by excited states of the nucleus.

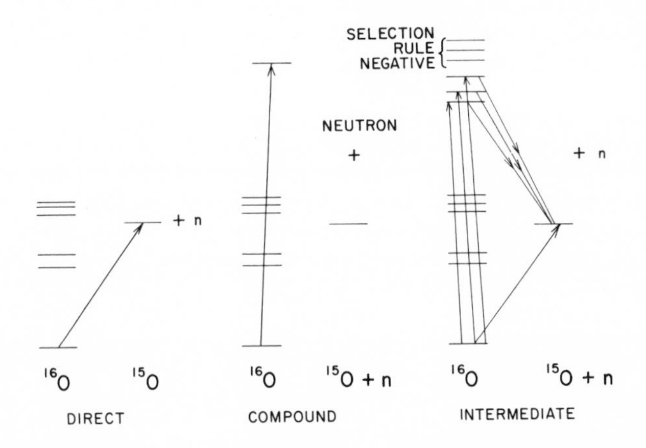

Fig. 1b. Columns marked Direct, Compound and Intermediate refer, respectively, to first, second and third rows of (a) in terms of nuclear levels.

characterized by resonances of narrow widths. The key difference between these two concepts is that in the case of the DR only one or at the most a few nucleons participate and the rest of the nucleus remains a spector, whereas in a CP, the whole nucleus participates. Moreover, in a DR, a nucleon is transferred directly from an initial to a final state without going through an intermediate state, whereas in a CP the formation of the intermediate compound nucleus plays a central role.

In actuality, a physical situation may occur which cannot be adequately idealized by either of these two extreme models and demands an intermediate approach blending salient features of both of them. Examples of such a situation are easy to visualize. In the case of photonucleon emission a fraction of the transition may be direct, whereas another fraction can go via an intermediate state to the continuum. In principle, it is not necessary to restrict the intermediate state to be a single one but there may be a number of them. This sharing of strength between a direct process and an intermediate process is even more likely, if there exists a relatively simple particle type of bound state embedded in the single nucleon continuum and the electromagnetic transition probabilities to this state are large. In that case, a fraction of the nucleons may be emitted directly and the remaining fraction may be ejected via an electromagnetic transition to this intermediate state.

A number of variations of this type of process can occur. There may be a bunch of intermediate states close to one another and an electromagnetic transition of a given multipolarity from the ground state to all of these states is allowed. By close, we mean that the mean separation of these states is of the order of their particle emission widths. In that case, the excitation function will have a series of overlapping resonances and exhibit a succession of peaks on a broad bump. In case the level spacing is shorter than the half widths of particle emission we may not see any peak but see only a broad bump in the excitation function even though there may be a number of intermediate states under this bump.

A somewhat more sophisticated variant of the same thing may also occur. This was first pointed out by Lane, Thomas and Wigner.[7] We may visualize that a very strong intermediate state is surrounded by a series of closely spaced states as shown on the left side of Figure 2. In the first approximation the selection rule does not permit any transition from the ground state to these surrounding states (let us call them satellite states or satellite doorway states). However, if this state is well described only in the first approximation by a shell model type or a particle-hole or any type of model, there is a residual interaction between this strong state and its surrounding weak states. As a result of this interaction a fraction of this strong state is mixed to its surrounding states and a situation similar to the one shown on the right side of Figure 2 can arise. This makes then a transition from the ground state to all these states permissible and nucleons then may be emitted from these states. The excitation function of such a process may again exhibit either a series of peaks on a background of a broad bump or a broad bump. The former will be the case if either the radiative width or the decay width or the combination of them associated with these satellite states is comparable to that of the strong state [primary intermediate (or doorway) states--PIS], and if it is not, the latter is the case.

Thus the key points associated with the occurrence of intermediate structures in the excitation function are:

a) The process is not entirely compound. This implies that a reasonable strength for a direct transition is available.

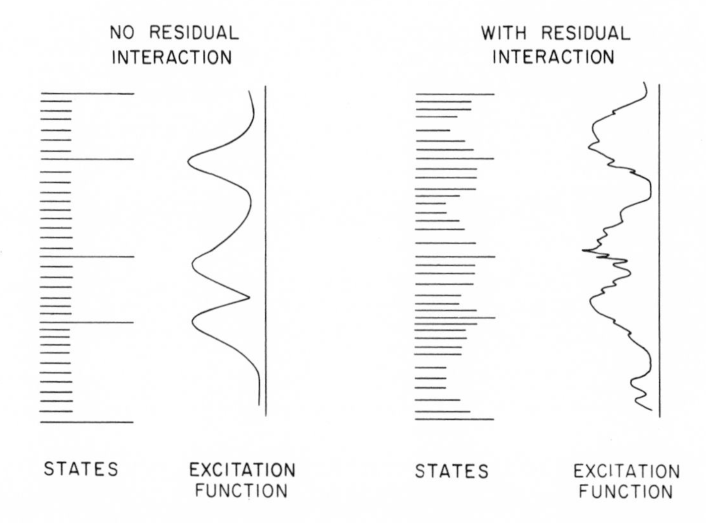

Fig. 2. A schematic representation of an intermediate reaction process. The first column represents unperturbed states generated by an approximate model Hamiltonian. Transitions from the ground state to a few states marked by long bars are allowed. These states then decay by particle emission yielding an excitation function shown in the second column. If the model Hamiltonian used to generate states shown in the first column is not an adequate approximation to the actual Hamiltonian, the states shown in column one can get mixed by a residual interaction and transitions from the ground state to nearby satellite states are allowed. The excitation function may look like the one sketched in the last column.

b) There may be one or many closely spaced intermediate states embedded in single nucleon emission continuum. A variant of this general process is the occurrence of one or more PIS states coupled to a series of satellite states.

c) The relative strength of *both* the radiative widths to these states and the decay widths from them determines the shape of the function.

d) The transition strength to a PIS state is shared or redistributed between a PIS state and its satellites. The total nucleon emission probability is again shared between a direct transition and the transition via one or more of these intermediate states.

Thus attention must be paid to the *normalization of the total wavefunction* in every stage of the analysis if one wants to obtain proper magnitude of the cross sections.

Once we have defined our physical picture, the next step is to find a suitable mathematical language to implement this. There are a variety of options starting from R-matrix theory to Green's function technology open, but I prefer to try the simplest one though this may not be very sophisticated. To this end, we start with the total Hamiltonian and proceed in a fashion analogous to Feshbach's approach[8,9] but avoid the explicit use of projection operators and adopt a theory based on Trefftz's treatment[10] of diaelectronic recombination. Furthermore, we cast the theory specifically for the case of photonuclear reactions but its application in other cases like fine structure in analog resonances is straight forward.

The present day experimental situation in the photonuclear reaction in light nuclei has recently been reviewed by Firk.[11] Considerable structure observed in the giant dipole excitation region led to the suggestion[12,13] that these are clear cases of intermediate coupling where a part of the transition occurs through intermediate states. These states are generated by the coupling of a strong single particle or particle-hole state to a series of other weak and more complicated states of boson-like character. This simple model using coupling constant and the decay width as free parameters was successful in correlating the observed structure in O^{16} and Si^{28}. Recently, Shakin and Wang[14] have applied a similar idea to the structure in O^{16} and using five parameters, namely the coupling constant, energy shifts and the decay widths of PIS and satellite states. They have been successful in generating enough structure in the excitation function. Rowe and Wang,[15] Goncharove *et al.*[16] and Kamimura *et al.*[17] have also shown that the strong particle-hole 1^- states in C^{12} can be successfully coupled to other nearby states and the strong radiative E1 transition strength to this state is, then, fragmented and shared by these nearby states via this coupling.

However, no attempt has been made to compute the actual excitation function.

II. THEORY

1. Equations

The total wave function for A nucleons in the initial or final state can be expanded in a complete orthonormal set

$$\Psi = \sum_c \Psi_c + \sum_b \Psi_b \equiv \Psi_I + \Psi_{II} \tag{1}$$

where Ψ_c refers to all continuum open and closed channels and Ψ_b refers to all channel wave functions which are bound. Thus

$$\begin{aligned} (\Psi_c, \Psi_{c'}) &= \int dE\, \delta(E_c - E_{c'})\, \delta_{cc'} \\ (\Psi_b, \Psi_{b'}) &= \delta_{bb'}, \quad (\Psi_c, \Psi_b) = 0. \end{aligned} \tag{2}$$

In order to specify Ψ_c and Ψ_b, we must specify the corresponding Hamiltonian. Since we shall be considering the photonuclear reaction of the type $O^{16}(\gamma,p)N^{15}$, it is appropriate to split the total Hamiltonian H(A) as follows

$$H(A) = H_o(A-1) + T(i) + V(i,A-1), \tag{3}$$

where $H_o(A-1)$ is the Hamiltonian of the (A-1) system, T(i) is the kinetic energy operator of the ith particle which is being scattered and V(i,A-1) is the interaction between the ith particle and (A-1) nucleons. Let $\Phi_c(A-1)$ and $\Phi_b(A-1)$ be the continuum and bound state wave functions of H_o satisfying the Schrödinger equations

$$H_o \Phi_\beta = E_\beta \Phi_\beta \quad ; \quad \beta = b,c. \tag{4}$$

Then the functions Ψ_b and Ψ_c are constructed from Φ_β's by taking an antisymmetic product

$$\Psi\,(A) = [\Phi_\beta(A-1)\;\Phi_j(i)]_a \quad ; \quad \beta,j = b,c. \tag{5}$$

To define Φ_j, we take the scalar product $<\Phi_\beta, (H-E)\psi> = 0$, and get the set of equations

$$[E - E_j - T(\vec{r}_i) - V_{jj}(\vec{r}_i)]\ \Phi_j(\vec{r}_i) = \sum_{j' \neq j} V_{jj'}(\vec{r}_i)\phi_{j'}(\vec{r}_i)$$

$$(j,j' = b,c\ ;\ b = 1 \ldots N,\ c = 1 \ldots N) \qquad (6)$$

$V_{jj'}$ are hermitian matrix elements

$$V_{jj'} = <\Phi_\beta(A-1),\ V(i,A-1)\ \Phi_{\beta'}(A-1)>. \qquad (7)$$

The form of (6) is obtained if the antisymmetrization is neglected. This coupled set of differential equations will have to be replaced by a different coupled set of integro-differential equations, if the Pauli Principle is incorporated, explicitly.[18]

Specializing now to those cases of Φ_j's having spherical symmetry, we may write

$$\Phi_j(i) = \xi_j(\alpha,\hat{r}_i)\ r_i^{-1}\chi_j(r_i), \qquad (8)$$

where α represents other quantum numbers necessary to define the state Φ_j. Displaying explicitly indices b and c, we can rewrite (6) to emphasize these two different classes of solutions distinguished by their asymptotic behaviors.

$$[E - E_c - T_c(r) - V_{cc}(r)]\ \chi_c(r) = \sum_{c' \neq c} V_{cc'}(r)\ \chi_{c'}(r)$$

$$+ \sum_b V_{cb}(r)\ \chi_b(r) \qquad (9a)$$

and

$$[E - E_b - T_b(r) - V_{bb}(r)]\ \chi_b(r) = \sum_{b' \neq b} V_{bb'}(r)\ \chi_{b'}(r)$$

$$+ \sum_c V_{bc}(r)\ \chi_c(r), \qquad (9b)$$

where $T_m = -\dfrac{\hbar^2}{2\mu}\left(\dfrac{d^2}{dr^2} - \ell_m(\ell_m+1)/r^2\right)$; m = c,b. and $V_{cc'}$ and

$V_{bb'}$ include an integration over α, and $\hat{r}_i$.

We now discuss different approaches to the study of photonuclear reactions done to date within the context of these sets of coupled equations. For $O^{16}(\gamma,n)O^{15}$ type reactions, initial bound states of O^{16} are usually taken to be the solutions of (9b) with V_{bc}, *i.e.*, the interaction of a bound state to a continuum state, to be zero. Futher discussion relates to different types of approximation used for the final state wave function.

2. Approximations

A. Direct Reaction

The extreme case of direct reaction is obtained by setting V_{cb}, V_{bb} and $V_{bb'}$ all equal to zero. In addition, the distorted wave approximation assumes $V_{cc'} = 0$. This corresponds to the physical picture that the outgoing particle is merely scattered by the average potential and there is no bound state in the continuum.

If, however, V_{cb} is not taken to be zero exactly but assumed to be small, following References 19 and 20, one can show that an evergy averaged contribution of these terms smooting out resonance structures leads to a complex potential. This conclusion can also be drawn within the context of Feshbach's formalism. Thus in a direct reaction model for the photonuclear reaction, the final state single nucleon wave function is a solution of the optical equation

$$[E - E_c - T_c(r) - \overline{V}_{cc} - i\overline{W}_{cc}]\ \chi_c(r) = 0, \qquad (10)$$

where $\overline{V}_{cc} + i\overline{W}_{cc} = V_{cc}$ + the complex potential originating from the coupling V_{bc}.

Actually, the first direct reaction model for the photonuclear reaction was used by Breit and Yost[1] to explain the $C^{12}(p,\gamma)N^{13}$ experiment of Hafstad and Tuve.[2] The fact that they achieved partial success in explaining these data lies in using only the real potential V_{cc} in obtaining $\chi_c(r)$ from (10). Since then, this model has been used in a number of cases. In particular, the photonuclear reaction in O^{16} and C^{12} above 30 MeV excitation energy has been successfully analyzed in References 4 and 21. Even in the giant dipole region, this process yields a substantial contribution. This point is, of course, important for the application of the intermediate coupling model discussed later because if the contribution from the direct process is negligible, the process is compound and can not be considered to be intermediate.

B. Bound State Calculations

The work of Elliot and Flowers,[22] Gillet and Vinh Mau[23] and Brown and his collaborators[24] is of this type. The wave function $\Psi_b(A)$ is a bound one particle - one hole state, *i.e.*, $\Phi_b(A-1)$ is the hole state and $\phi_b(i)$ is the bound excited single particle state. For example, in O^{16}, $\Phi_b(A-1)$ is a $1p_{1/2}$ or $1p_{3/2}$ hole state and $\phi_b(i)$ is a $1d_{5/2}$, or $1d_{3/2}$, or $2s_{1/2}$ single particle state. Since no continuum state is present, the interactions are

$$V_{cc} = 0 = V_{cc'}$$

$$V_{cb} = 0 = V_{bc}.$$

The relevant coupled equation is the following:

$$\left[E - E_b - T_b(r) - V_{bb}(r)\right] \chi_b(r) = \sum_{b' \neq b} V_{bb'}(r)\, \chi_{b'}(r).$$

Using an appropriate residual interaction, the final wave function and energies are obtained after diagonalization in the one particle-one hole basis. These calculations showed a 2-4 MeV shift in the energies of the unperturbed one particle-one hole states and also showed that the dipole transition strength between these states and the ground state is concentrated in one or two transitions.

The study of the structure[15-17] in giant dipole region of C^{12} also falls in this category. In essence, all these calculations calculate discrete lines, which constitute only the first step towards calculating cross sections.

Locations of particle-hole states within the standard shell model scheme have also been carried out in Si^{28}. The E1 transition rates from the ground state to these 1^- states exceeds substantially the observed cross sections and fails to account for the fine structure. This is, indeed, indicative that these states may be sharing a part of their strength with other states.

C. Continuum State Calculations

These calculations attempt to solve the set of equations (9a) by incorporating the $(V_{cc} + V_{cb})$ term in terms of a smooth optical potential and the $\chi_{c'}$ are expanded in terms of 1 particle-1 hole (1p-1h) shell model type states having bound state properties.[25] The antisymmetrization has been incorporated and it is found that the antisymmetrization primarily affects the magnitude of the dipole transition but does not change the structure of the cross section. The main feature of the two major peaks in O^{16} at 22.3 and 24.3 MeV is well accounted for by these calculations. However, the observed structure could not be reproduced and the computed cross section remains 2 to 3 times larger than observation. Incorporation of 2p-2h states in the expansion of $\chi_{c'}$ does not improve the situation adequately.[26] A series of studies involving interactions between a number of continuum states has also been reported.[27]

D. Intermediate Coupling Approximation

While the continuum state calculations focus primarily on the interaction between the continuum states and incorporate the interaction between a continuum and a bound state either as a smooth background in terms of the optical potential or as a well separated resonance, the intermediate coupling model demands that the strong 1p-1h states are coupled to a number of other bound states by a residual interaction and thereby shares its strength with them, and the E1 transition occurs in two steps: first the electromagnetic transition takes place from the ground state to these intermediate states in the continuum. These then decay by particle emission very much in the same way as in the autoionization process in atomic physics.

Duke, Malik, and Firk (DMF)[13] have described the giant dipole resonances in light nuclei by an intermediate coupling model which attributes their fine structure and widths to the coupling between single-fermion states and collective bosons associated with the quasi-stationary stage of the reaction. In their analysis, they assumed a linear coupling in momentum space between these fermion and boson states, and they have approximated these states by harmonic oscillator type wave functions. With such wave functions the diagonalization of the Hamiltonian has been achieved. It is natural that such wave functions will only produce discrete lines in the excitation function corresponding to different transition matrix elements which represent only the first phase of the intermediate coupling. The next phase involves the computation of the decay probability.

The final cross section will be discrete if the particle decay width is assumed to be a δ-function. However, this is obviously unphysical and DMF used a normalized Lorentzian width of Γ_s to compute the final cross section. This Γ_s is related to the lifetime of the particle-hole (or rather PIS) state in the absence of the coupling term.

In connection with our two sets of coupled differential equations (9), the DMF approximation consists of

a) neglecting $V_{cc'}$ $(c \neq c')$,

b) use $V_{bb'} = g_\lambda f(r)$ where g_λ is some simple coupling constant. The subscript λ identifies the different single particle excited states the combination of which gives the giant resonance cross section,

c) take $\chi_b(r)$ as harmonic oscillator wave functions, and

d) perform a diagonalization to account for $V_{bb'}$ and obtain a new set of basis $\overline{\chi}_b$ and express cross section in terms of the resonant amplitudes and decay widths of these $\overline{\chi}_b$.

Using this simple model the structure of giant dipole resonances in O^{16} and Si^{28} has been fitted using the coupling constant and decay widths as parameters. The purpose here is to investigate this intermediate coupling model within the context of reaction theory and to relate energy shifts, widths, and cross sections to two body matrix elements. This will, then, throw more light on the resonant nature of structures in the giant dipole region.

III. FORMAL SOLUTION OF THE COUPLED EQUATIONS IN THE INTERMEDIATE COUPLING APPROXIMATION

To solve the system of equations (9), we neglect coupling between different continuum channels $V_{cc'}$ ($c \neq c'$) in (9a). Although there is *a priori* no justification in neglecting these terms, it is hoped that its effect can be partly incorporated by modifying the diagonal interaction. Continuum coupled channel calculations emphasizing these terms already indicate that they do not account for fine structures, although they influence the width of the resonances and may shift their energies. Their neglect, on the other hand, helps us to obtain a simple algebraic solution and to see clearly how the resonating part of the cross section influences the non-resonating background or direct component and vice versa.

Using

$$t_m(r) = d^2/dr^2 - \ell_m(\ell_m + 1)/r^2,$$

$$k_c^2 = (2\mu/\hbar^2)(E - E_c),$$

$$k_b^2 = (2\mu/\hbar^2)(E_b - E),$$

and

$$U_{mn} = (2\mu/\hbar^2)\ V_{mn},$$

we may rewrite (9a) and (9b) in this approximation as

$$[t_c(r) + k_c^2 - U_{cc}(r)]\ \chi_c(r) = \sum_b U_{cb}\chi_b \tag{11a}$$

and

$$[t_b(r) - k_b^2 - U_{bb}(r)]\ \chi_b = \sum_{b'\neq b} U_{bb'}\chi_{b'}(r) + \sum_c U_{bc}\chi_c(r). \tag{11b}$$

Following Mott and Massey[39] the formal solution of (11a) is

$$\chi_c(r) = \chi_c^{(o)}(r)\ \delta_{cc'} + \int K(r,r')\ \sum_b U_{cb}(r')\ \chi_b(r')\ dr', \tag{12}$$

where

$$K(r,r') = (-k_c)^{-1}[\chi_c^{(o)}(r_<)\ \chi_c^{(1)}(r_>) + i\chi_c^{(o)}(r_<)\chi_c^{(o)}(r_>)]. \tag{13}$$

Here $r_<(r_>)$ is the smaller (larger) of r and r', $\chi_c^{(o)}$ and $\chi_c^{(1)}$ are respectively, the regular and irregular solutions of the homogeneous part of (11a) (*i.e.*, the right hand side of (11a) is set equal to zero) having the following asymptotic forms:

$$\chi_c^{(o)} \xrightarrow[r\to\infty]{} \sin\left(k_c r - \eta_c \ln(2k_c r) - \pi\ell_c/2 + \lambda_c + \delta_c\right)$$

and

$$\chi_c^{(1)} \xrightarrow[r\to\infty]{} \cos\left(k_c r - \eta_c \ln(2k_c r) - \pi\ell_c/2 + \lambda_c + \delta_c\right),$$

where $\eta_c = Z_1 Z_2 e^2/\hbar v$; Z_1 is the charge of the residual nucleus, Z_2, the charge of the outgoing particle and **v**, the velocity of the particle in the continuum. δ_c and λ_c are respectively, Coulomb and non-Coulomb phase shifts.

Two points may be noted: (i) one can add or subtract any potential in defining the homogeneous part of (11a), *e.g.*, the imaginary smooth potential simulating the nonresonant effect of the U_{bc} can be included in defining homogeneous solution and (ii) formerly, continuum-continuum coupling terms $\sum_{c'} U_{cc'}\chi_{c'}$ could have been incorporated in (12) by replacing the single sum over b by a double sum over b and c', provided all matrix elements $U_{cc'}\chi_{c'} \to 0$ at least exponentially. We also note that solution (12) along with appropriate boundary conditions and (11) are equivalent to defining a scattering matrix.

Although, formally, (12) appears to be a solution, in principle it is nothing but an integral representation of (11), since the χ_b's are still unknown. Following Trefftz,[10] we write a general solution of χ_b in terms of the orthonormal set set

$$\chi_b(r) = \sum_n a_{nb}\chi_{nb}(r) + \int dq\, a_{qb}\chi_{qb}(r) \tag{14}$$

with

$$\int \chi_{nb}^2 dr = 1, \int \chi_{qb}\chi_{q'b} dr = \delta(q - q')$$

and

$$\int \chi_{nb}\chi_{nb'} dr = 0.$$

Coefficients a_{nb} and a_{qb} can now be obtained from (11b), noting that χ_{nb} and χ_{qb} satisfy the homogeneous part of these equations.

$$(k_{nb}^2 - k_b^2)a_{nb} = \sum_{b'\neq b} \int \chi_{nb}U_{bb'}\chi_{b'}dr + \sum_c \int \chi_{nb}U_{bc}\chi_c dr \tag{15}$$

$$-(q^2 + k_b^2)a_{qb} = \sum_{b'\neq b} \int \chi_{qb}U_{bb'}\chi_{b'}dr + \sum_c \int \chi_{qb}U_{bc}\chi_c dr. \tag{16}$$

The a_{nb} in (15) have clearly a resonant behavior when k_b^2 is very close to k_{nb}^2. Near such a resonance all the χ_b may be approximated as (note that *this is not the case of isolated resonance but a series of successive resonances*)

$$\chi_b(r) \simeq a_{nb}\chi_{nb}(r) \quad ; \quad b = 1....N. \tag{17}$$

Putting (12) and (17) into (15) we get

$$\sum_{b'} \left[(k_{b'}^2 - k_{nb'}^2)\delta_{bb'} - \Delta k_{bb'}^2 + i\Gamma_{bb'}\right] a_{nb'}$$

$$= - \sum_c \int_0^\infty \chi_{nb}(r)\, U_{bc}(r)\, \chi_c^{(o)}(r)\, dr, \tag{18}$$

where

$$\Delta k^2_{bb'} = - \int_0^\infty \chi_{nb}(r)\ U_{bb'}(r)\ \chi_{nb'}(r)\ dr$$

$$+ \sum_c k_c^{-1} \iint \chi_{nb}(r)\ U_{bc}(r)\ \chi_c^{(o)}(r_<)\ \chi_c^{(1)}(r_>)$$

$$\times\ U_{cb'}(r')\ \chi_{nb'}(r')\ dr'\ dr \qquad (19)$$

and

$$\Gamma_{bb'} = -\sum_c k_c^{-1} \int_0^\infty \chi_{nb}(r)\ U_{bc}(r)\ \chi_c^{(o)}(r)$$

$$\times \int_0^\infty \chi_c^{(o)}(r)\ U_{cb'}(r)\ \chi_{nb'}(r)\ dr. \qquad (20)$$

Here $\Delta k^2_{bb'}$ is the energy shift matrix. The first term on the right hand side of (19) is the energy shift due to the interactions between different bound states and the second term represents the shift due to the interaction of the bound states with the scattering states. The matrix $\Gamma_{bb'}$ is connected with the width of the resonance. Both $\Delta k^2_{bb'}$ and $\Gamma_{bb'}$ depend on the nature of two-body interaction.

Now (12) along with (17) and (18) represent a complete set of equations defining the theory of intermediate structure. Clearly for all large $a_{nb'}$ the wave function (12) has two distinct parts: one is $\chi_c^{(o)}$ corresponding to the scattering by a standard potential (or optical model), and the other connects a bound state to a bound state like structure embedded in the continuum which, finally, decays to the continuum. Thus intermediate structures are clear cut resonances but unlike the cases of isolated resonances, they can often be overlapping because if b and b' are far apart, the first term in (19) and off diagonal $\Gamma_{bb'}$ may be neglected leading to well separated resonance like behavior.[20]

IV. CROSS SECTIONS

For completeness we note that the above formalism is suitable for the study of intermediate structure in elastic and inelastic scattering. For these cases, the scattering amplitude $f_{cc'}$ is defined by the asymptotic condition on χ_c as

$$\chi_c \xrightarrow[r\to\infty]{} \delta_{cc'}\left(\sin k_c r - \eta_c \ln 2k_c r - \pi\ell_c/2\right)$$

$$+ f_{cc'} \exp i(k_c r - \eta_c \ln 2k_c r - \pi\ell_c/2) \tag{21}$$

with

$$\phi_b \xrightarrow[r\to\infty]{} 0(r^{-2}),$$

$$\phi_b \xrightarrow[r\to 0]{} 0$$

and

$$\phi_c \xrightarrow[r\to 0]{} 0.$$

Then the differential cross section for inelastic processes ($c \neq c'$) is proportional to $|f_{cc'}|^2$

$$|f_{cc'}|^2 = |\sum_b a_{nb} \int \chi_c^{(o)} U_{cb}\, \chi_{nb}\, dr|^2$$

$$= \sum_{bb'} a^*_{nb} a_{nb'} \int \chi_{nb} U_{bc} \chi_c^{(o)} dr$$

$$\times \int \chi_c^{(o)} U_{cb'} \chi_{nb'} dr. \tag{22}$$

For elastic scattering f_{cc} is given by (22) plus a contribution coming from the potential scattering U_{cc}. If a_{nb} and $a_{nb'}$ are smooth functions, there will be no resonance-like structure in the cross section but otherwise one may obtain a series of resonance-like structures which will look like isolated resonances only if $\Gamma_{bb'}$ and $\Delta k^2_{bb'}$ (for $b \neq b'$) are zero.

The differential photonuclear reaction cross section is given by

$$\frac{d\delta}{d\Omega} = (2\pi/\hbar)(\rho_f/\text{Photon Flux})\ \Sigma|\langle\psi_f|H_{Int}|\psi_i\rangle|^2, \tag{23}$$

where H_{Int} is the interaction between photon and nuclei; ψ_f and ψ_i are eigenfunctions of H. Here Σ refers to an average over initial and sum over the final spin orientations. For cases like $O^{16}(\gamma,n)O^{15}$, the initial ground state is far away from the continuum states, and one can approximate $\psi_i \simeq \sum_b \psi_b$ where ψ_b are constructed from shell model wave function solving (9b). This corresponds to incorporating correlations in the initial state. In the intermediate coupling model $\psi_f = \sum_c \psi_c + \sum_b \psi_b$ where ψ_c and ψ_b are constructed from (5) using the following approximations

$$\chi_c(r) \simeq \chi_c^{(o)} + \sum_b a_{nb} \int K(r,r')\, U_{cb}(r')\, \chi_{nb}(r')\, dr' \qquad (24)$$

and

$$\chi_b(r) \simeq a_{nb} . \chi_{nb}(r), \qquad (25)$$

where $K(r,r')$ is given by (13), χ_{nb} are solutions of homogeneous equations defining χ_b, and a_{nb} are obtained by solving the set of coupled equations (18) exactly or in a suitable approximation.

V. CASE OF ONE INTERMEDIATE STATE

The matrix element for the transition from the ground state to a final state ψ_f via an *isolated* single intermediate state involves a simple expression for a_b (if there is no other intermediate state).

$$a_b = - \frac{\int \chi_b(r)\, U_{bf}(r)\, \chi_f^{(o)}(r)\, dr}{k^2 - (k_b^2 + \Delta k_{bb}^2) + i\, \Gamma_{bb}} \qquad (26)$$

$$\equiv \frac{\gamma_{bf}}{E - (E_b + \Delta E_b) + i\, \Gamma_{bf}/2} \qquad (27)$$

where k_b in (18) is simply written as k and the index n in other quantities has been dropped for convenience. The γ_{bf} is the reduced width and is related to the partial width by

$$\gamma_{bf} \equiv - \int \chi_b(r)\, V_{bf}(r)\, \chi_f^{(o)}(r)\, dr. \qquad (28)$$

and

$$\Delta E_b = (2/\hbar v) \iint \chi_b(r)\ V_{bf}(r) \chi\chi_f^{(o)}(r_<)\ \chi_f^{(1)}(r_>)$$

$$\times\ V_{fb}(r')\ \chi_b(r')\ dr'\ dr. \tag{29}$$

This is the expression derived by Fano[25] and leads to the expression for the Breit-Wigner resonance.[26] This is evident because the contribution to the matrix element M in (23) due to the resonant part of the wave function is

$$M = \frac{\langle\psi_i|H_{Int}|\psi_b\rangle\gamma_{bf}}{E - (E_b + \Delta E_b) + i\ \Gamma_{bf}/2}\ . \tag{30}$$

The cross section is then proportional to

$$\frac{\Gamma_{ib}\ \Gamma_{bf}}{\left(E - (E_b + E_b)\right)^2 + \Gamma_{bf}^2/4} \tag{31}$$

with

$$\Gamma_{ib} = |\langle\psi_i|H_{Int}|\psi_b\rangle|^2 \tag{32a}$$

and

$$\Gamma_{bf} = \left(4/hv\right)|\int \chi_b(r)\ V_{bf}\ \chi_f^{(o)}(r)\ dr|^2. \tag{32b}$$

This is the Breit-Wigner resonance relation.

In principle however M also contains a term proportional to $\langle\psi_i|H_{Int}|\chi_c^{(o)}\rangle$ which is the contribution of a direct transition and there is an interference between these two terms in the expression for the cross section. This is important for the intermediate coupling case.

VI. CASE OF OVERLAPPING RESONANCES AND MANY INTERMEDIATE STATES

In this case, one has to diagonalize the following equation

$$\sum_{b'} [(k^2 - k_b^2)\, \delta_{bb'} - \Delta k_{bb'}^2 + i\Gamma_{bb'}]\, a_{b'}$$

$$= - \sum_c \int \chi_b(r)\, U_{bc}\, \chi_c^{(o)}(r)\, dr$$

or symbolically

$$\sum_{b'} [k^2\, \delta_{bb'} - (W_{bb'} - i\Gamma_{bb'})]\, a_{b'} = \sum_c X_c . \tag{33}$$

That is one has to solve the matrix equation

$$(k^2 \cdot \underline{1} - \underline{W} + \underline{\Gamma})\underline{a} = \underline{X}. \tag{34}$$

This can simply be achieved by a similarity transformation which diagonalizes $\underline{W} - i\underline{\Gamma}$, *i.e.*, to find a $\underline{t}$ such that

$$(\underline{W} - i\underline{\Gamma})\underline{t} = \underline{t}\ \underline{\Lambda} \tag{35}$$

and if $t_{\alpha\beta}$ and $\Lambda_{\alpha\beta}$ are elements of $\underline{t}$ and $\underline{\Lambda}$ then we have

$$\sum_\beta (W_{\alpha\beta} - i\Gamma_{\alpha\beta})\, t_{\beta\nu} = t_{\alpha\nu}\, \Lambda_\nu$$

with $\Lambda_\nu = \delta_{\nu\nu'}(k_\nu^2 - i\gamma_\nu)$, $\underline{W} - i\underline{\Gamma}$ is a symmetric matrix, and $\underline{t}$ is an orthogonal matrix, *i.e.*, $\sum_\nu t_{\alpha\nu}\, t_{\alpha\nu'} = \delta_{\nu\nu'}$ and $\sum_\nu t_{\alpha\nu}\, t_{\beta\nu} = \delta_{\alpha\beta}$ and hence $\underline{t}^{-1} = \underline{t}^T$ (its transpose). Since $\underline{W} - i\underline{\Gamma} = \underline{t}\ \underline{\Lambda}\ \underline{t}^T$, we get

$$(k^2 \cdot \underline{1} - \underline{W} + \underline{\Gamma})^{-1} = t(k^2 \cdot \underline{1} - \underline{\Lambda})^{-1} t^T$$

yielding

$$\underline{a} = \underline{t}(k^2 \cdot \underline{1} - \underline{\Lambda})^{-1}\, \underline{t}^T\, \underline{X} \tag{36}$$

or

$$a_b = \sum_{\nu b'} \frac{t_{b\nu} t_{b'\nu} x_{b'}}{(k^2 - k_\nu^2) + i\gamma_\nu} \tag{37}$$

and it again shows resonance like behavior at an energy k_ν^2 which is obtained by diagonalizing $\underline{W}$ and a width γ_ν obtained by diagonalizing $\underline{\Gamma}$.

VII. MODEL OF DMF

Instead of performing the complete transformation (35), one can do a partial diagonalization which is good if $U_{bc} << U_{bb'}$, as it is visualized by Lane, Thomas and Wigner.[27] In that case $\Gamma_{bb'}$ defined in (20) can be neglected for all off diagonal elements and $\Delta k^2_{bb'}$ is given by the first term of (19).

In this particular case, one can first diagonalize (11b) and obtain a new basis χ_{nb} and expand χ_b in terms of these χ_{nb}, *i.e.*, (14) is replaced by the basis set

$$\chi_b(r) = \sum_n \bar{a}_{nb}\bar{\chi}_{nb} + \int dq\ \bar{a}_{qb}\bar{\chi}_{qb}(r). \tag{38}$$

For this basis set, $\Delta k^2_{bb'}$ is given by the second term of (19) only with χ_{nb} replaced by $\overline{\chi_{nb}}$ (we call it $\overline{\Delta k^2_{bb'}}$) and $\Gamma_{bb'}$ is given by (20) with χ_{nb} replaced by $\overline{\chi_{nb}}$ (we call it $\overline{\Gamma_{bb'}}$). If, therefore, one of the $U_{cb'}$ in (19) and (20) is small, these new $\overline{\Delta k^2_{bb'}}$ and $\overline{\Gamma_{bb'}}$ can be neglected. This is the model used by DMF under the assumption that $U_{bb'} = g_\lambda f(r)$, and Γ_{bf} of (32b) is taken as a parameter.

Since in this case $\overline{k^2_{bb'}} \simeq 0$ and $\overline{\Gamma_{bb'}} \simeq 0$, the coupled equations (18) (defined by basis χ_{nb}) decouple and $\overline{a_b}$'s can be found easily. To illucidate this point we write (18) for the case of two amplitudes $\overline{a_b}$ and $\overline{a_{b'}}$:

$$\begin{pmatrix} k^2 - \bar{k}_b^2 - \overline{\Delta k^2_{bb}} + i\bar{\Gamma}_{bb} & -\overline{\Delta k^2_{bb'}} + i\bar{\Gamma}_{bb'} \\ -\overline{\Delta k^2_{bb'}} + i\bar{\Gamma}_{bb'} & k^2 - \bar{k}^2_{b'} - \overline{\Delta k^2_{b'b'}} + i\bar{\Gamma}_{b'b'} \end{pmatrix} \begin{pmatrix} \bar{a}_b \\ \bar{a}_{b'} \end{pmatrix} = \begin{pmatrix} \bar{\gamma}_{bf} \\ \bar{\gamma}_{b'f} \end{pmatrix},$$

if $\overline{\Delta k}_{bb'} \simeq 0$ and $\overline{\Gamma}_{bb'} \simeq 0$ then both $\overline{a}_b$ and $\overline{a}_{b'}$ are given by

$$\overline{a}_i \simeq \frac{\overline{\gamma}_{if}}{k^2 - (\overline{k_i^2} + \overline{\Delta k_i^2}) + i\overline{\Gamma}_{ii}} \quad ; \quad i = b,b'. \tag{39}$$

Obviously, (39) is true for an arbitrary number of intermediate states.

Actually, one may note that for (39) to hold, it is not essential that $\overline{\Delta k^2_{bb'}} \simeq 0$ and $\overline{\Gamma}_{bb'} \simeq 0$. They hold for the more general case of $\overline{\Delta k^2_{bb'}}$ and $\overline{\Gamma^2_{bb'}}$ being independent of either b or b', *e.g.*, large but constant or $\Gamma_{bb'}/\Gamma_b = f_b$.

Even in the approximation (39) generalized to contain an arbitrary number of resonances, the cross section is not just proportional to

$$\sum_b \frac{\overline{\Gamma}_{ib}\overline{\Gamma}_{bf}}{\left(E - (\overline{E}_b + \overline{\Delta E}_b)\right)^2 + \overline{\Gamma}^2_{bb}/4} \tag{40}$$

but contains an interference term between different a_b's.

We note once more that the key difference between intermediate coupling with overlapping resonances and a simple case of overlapping resonances leading to either a background imaginary potential and/or fluctuations is that $U_{bb'}$ in (11b) is non-negligible. In the limit where (37) or (39) is valid the structures accounted for in the intermediate coupling model are *actual resonances*. However, if the density of intermediate states is very large in comparison to their widths, these resonances can loose their own identities leading to fluctuations in cross sections.

This completes our formalism of the intermediate structure and overlapping resonances, and a microscopic justification and an analysis of assumptions involved in the DMF model. Obviously, the theory is a general one and not restricted to the photonuclear case.

VIII. APPLICATIONS

We shall discuss in detail the situation in the photonuclear reaction in O^{16} within the context of an intermediate structure treatment and mention progress made in Si^{28} and C^{12}.

A. O^{16}

An important clue suggesting that an intermediate structure type of situation may be accompanying a nuclear reaction is to estimate the direct transition probabilities and compare them with observation. If the direct transition probabilities form a significant fraction of the observed cross section, the entire process is not clearly a compound nuclear one in the classic sense and may admit an intermediate structure type of analysis.

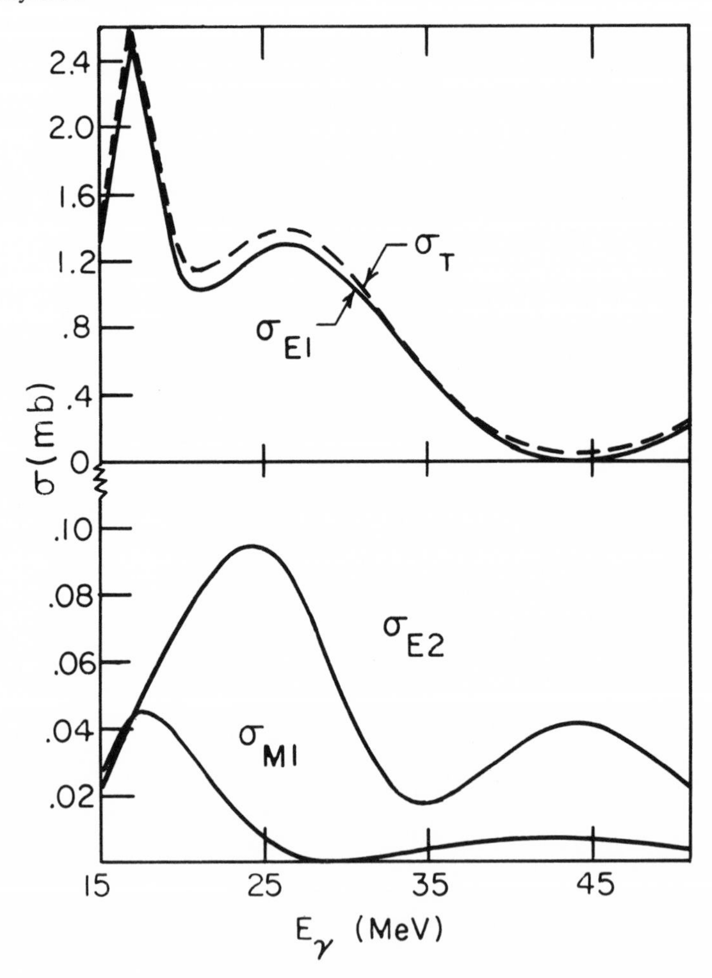

Fig. 3. Calculated E1, E2, and M1 integrated cross sections in a direct reaction model for the reaction $O^{16}(\gamma,p_o)N^{15}$ as a function of γ ray energy E_γ. σ_{E1}, σ_{E2} and σ_{M1} are respectively the E1, E2 and M1 integrated cross sections, and σ_T is the sum of these three. (From Ref. 21)

The work of Buck and Hill[25] indicates already the presence of a reasonable direct component in the giant dipole energy region of O^{16}. A detailed analysis done in Reference 21 shows that the direct transition probabilities form a reasonable fraction of the excitation function. This is shown in Figure 3. The strength of the direct transition probabilities is roughly 10% of the observed data.

For further analysis, we discuss equations (11) through (16). The implications of these equations are

1) The interaction between bound states χ_{nb} has not been completely diagonalized. If a complete diagonalization of the Hamiltonian with respect to χ_{nb} is achieved, $U_{bb'}$'s is on the right sides of (15) and (16) are zero and no further fragmentation is possible. In that case, (15) and (16) define the problem of *overlapping resonances only*. Mathematically, it is immaterial how the diagonal part of the interaction is defined. The important point is to include in off-diagonal matrix elements all interactions, which are not included in the diagonal part.

2) In (15) k^2_{nb} refers to entire bound state spectrum. Thus a model Hamiltonian which describes low lying excitation spectrum should also be used to specify k^2_{nb} lying in single particle continuum. One cannot consider states in the giant dipole region with total disregard of the low lying spectrum.

3) Total normalization of the wavefunction is to be maintained. If an El transition rate from the ground state to a single unperturbed principal state k_{nb} is strong, and very weak to its satellite states, the consideration of the coupling terms $U_{bb'}$ between this original PIS state and its satellite redistributes the strength of that transition. Consequently, the original strength of the transition is *reduced* and gets *redistributed* giving rise to *broadenings*.

4) The correlation in the ground state (or the initial state) is to be incorporated. As shown in Reference 4, the inclusion of the ground state correlation influences the magnitude of the cross section. In the case of O^{16}, the magnitude is reduced.

5) Actual location of states depends upon ΔE_{bb}, and the energy shift while the shape depends upon two widths, the radiative width Γ_{ib} and the decay width Γ_{bf}. This is evident from (31).

6) Since resonances corresponding to states of intermediate coupling are closely spaced the total width is not a simple sum of their individual widths and one must diagonalize (34) and find the new widths.

In actual computation all these points have not yet been fully incorporated.

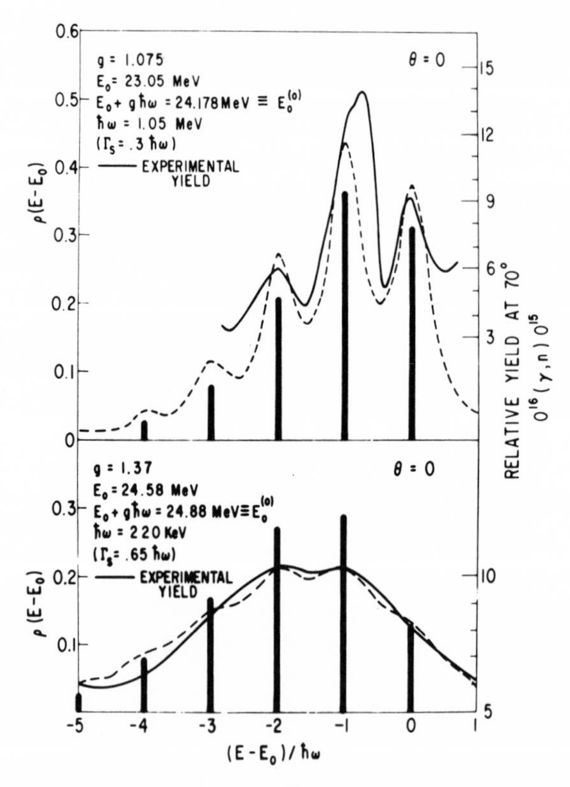

Fig. 4. Heavy solid vertical lines are computed intensities ρ of the spectral lines in a photonuclear process in O^{16} calculated using the model of Ref. 13. The dashed lines are the expected cross section if the principal state has a width Γ_s. The light solid line represents the experimental points in the $O^{16}(\gamma,n)O^{15}$ reaction (Ref. 35) obtained by Firk *et al.* (From Ref. 13)

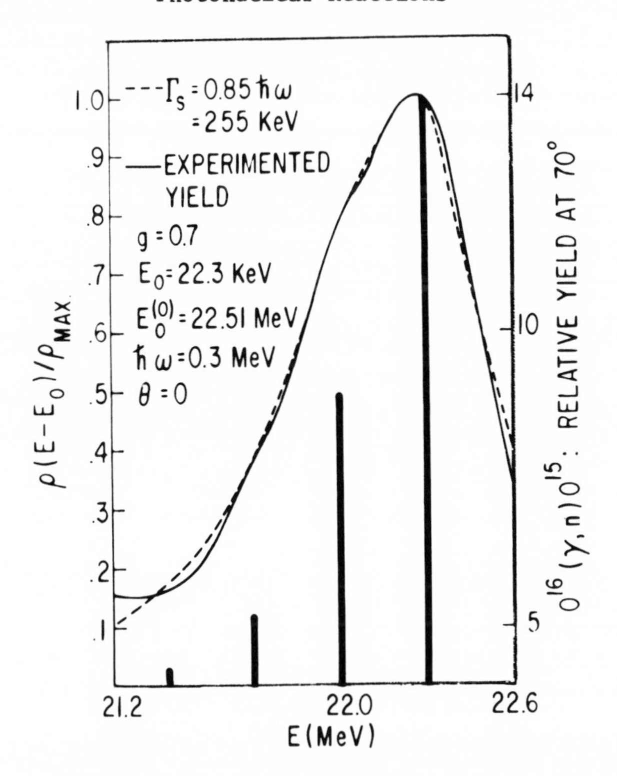

Fig. 5. The same as in Figure 4 but for a different set of parameters. This demonstrates that if the particle emission width is large in comparison to level spacings, the underlying states may be difficult to detect.

The model described in References 12 and 13 uses an arbitrary coupling parameter which is linear in momentum to couple strong particle-hole states to nearly boson-like states. The decay width associated with the PIS is taken to be of Lorentzian type and its width is used as a free parameter. The computed cross section is shown in Figure 4. The interesting part is to note that because of the coupling E1 transitions can take place to a number of adjacent states and the whole width can take a structure observed in the $O^{16}(\gamma,n_o)O^{15}$ experiment.

In Figure 5, we present an alternative fit to the broad resonance at 22.3 MeV. It clearly indicates that if the level spacing of satellite states is further shortened but the level width is nearly kept the same, the line shape can again be fitted and although there are a number of strong lines folded in the resonance, the fine structure is not pronounced.

Shakin and Wang[14] adopt basically the same approach and idea and try to generate these boson states from particle-hole states. These are then folded in a Lorentzian function to generate the final cross section. The calculated result is compared with the experimental yield curve in Figure 6. The absolute value is off by a factor of two. However, if the integrated cross section is normalized to the experimental data and then the decay width is used as a fitting parameter, the theoretical calculation fits the experimental data.

The most likely reason for Shakin and Wang's inability to account for the absolute magnitude of the cross section lies in their not considering points (2) and (3) discussed earlier. They do not consider ground state correlations and investigate the extent of compatibility of their states with the low lying spectrum of O^{16}. This point is discussed below in a schematic fashion.

One of the best O^{16} theoretical spectra is due to Zucker Buck and McGrory.[28] Here O^{16} is considered by them as a C^{12} core plus four particles sharing $1p_{1/2}$, $2s_{1/2}$, and $1d_{5/2}$ shell model orbitals. E1 transition rates from their ground states to all possible 1^- states in 10 to 25 MeV excitation energy generated by their wave function is shown in Figure 7. Clearly, their wave function does not explain strong dipole transitions. This is, of course, not at all surprising because the main E1 strength originates from the transition of a $1p_{3/2}$ particle to 2s-1d shell and ZBM wave functions do not have a $1p_{3/2}$ component. However a simple shell model wave function for $1p_{3/2}$ and 2s-1d states does not produce enough strength at correct energies. This is shown in Figure 8a. As it has been pointed out by Brown and his collaborators[24] a particle-hole type of residual interaction is able to place them in a proper position: this is shown in Figure 8b. The main problem after this is

a) to reduce the strength of particle-hole states, and

b) to generate additional fragmentation in the giant dipole excitation energy region keeping the basic structure of the ZBM wave function in tact.

Obviously, a good way to solve the problem is to open up the C^{12} core and add a $1p_{3/2}$ component to the ZBM wave function and then perform the diagonalization. However, this is a very formidable task and will not be discussed here. A simple way to realize the effect of such a procedure is to weak couple a $1p_{3/2}$ hole to the ZBM wave function. The unperturbed energies of 1^- states obtained by weak coupling a $1p_{3/2}$ hole respectively, with a $1d_{5/2}$ and $2s_{1/2}$ particle states of the ZBM spectrum is shown in Figures 10a and 10b, respectively. A 6.00 MeV spin orbit splitting between $1p_{1/2}$ and $1p_{3/2}$ has been used

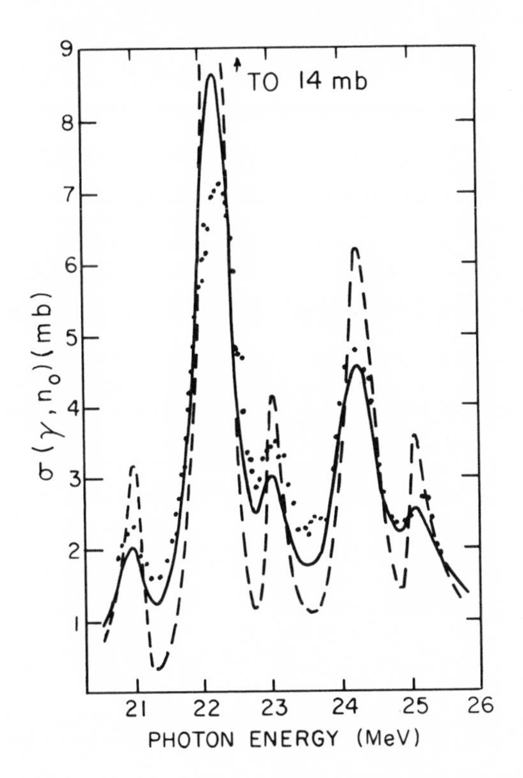

Fig. 6. Experimental cross section σ for the reaction $O^{16}(\gamma,n_o)O^{15}$ (dots) are compared to theoretical calculation of Ref. 14. The dashed curve is the actual calculation using the decay width as a parameter. The solid line is the theoretical cross section normalized to yield total integrated experimental value.

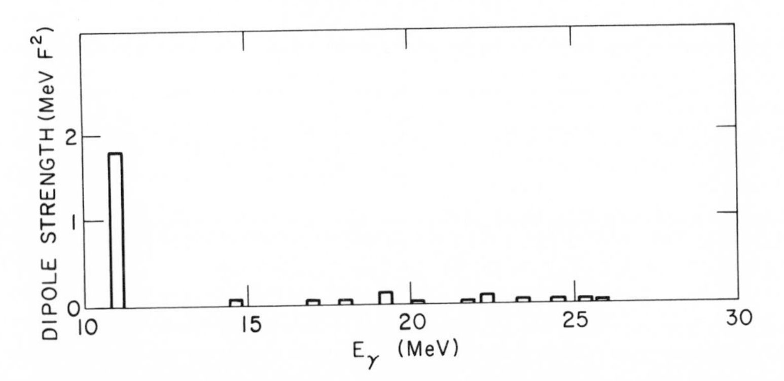

Fig. 7. E1 transition rates from the ground 1^- states in the giant dipole region of O^{16} calculated from the ZBM wave-functions (Ref. 28).

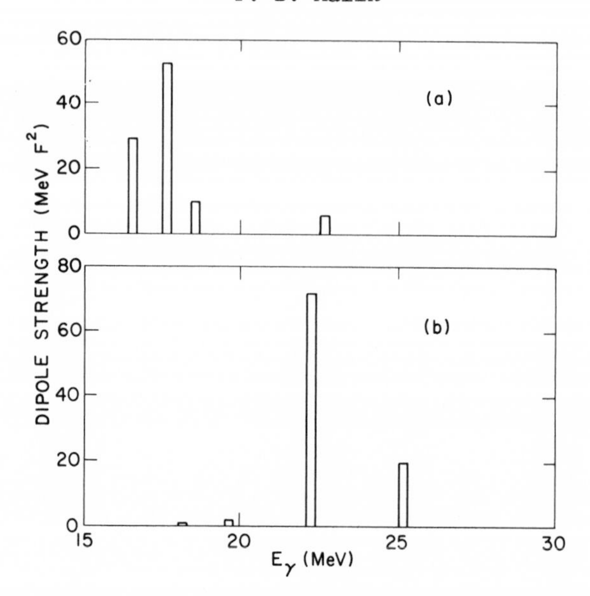

Fig. 8. (a) E1 transition rates from the ground state of O^{16} to 1^- states in a single particle elementary shell model where a $1p_{3/2}$ nucleon is transferred to the 2s-1d shell. (b) E1 transition rates from the ground state of O^{16} to 1^- particle-hole states generated by coupling a $1p_{3/2}$ nucleon hole state of an elementary shell model to the 2s-1d particle states.

to generate these states. From the actual particle-hole calculation of Brown and others,[23,24] it is known that the 18.31 MeV state of Figure 10a is shifted by about 4 MeV to 22.3 MeV, and the rest of the states involving $1d_{5/2}$ particle orbital are shifted by the amounts mentioned in Figure 9. The energy shifts of coupling a $1p_{3/2}$ hole to a $1d_{5/2}$ and $2s_{1/2}$ particle orbitals are given in Figure 9 . Using these energy shifts but generating particle-hole states from the ZBM we get the observed strengths shown. This figure also includes calculated dipole strengths for the 1^- states generated from $(1p_{1/2})^{-1}1d_{3/2}$ and $(1p_{3/2})^{-1}1d_{3/2}$. These yield additional strengths around 18 and 25 MeV. Thus the expected dipole strength for a transition from the ground state of O^{16} within the context of the ZBM model is the sum of Figures 7 and 9 . The most interesting point to observe is that dipole strengths have now correct magnitudes. The consideration of the correlation in the ground state configuration and a proper normalization of the total wave function are primarily responsible for this. This was not the case in the work of Shakin and Wang, who, therefore, failed to get correct magnitudes.

It is to be emphasized that ours is only a simple schematic analysis of the problem. To complete this analysis, one would like to estimate properly the actual energy shifts in a particle-hole calculation based on the ZBM wave function, and then either to fold the discrete dipole strength in a Lorentzian decay function as it has been done in References 13 and 14, or a more thorough analysis can calculate the decay widths and additional energy shifts themselves from a two nucleon potential using (19) and (20) or (37). We should also keep in mind that actual location of a state need not be at a maximum of a peak if $\Delta k_{bb'}^2$ in (19) does not equal 0.

Another complexity in comparing a theoretical calculation with an experiment arises from the fact that experimental results in photoemission involves a radiative transition of all multipoles, whereas theoretical results, discussed so far, consider only dipole transitions. Experimental results on the study of angular distribution indicate an asymmetry about 90°

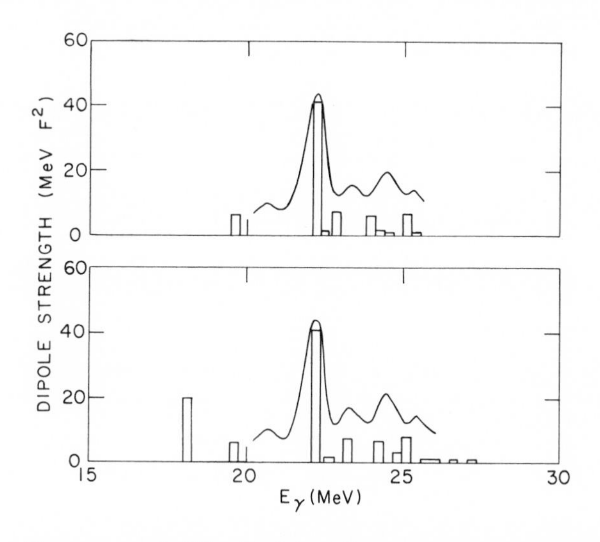

Fig. 9. The upper half of this figure represents the computed (vertical bars) E1 dipole strength from the ground state of O^{16} to 1^- states generated by weak coupling a $1p_{3/2}$ hole to $2s_{1/2}$ and $1d_{5/2}$ states of ZBM (Ref. 28). The actual energy shifts ΔE calculated using a δ-function force discussed in Ref. 14 are 3.89, 0.21, 0.58 and 0.15 MeV respectively, for states at 18.31, 22.25, 23.34 and 24.00 MeV in Figure 10(a) and are 1.07, 0.20, 0.15, and 0.11 MeV for states at 18.66, 22.10, 24.68, and 25.20 MeV, respectively, of Figure 10(b). Vertical bars in the lower half of this figure represent the same E1 transitions if an average shift of 1.0 MeV is assumed for all except the 18.31 MeV state of Figure 10(a) and an average shift of 0.5 MeV is assumed for all except 18.66 state of Figure 10(b). Solid curve in each figure represents experimental results (Ref. 35).

which may arise from other multipole transitions. A simple theoretical study of the excitation function due to M1 and E2 direct transitions clearly shows that the net contribution of such processes is only a few per cent of an E1 transition. This is shown in Figure 3. Thus we do not expect a large E2 or M1 strength in this excitation energy. It is, however, interesting to note that this slight admixture of E2 and M1 transitions can reproduce the observed angular distribution and polarization data. This is shown in Figures 11 and 12.

A note of disparity hangs over the entire analysis of the photonuclear process. Because neutrons and protons have been treated on the same footing, *i.e.*, an effective charge has been used for all neutrons. This may be indicative of a serious drawback of using spherical shell model states as a basis. One should ponder legitimately about this effective charge and open one's mind to the use of a deformed basis. Such an attempt by Kluge[29] has already produced encouraging results. In particular, about six K = 1 and three K = 0 states can be generated between 19 and 26 MeV and the dipole strength to these states from the ground state is rather large. This is shown in Figure 13.

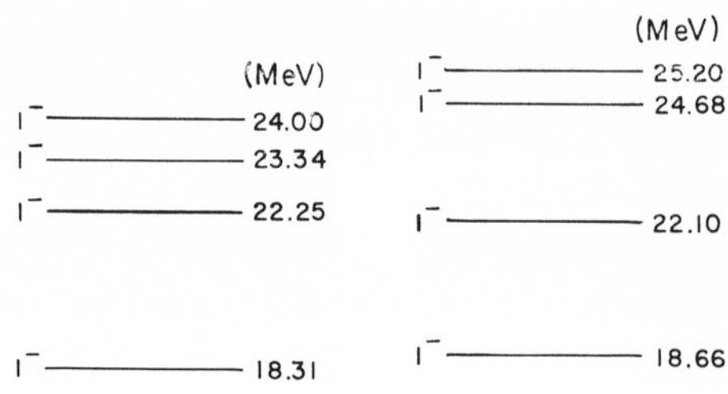

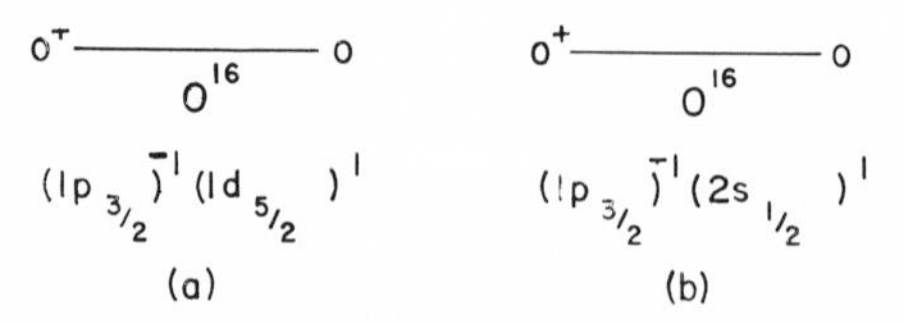

Fig. 10. (a) Unperturbed energies of 1^- states generated by coupling a $1p_{3/2}$ hole to a $1d_{5/2}$ particle state of the ZBM (Ref. 28). A 6 MeV spin-orbit splitting between the $1p_{3/2}$ and the $1p_{1/2}$ orbitals has been used. (b) The same as (a) for the coupling of a $1p_{3/2}$ hole to a $2s_{1/2}$ particle state.

B. C^{12}

That the ground state correlation plays an important role in determining the correct magnitude of the dipole strength is also demonstrated in the case of C^{12}. In Figure 14, the dipole strength is calculated in a particle-hole approximation using no correlation in the ground state. Figure 14 shows also transitions to the same particle-hole states but with ground state correlations incorporated by the open shell Tamm-Dancoff method formulated by Rowe.[30] Although the incorporation of this ground state correlation yields a reduced strength, it cannot account for the fine structure. To obtain further fragmentation Wong and Rowe[15] performed an intermediate coupling calculation using $(1s)^{-1}(1p)^9$ and $(1p)^7(2s\text{-}1d)^1$ configurations and Oak Ridge-Rochester shell model code.[31] This generated 41 $J = 1^-$, $T = 1$ states and the fragmentation of the dipole strength is also shown. A similar calculation with a much more restrictive space and yielding somewhat less fragmentation has also been performed by Goncharova and Yudin.[16] A comparison of the center part of Figure 14 with its lower part demonstrates that the fine structure originates from a redistribution of 1p-1h strengths among a series of satellite states by an intermediate coupling.

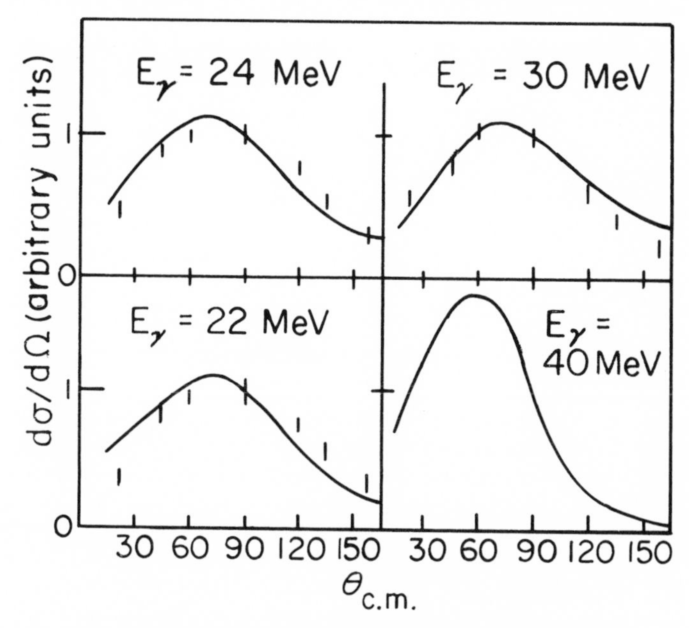

Fig. 11. The measured angular distributions in the $O^{16}(\gamma,p_0)N^{15}$ reaction normalized to unity at 90° (c.m.) (Ref. 36) marked as vertical bars are compared with theoretical results (solid curves) of direct emission (Ref.21).

For a proper comparison of the observed total photonucleon cross section, these discrete dipole strengths are to be folded into particle decay widths. Since the particle decay width in C^{12} is a few MeV, many of these dipole strengths can be buried underneath without showing a great deal of fine structure. However, this results in broad bumps. Moreover, the problem of overlapping resonances must now be solved more accurately using (37).

In C^{12}, as in the case of O^{16}, the photonucleon emission of a neutron poses the intriguing question of the "effective charge". Once again a deformed orbital approach to generate intermediate structure may provide a solution to this question. The preliminary work of Nilsson, Sawicki, and Glendenning[32] looks promising.

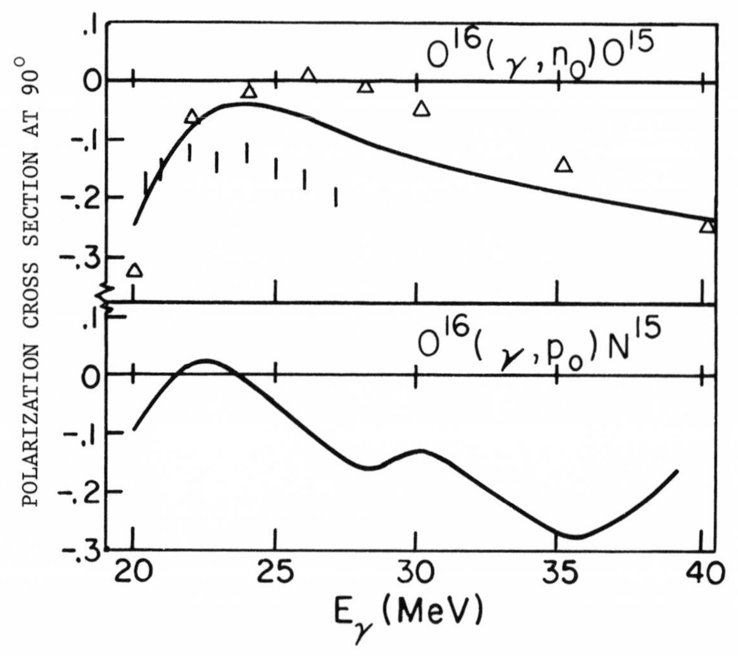

Fig. 12. In the upper part of the figure, observed $O^{16}(\gamma,n_o)O^{15}$ polarization cross section at 90° (Refs. 37 and 38) are plotted as vertical bars and compared to the theoretical prediction of a direct emission used in Ref. 21 . The solid line in the upper half represents calculated results using a single effective charge of ±0.5e for E1 and E2 transitions. Triangles are theoretical computations using neutron effective charges of (-Z/A) and Z/A^2 for E1 and E2 respectively. The curve in the lower half of the figure is the theoretical proton polarization at 90° in the $O^{16}(\gamma,p_o)N^{15}$ reaction using a single proton effective charge of +0.5e for E1 and E2 transitions.

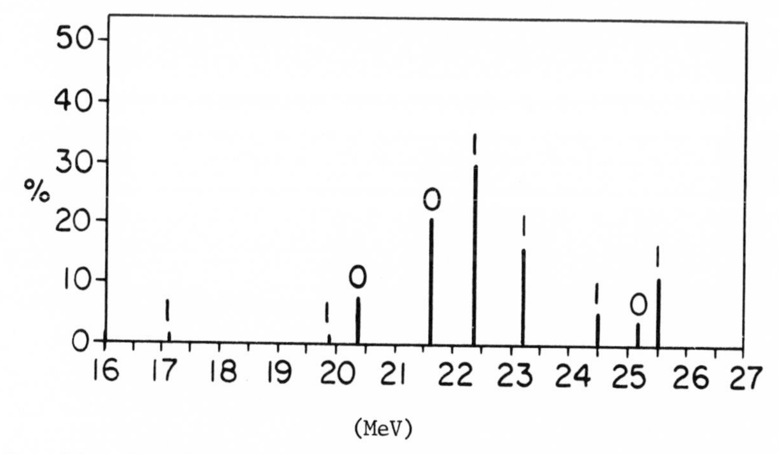

Fig. 13. The dipole strength from the ground state rotational band to rotational states in the giant dipole region (From Ref. 29).

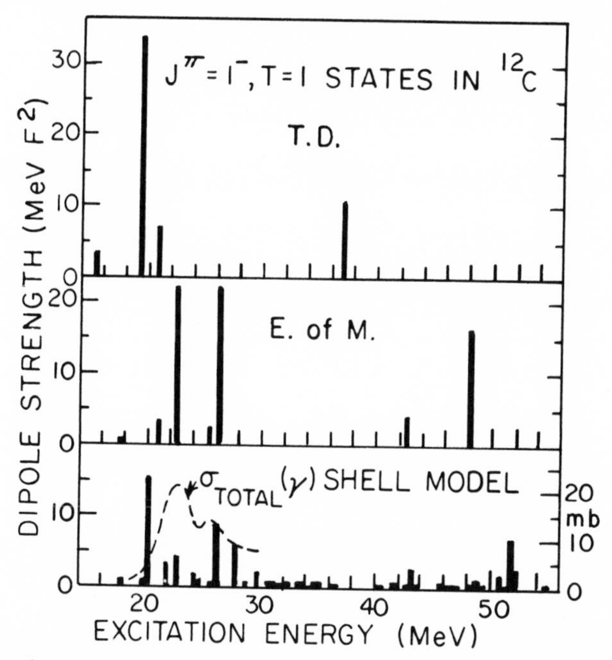

Fig. 14. The uppermost part represents the theoretical calculation of dipole strength from the ground state of C^{12} to 1^{-} states generated by a conventional 1 particle-1 hole calculation. The center represents the dipole strengths to 1^{-} states generated by an open shell Tamm-Dancoff calculation suggested in Ref. 15 . The bottom respresents E1 rates from the ground state of C^{12} to all 1^{-} states generated by an intermediate coupling shell model calculation incorporating $(1s)^{-1}(1p)^{9}$ and $(1p)^{7}(2s\text{-}1d)^{1}$ configuration space (From Ref. 15). Dashed curve represents experimental cross section.

C. Si^{28}

Figure 15 indicates the experimental results[33] of the neutron yields from $Si^{28}(\gamma,n)Si^{27}$. The experiment seems to contain three well separated broad major resonances, the envelope of each of which contains small resonances. The spacing of these small or minor peaks is fairly uniform for a particular major resonance, which is analogous to the typical feature of phonon broadened electronic transition in solids due to low energy vibration.

The particle-hole calculations[34] yield three 1^- states in the 17-21 MeV excitation energy region but fail in magnitude and cannot explain this fine structure.

Although Si^{28} is a complicated nucleus, it is reasonable to visualize that a large number of collective states of boson like character are present at such a high excitation energy. In fact a simple estimate based on a classical liquid drop model indicates a level spacing of 30 to 80 keV between these states at 17-21 MeV. In that case, these strong particle-hole states can be coupled to the boson-like states and this generates a situation which is very similar to the phonon-broadened electronic transition in solids. The intermediate coupling model used in References 12 and 13 is particularly suitable for this. The computed excitation function is shown in Figure 16, and can account for the magnitude and the broadening of observed resonances.

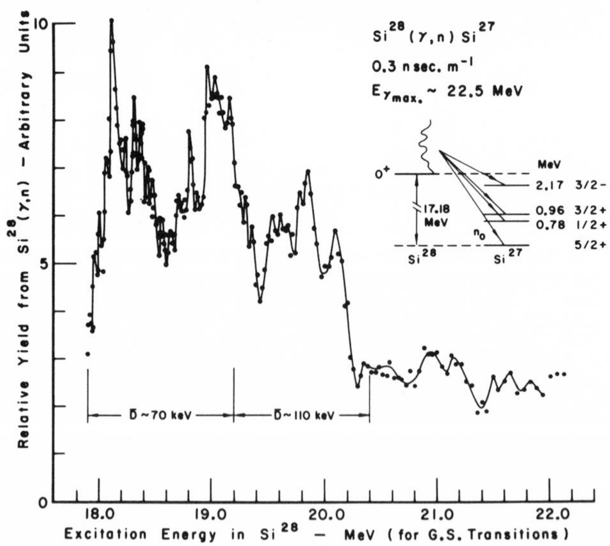

Fig. 15. The experimental neutron yield obtained for the reaction $Si^{28}(\gamma,n)Si^{27}$ by Firk (unpublished). The yield below 19 MeV may be artificially enhanced by the inclusion of non-ground state transitions. (From Ref. 13).

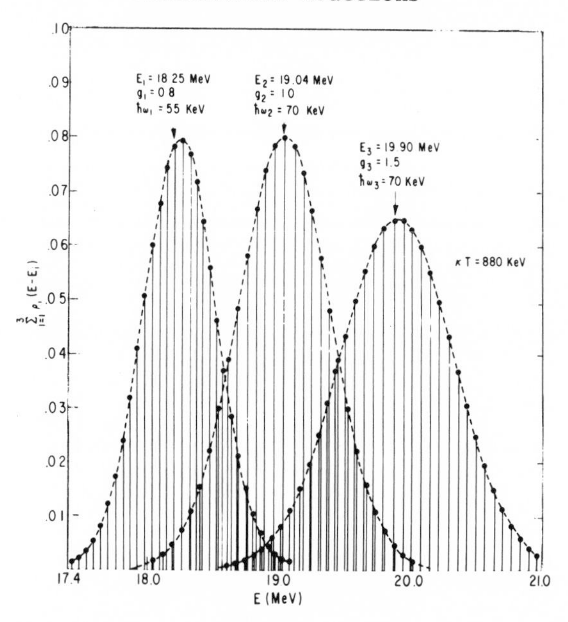

Fig. 16. The vertical lines are E1 transition probabilities from the ground state of Si^{28} to a series of 1^- states generated by coupling three strong particle-hole states to its satellites by a linear coupling (from Ref. 30). The lightly dashed line is inserted as a visual aid to locate the envelope of the sharp line-spectra. Parameters used are noted in the figure.

IX. CONCLUDING REMARKS

The emerging picture associated with excitation functions in photonuclear reactions indicated that these comprise a direct emission and a transition via a series of closely lying intermediate states. Thus these are clear cases of overlapping resonances where each resonance has not yet completely lost its individual identity to the extent that they can be termed as fluctuations. On the other hand, we may learn more about a sophisticated reaction mechanism by trying to understand in greater depth some of the basic parameters like particle width from a two-body potential.

This analysis also points towards a systematic change in the character of the observed structure across the periodic table. Whereas the large nucleon emission width in light nuclei can smear out resonances associated with complicated states, as in C^{12}, the situation changes with heavier nuclei because of the systematic variation of two basic parameters with increasing mass number A. First the single particle decay width decreases rapidly with increasing A and, second, the density of collective states of boson-like character increases with A. Because of this and of the finite experimental resolution the fine structure may no longer be observed beyond A ∿ 40 or 50 and instead one would see a single large bump.

Clearly, this intermediate structure is not restricted to the photonuclear reaction only. In the case of (p,n) reactions through analog states in light and medium weight nuclei, a similar physical situation persists. The strong analogue state can be fragmented by a coupling to its satellite state. Apart from the usual nuclear residual interaction, the Coulomb interaction couples such analogue states with other nearby complex particle or collective states.

REFERENCES

1. G. Breit and F. L. Yost, Phys. Rev. 48, 203 (1935).
2. L. R. Hafstad and M. A. Tuve, Phys. Rev. 47, 506 (1935); C. C. Lautrisen and H. R. Crane, private communication to G. Breit and F. L. Yost (1935).
3. See *e.g.*, G. A. Bartholomew, Ann. Rev. Nucl. Sc. 11, 259 (1961); or R. R. Christy and I. Duck, Nucl. Phys. 24, 89 (1961).
4. M. G. Mustafa and F. B. Malik, Phys. Rev. C1, 753 (1970).
5. N. Bohr and F. Kalckar, Kgl. Danske Videnskab. Selskab. Mat-fys. Medd. 14, No. 10 (1937).
6. E. Amaldi, O. D'Agostino, E. Fermi, B. Pontecorvo, F. Rosetti and E. Segre, Proc. Roy. Soc. (London) A149, 387 (1935).
7. A. M. Lane, R. G. Thomas and E. P. Wigner, Phys. Rev. 98, 693 (1955).
8. H. Feshbach, Ann. Phys. (N.Y.) 5, 357 (1958).
9. H. Feshbach, Ann. Phys. (N.Y.) 19, 287 (1962).
10. E. E. Trefftz, Z. Astrophysik, 65, 299 (1967).
11. F. W. K. Firk, Ann. Rev. Nucl. Sc. 20, 39 (1970).
12. C. B. Duke and F. B. Malik, *Proceedings of the International Nuclear Physics Conference*, Richard L. Becker, ed., Academic Press (New York) 1967.
13. C. B. Duke, F. B. Malik and F. W. K. Firk, Phys. Rev. 157, 879 (1967).
14. C. M. Shakin and W. L. Wang, Phys. Rev. Lett. 26, 902 (1971); Phys. Rev. C5, 1898 (1972).
15. D. J. Rowe and S. S. M. Wong, Phys. Lett. 30B, 147, 150 (1969).
16. N. G. Goncharova and N. P. Yudin, Phys. Lett. 29B, 272 (1969).
17. K. Kamimura, K. Ikeda and A. Arima, Nucl. Phys. A95, 129 (1967).
18. M. G. Mustafa, doctoral thesis, Yale University, 1970, (unpublished).

19. N. Austern, Ann. Phys. (N.Y.) 45, 113 (1967).
20. F. M. Nicolau, Z. Naturforsh, 16a, 603 (1961).
21. M. G. Mustafa, and F. B. Malik, Bull. Am. Phys. Soc. 14, 607 (1969); Phys. Rev. C2, 2068 (1970).
22. J. P. Elliot and B. H. Flowers, Proc. Roy. Soc. (London) A242, 57 (1957).
23. V. Gillet and N. Vinh Mau, Nucl. Phys. 54, 321 (1964).
24. G. E. Brown and M. Bolsterli, Phys. Rev. Lett. 3, 472 (1959); G. E. Brown, L. Castillejo and J. A. Evans, Nucl. Phys. 22, 1 (1961).
25. C. Mahaux and H. A. Weidenmuller, Phys. Lett. 19, 408 (1965); B. Buck and A. D. Hill, Nucl. Phys. A95, 271 (1967); J. D. Perez and W. M. MacDonald, Phys. Rev. 182, 1066 (1969); W. P. Beres and W. M. MacDonald, Nucl. Phys. A91, 529 (1967); J. Raynal, M. A. Melkanoff, and T. Sawada, Nucl. Phys. A101, 369 (1967); A. M. Saruis and M. Marangoni, Nucl. Phys. A132, 433 (1969).
26. V. Gillet, M. A. Malkanoff and J. Raynal, Nucl. Phys. A97, 631 (1967).
27. C. Mahaux and H. A. Weidenmuller, *Shell Model Approach to Nuclear Reactions*, North Holland Publishing Co., Amsterdam, 1969; W. Glockle, J. Hufner and H. A. Weidenmuller, Nucl. Phys. A90, 481 (1967).
28. A. P. Zuker, B. Buck and J. B. McGrory, Phys. Rev. Lett. 21, 39 (1968); Brookhaven National Laboratory Report, BNL 14085 (1968).
29. G. Kluge, Z. Physik, 197, 288 (1966).
30. D. J. Rowe, Rev. Mod. Phys. 40, 153 (1968); Nucl. Phys. A107, 99 (1968).
31. J. B. French, E. C. Halbert, J. B. McGrory and S. S. M. Wong, *Advances in Nuclear Physics, Vol. III*, M. Baranger and E. Vogt, eds. Plenum Press, New York, 1969.
32. S. G. Nilsson, J. Sawicki and N. K. Glendenning, Nucl. Phys. 33, 239 (1962).
33. F. W. K. Firk, Bull. Am. Phys. Soc. 11, 367 (1966).
34. J. B. Seaborn and J. M. Eisenberg, Nucl. Phys. 63, 496 (1965).
35. F. W. K. Firk and K. H. Lokan, Phys. Rev. Lett. 8, 321 (1962); F. W. K. Firk, Nucl. Phys. 52, 437 (1964).
36. J. E. E. Baglin and M. N. Thompson, Nucl. Phys. A138, 73 (1969).
37. G. W. Cole, Jr., F. W. K. Firk and T. W. Phillips, Phys. Lett. 30B, 91 (1969).
38. G. W. Cole, Jr., Ph.D. Thesis, Yale University, 1970 (unpublished).
39. N. F. Mott and H. S. W. Massey, *The Theory of Atomic Collisions*, Oxford University Press, London, 1949.

DISCUSSION

KOSHEL: The approximation of setting the continuum-continuum coupling to zero has been used quite a few times. It was first introduced by Fano in the study of autoionization of electron-atom scattering. It was later used by Bloch, Weidenmüller and others and they found that it is a poor approximation if you have narrow resonances in the region of interest. This is true in particular for single particle resonances. Now, you seem to have that here, so that this may really be a poor approximation for O^{16}. In fact, calculations not setting the continuum-continuum coupling to zero have been done by Melkanoff and Raynal. How does your calculation differ from these, other than setting the continuum-continuum coupling term equal to zero?

MALIK: May I comment on that? The original work of Fano was only done in the case of two levels which is nice but simple. It doesn't consider the problem of overlapping resonances which was really done by Trefftz. The work of Weidenmüller and others completely neglected that interaction between many bound states, and the continuum states to the continuum. Well, I had to read that paper because somebody else made the same comment some time ago and we agreed that they considered one bound state and the continuum-continuum coupling. Melkanoff and Raynal said that people cannot reproduce either the fine structure or the magnitudes. This is just going to the other direction. They neglected the resonances which are *not* narrow. We have shown that the bound state coupling is important and that their interaction with the continuum and normalization are also important.

CRAWLEY: In the early part of your talk you made the point that the direct component of the giant dipole resonance was about 10%. You made this argument from the fact that the theoretical calculation gave about 10%. How reliable do you feel this calculation is for a direct reaction? We know it is really very difficult to obtain the absolute magnitudes of cross sections in direct reactions. I think you've shown later on that magnitudes are a real problem.

MALIK: Oh, we played with that thing. It is reliable to within about 50%, *i.e.*, if one is pessimistic, up to 15% depending on the nature of the optical model. In the silicon case, of course, there is another piece of evidence because it was analyzed in terms of fluctuations and it was found that it can only be fitted with a parameter indicating that the direct component is large. This is the parameter which is the ratio of the direct to compound process.

GLAVISH: I was interested in your predictions for the polarization in the O^{16} giant dipole resonance. At Stanford we have made very accurate measurements of the equivalent thing in the inverse capture reactions. We have measured the analyzing power using incident polarized protons. The point is that the polarization is near enough to being identically zero at 90° throughout the whole giant dipole resonance but quite large at 45° and 135°, which implies that the radiation is almost pure E1 and this feature persists at every giant dipole resonance we studied from He^4 through C^{12}, O^{16}, Ne^{20} and Si^{28}. It is only in Si^{28} that we see evidence for a polarization that is significantly different from zero at 90°. Even through the intermediate structure the polarization remains almost constant.

MALIK: There is a very interesting thing I would like to show on slide 3 (Fig. 3), please. There is an interesting question there: can you reconcile that picture with the angular distribution?

GLAVISH: Yes, we have in fact done this for each reaction.

MALIK: In this calculation, of course, we have only calculated the direct part. The primary component of the total cross-section is always E1. E2 and M1 transition probabilities form only a fraction of the total cross section. It is about 1% or 2% but in the next slide (Fig. 4) we have experimental points at 22 MeV and 24 MeV. If it were a pure E1 direct transition, the angular distribution would be symmetric about 90°. These experimental points were taken about 2 years ago and I can give you the reference. This slight mixture of E2 and M1 changes the shape of the theoretical curve and we get an agreement with the experimental data. Of course, in this calculation we have not calculated the contribution to the cross sections for transitions through intermediate states.

GLAVISH: We have, in fact, simultaneously considered the angular distribution data and the polarization data and extracted unique values for the reaction amplitudes and phases for the $N^{15}(p,\gamma)O^{16}$ giant resonances. It is true that the unpolarized angular distribution reflects E2 interference. But my real point was that the polarized angular distribution certainly doesn't show this and one of your slides (Fig. 12) indicated that there was a large 90° polarization which is entirely inconsistent with our results.

MALIK: Well, I would be very happy if it could be shown that it's a pure E1 transition. In that case, we can simplify the calculation and not worry about E2 and M1. But I can give you those experimental data. Probably you know it yourself. If these data can be reconciled with your data, we'd be very happy and if it is a pure E1 we'd be even happier.

TEMMER: Just as a point of information: did I understand you correctly to say that you coupled 0^+ and 2^+ states in the vicinity of that strong state, to generate your intermediate structure?

MALIK: That is what Shakin and Wang did and they also coupled to some of the 3^+ and 3^- states but primarily to 0^+ and 2^+ states in that vicinity. These 0^+ and 2^+ states which they could generate from a particle-hole type of interaction, in the model which we use for O^{16}, we generate by weak-coupling a 1p hole with the 2s-1d particle states of Zuker, Buck, and McGrory.

TEMMER: At the risk of sounding very ignorant I thought that these primary privileged states would couple only to states having spin one minus.

MALIK: I think if I understand correctly, that one can couple 0^+ and 1^- and get 1^-.

TEMMER: I believe that I begin to see what you are trying to say. The primary 1^- states couple to various more complicated states, also having spin 1^-, via the residual nucleon-nucleon interaction. These more complicated states, in turn, are built by coupling 1^- states to other 0^+ or 2^+ "boson" states, via the electromagnetic E1 operator. You have simply, by force of habit and familiarity, referred to those more complicated states as "0^+ and 2^+ states." They have in fact, spin 1^-.

III.B. FINE STRUCTURE OF ANALOG STATES*

G. E. Mitchell
North Carolina State University
Raleigh, North Carolina
and
Triangle Universities Nuclear Laboratory
Durham, North Carolina

I. INTRODUCTION

The discovery of analog states in medium and heavy nuclei and the subsequent renaissance of interest in isospin is well known and well documented.[1-3] In this talk it is impractical to give more than cursory attention to the history of analog states, but we should give credit to the pioneering work of Anderson[4] and collaborators (the observation of analog states via the (p,n) charge exchange reaction), of Lane[5] (the Lane potential), of Fox, Moore and Robson[6] (observation of analog states in the compound nucleus), and of Robson[7] (first theoretical description of compound nuclear analog states).

The aspect of analog states which we wish to emphasize today is fine structure. Considered as a special case of the general problem of line broadening, the fine structure of analog states is an old problem revisited. There are analogies with the fragmentation of single particle states in nuclei (Ref. 8) and with autoionizing states in atoms (Ref. 9). One expects "giant resonance" type of behavior, with the special (analog) state fragmented. This is illustrated schematically in Figure 1. The first direct evidence of this fragmentation was obtained by Richard, Fox, Moore and Robson[10] on Mo^{92}(p,p). Their work indicated fragmentation of the analog, although the individual states were not completely resolved. The question was answered conclusively by Keyworth, Kyker, Bilpuch and Newson[11] in their classic experiment on Ar^{40}. In this experiment, involving a cryogenically pumped gas target and a feedback system to correct for beam energy fluctuations, an overall proton energy resolution of <200 eV was obtained. The individual states were resolved, and the existence of fine structure for analog states established.

Our efforts towards studying fine structure in depth form the basis of this talk. This work was made possible by the beautiful high resolution system developed over many years by Professors H. W. Newson and E. G. Bilpuch of Duke University. The work described today is the result of a collaborative effort by Professors Bilpuch, Newson, and myself, helped

---

*Supported in part by the U.S. Atomic Energy Commission.

immeasurably by many dedicated graduate students (see acknowledgement).

The scope of the project is suggested by the following numbers--over twenty isotopes have been studied and several thousand (∿3,000) resonances have been analyzed. Resolution and energy considerations limit the experiments to A $\lesssim$ 64 (with our 3.3 MV accelerator). Although we have succeeded in extending these high resolution techniques to higher energies on our other single-ended Van de Graaff accelerator and on our tandem Van de Graaff accelerator, the results presented today were all obtained with the smaller accelerator. Specifically we have studied (with solid targets) elastic scattering on Si^{30}, $Ca^{40,42,44,48}$, $Ti^{46,48,50}$, $Cr^{50,52,54}$, $Fe^{54,56,58}$, $Ni^{58,60,62,64}$, and $Zn^{64,66}$. Inelastic scattering was also measured where the cross sections permitted. We studied not just the immediate vicinity of the analog states,

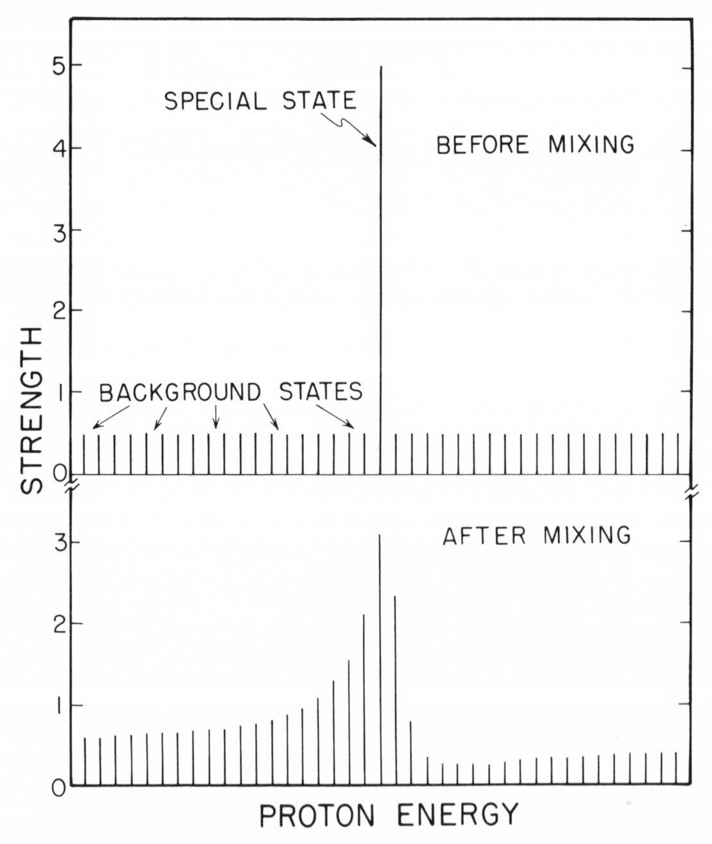

Fig. 1. Mixing of the special state with the background states, assuming a Robson distribution. A "picket fence" background is assumed for simplicity.

but also the region between the analog states. Thus a large amount of information of a statistical nature (*e.g.*, proton strength functions and nuclear level densities) has been obtained. This is not really a completely separate topic, for it is essential to understand the background states in order to understand the analog state fine structure. The (p,n) reaction has been studied on several isotopes in order to examine the important question of possible enhancement of neutron widths by the analog state. The capture reaction has been studied on several isotopes.

Our governing philosophy should be made explicit. We are of course interested in spectroscopic factors and Coulomb energies, in inferring the structure of parent states via the unique method of inelastic scattering from analog states, in antianalog states and in comparisons with beta decay. However, all of these topics may be explored without recourse to fine structure measurements (although our prejudice says--often not as well). To paraphrase Mahaux,[12] the value of fine structure experiments lies in examining analog states with a better "microscope" and therefore subjecting the various theories to closer, more detailed scrutiny. We therefore emphasize topics such as the fine structure distributions and correlations between the partial widths in various channels. A comprehensive treatment of these topics is given by Lane.[13]

The spirit of the presentation is to present illustrative examples, with no pretense of being either complete or definitive. After a discussion of experimental matters, examples of elastic scattering data are presented. After discussing Coulomb energies, spectroscopic factors and fine structure distributions, results for the (p,n), (p,γ) and (p,p') reactions from analog states are presented. Following a brief discussion of statistical considerations, the talk concludes with brief comments on projected high resolution experiments.

II. EXPERIMENTAL APPARATUS AND PROCEDURE

The high resolution system[14] is illustrated in Figure 2. One beam (H^+) is used to perform the experiment, while the associated beam (HH^+) passes through an electrostatic analyzer. A signal is obtained from the difference in current on the exit slits of the analyzer and a correction voltage applied to one plate of the analyzer in order to center the beam on the exit slits. The feedback system (denoted in Figure 2 by homogenizer) also applies a voltage to the target to correct for beam fluctuations. Although some phase lag is inherent in the system, the primary terminal voltage fluctuations are of rather low frequency and in practice the system works very well.

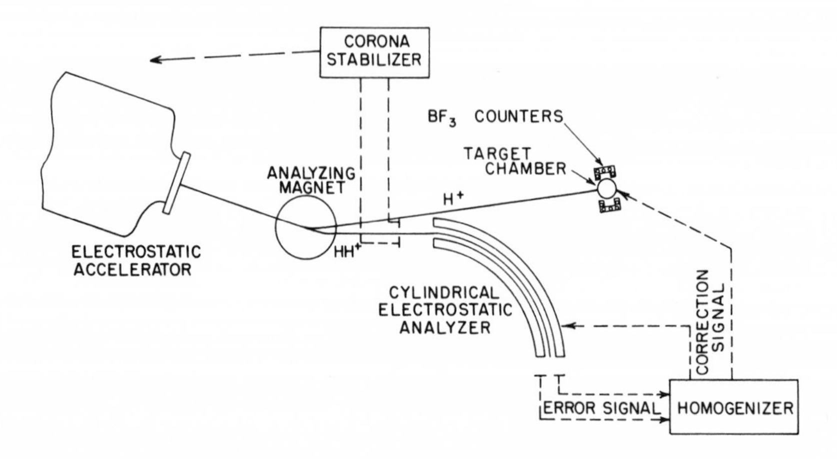

Fig. 2. Schematic of the high resolution system. The feedback system is denoted by "homogenizer."

A key feature of the system is the use of separate control and experimental beams. There is no need to restrict the proton beam in order to obtain high resolution. Over 100 μA of protons have been placed on a gas target; for solid targets typical beam intensities are ∿5 μA, but this is a target limitation. Although the system is capable of 100 eV resolution, for solid targets the best overall resolution is about 275 eV. The limiting factor is the target Doppler broadening. Typical targets consist of 1-2 μg/cm^2 of target material on 10 μg/cm^2 carbon backings.

A word about "resolutionmanship". The resolution that we quote is the width of a Gaussian resolution function which is used in fitting the resonance data. The resolution is determined empirically by fitting narrow resonances. Since it is the overall resolution that determines what is actually observed, usage of other quantities as a measure of resolution is misleading at best.

The experimental procedure followed is described in a number of our papers and will not be repeated here in detail. Briefly, we measure at four angles simultaneously for charged particle reactions, with conventional solid state detectors. A semiautomated data acquisition system was fabricated for these elastic scattering experiments. The data are recorded on punched cards for later processing and analysis. Gamma ray experiments were performed using NaI(Tl) and Ge(Li) detectors. The γ ray spectra were recorded with a DDP-224 online computer.

III. ELASTIC SCATTERING DATA

In order to simplify the analysis, our experiments have been limited to spin zero targets. The proton bombarding energies are 2 to 3 MeV and the targets have a Z value in the range 20 to 30. Since the proton energy is so low compared to the Coulomb barrier, the background is adequately represented by Coulomb plus hard sphere phase shifts. Normally only s-, p- and d-wave resonances are observed. Resonance parameters are extracted by fitting the data with a computer program based on the multi-level, multi-channel R matrix formalism.[15]

In practice, for a fixed bombarding energy, the proton separation energy (and thus the excitation energy in the compound nucleus) changes appreciably from isotope to isotope. The observed level densities therefore vary dramatically. Whether analog states are fragmented or single levels is primarily determined by the character of the background -- that is, by the level density and the strength function.

Figures 3 and 4 illustrate the change of the level density in the iron isotopes.[16,17] Figure 3 shows elastic scattering from Fe^{54}. Resonances are very well isolated and the analysis is straightforward. Note the increase in observed level density as a function of bombarding energy. This is due in part to a real increase in the level density with in-

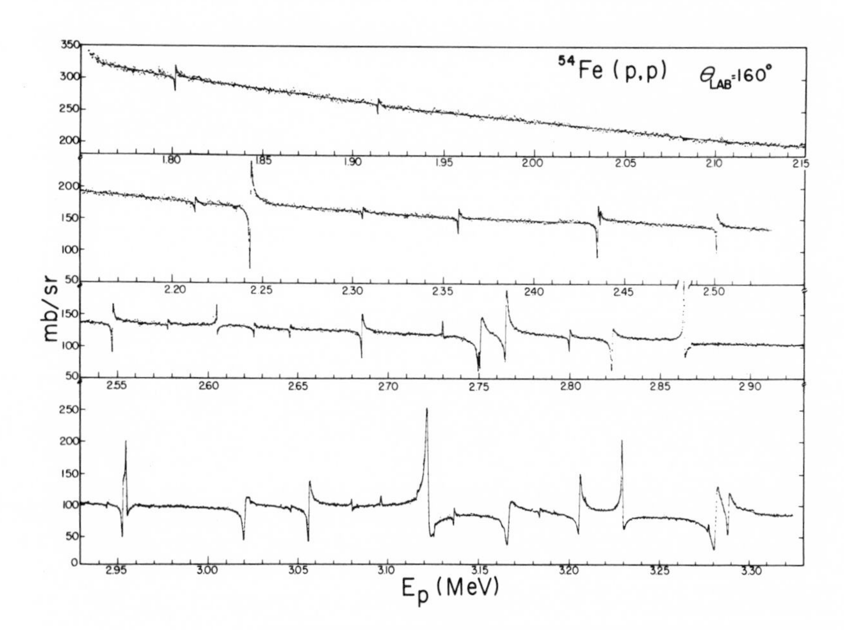

Fig. 3. Excitation function for the Fe^{54}(p,p) reaction at 160°.

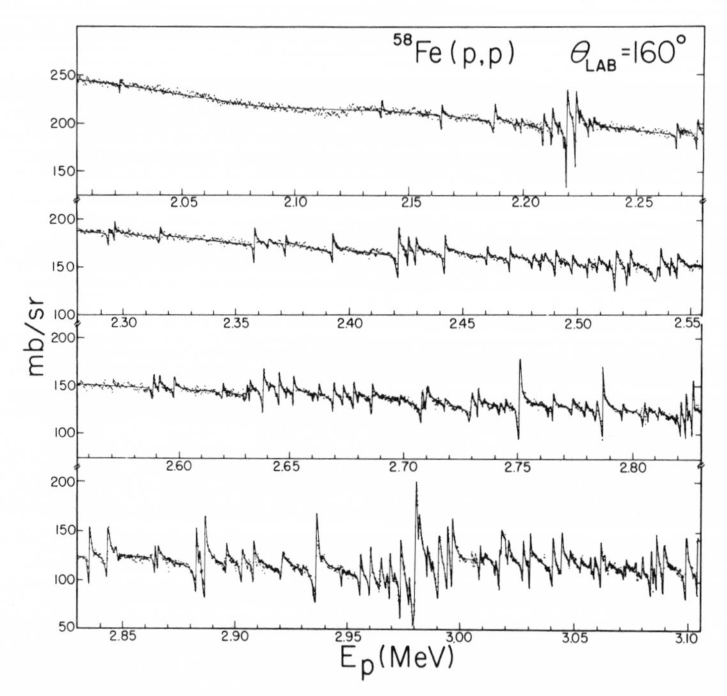

Fig. 4. Excitation function for the Fe^{58}(p,p) reactions at 160°. Analog states occur at $E_p \sim$ 2.22-, 2.5- and 2.98-MeV.

creasing excitation energy and in part to the increase in penetrability with bombarding energy. The latter effect dominates for higher values of orbital angular momenta. Figure 4 shows elastic data for Fe^{58}. With such a high density, fragmentation of the analog states is expected. For example, the clusters of resonances at $E_p \simeq$ 2.2 and 3.0 MeV are analog states. The lower energy analog is apparent by inspection.

Figure 5 shows the overall results[18] on Ti^{48}, with a more compressed energy scale. The lines through the data are fits; about 300 resonances are observed in the elastic channel. This example suggests the difficulty of interpreting such data without ultrahigh resolution. Even the resolution achieved for this experiment (about 300 eV) is perhaps marginal. Ti^{48} was notable in several respects -- at the time this was the most complicated set of data that we had attempted to analyze. A 1 eV f-wave resonance was analyzed in Ti^{48}(p,p); this is the smallest elastic resonance that we have fit. In the lower part of Figure 5 some of the data is shown on an expanded scale, illustrating the quality of fit obtained for a complicated excitation function. Inelastic data are also shown; the line through the inelastic data is again a fit.

Figure 6 shows Ti^{50}(p,p) elastic data[18] at four angles in a limited energy region. This figure illustrates that the ℓ-value of the resonance may be determined almost by inspection. The 90° data indicate the parity immediately, since positive parity states show dips at 90° while negative parity states show rises at 90°. States with ℓ = 0 and 2 have strikingly different patterns at back angles. Differentiating between resonances with $j = \ell \pm \frac{1}{2}$ is more difficult. Whether the j value can be reliably assigned for p-wave resonances depends upon the strength of the resonance, on the amount of interference from neighboring resonances and on the resolution. Information from inelastic scattering is quite valuable in determining the j-value. Often there is some ambiguity in

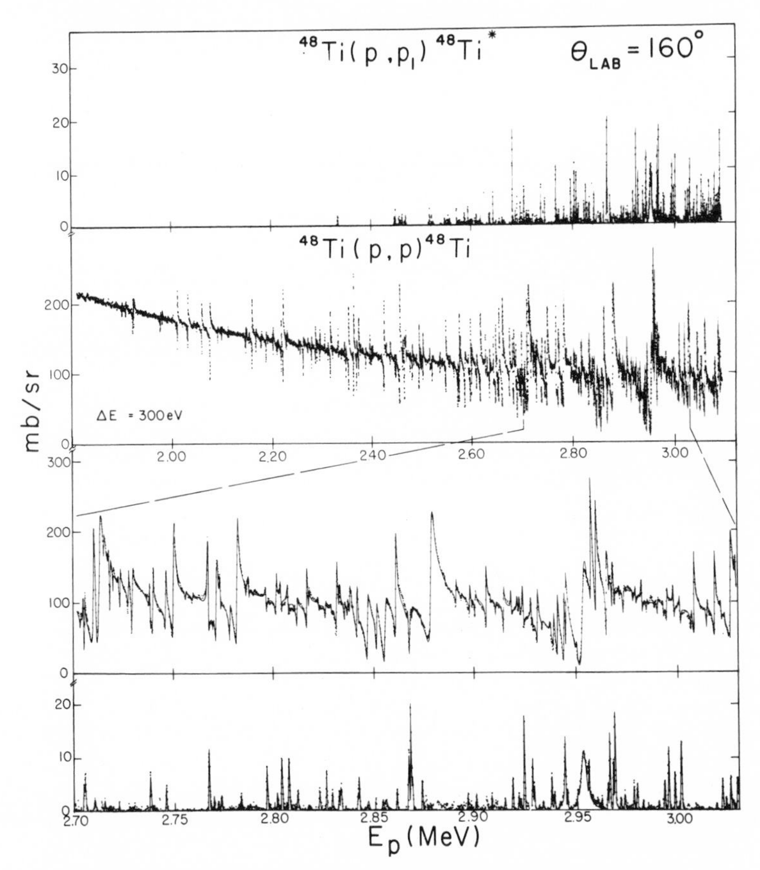

Fig. 5. Elastic and inelastic proton scattering from Ti^{48} on a compressed scale. The lower part of the figure shows a subset of the data on an expanded scale in order to illustrate quality of fit. All lines through the data represent fits.

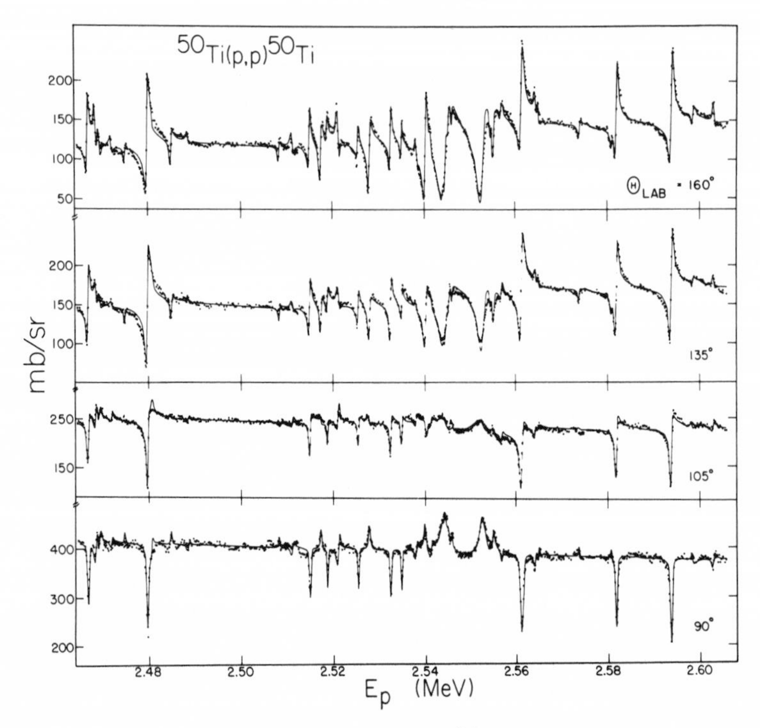

Fig. 6. Elastic scattering from Ti^{50} at four angles. Note that at 90^0 positive parity states show dips, while negative parity states show rises. There is a fragmented p-wave analog centered near $E_p \sim 2.54$ MeV. Lines through the data are fits.

the spin assignments for weak p-wave resonances. The situation is much worse for d-wave resonances, where it is usually not possible to distinguish the j-value without information from the inelastic channel.

Interpretation of the data is facilitated by sorting the reduced widths according to spin and parity. In Figure 7 the elastic reduced widths for $3/2^-$ states in Co^{55} are plotted both individually and as cumulative sums. The first, second, fourth and last states are analogs of low lying states in Fe^{55} and appear as single levels. Since there are no background states with which to mix, there is no fragmentation. As a consolation, there is no difficulty in identification of the analog states.

The situation in Fe^{58}(p,p) is quite different, as is shown in Figure 8. The sum of the s-wave strength varies smoothly with energy, while anomalies are observed in the p-wave reduced widths. After subtracting the analog strengths the sum of reduced widths for $p_{1/2}$ and $p_{3/2}$ states is much smaller than the sum of reduced widths for the $s_{1/2}$ states. There are two striking $3/2^-$ analogs at $E_p \simeq 2.2$ and 3.0 MeV;

they are highly fragmented. Notice that there are almost no $p_{3/2}$ resonances away from the analog states. In this case, identification of the analog states is as easy as in the single level case illustrated by Fe^{54}. There are unfortunately intermediate situations where the background states have strengths comparable to the analog strength, and the identification of analog states is more difficult. The relative sizes of the $p_{3/2}$ and $p_{1/2}$ strength functions are apparent in Figure 8. The $p_{1/2}$ strength function is much larger than the $p_{3/2}$ strength function, but not large in absolute magnitude. This is reasonable, since mass 58 should be well above the 2p giant resonance, but (due to the spin-orbit

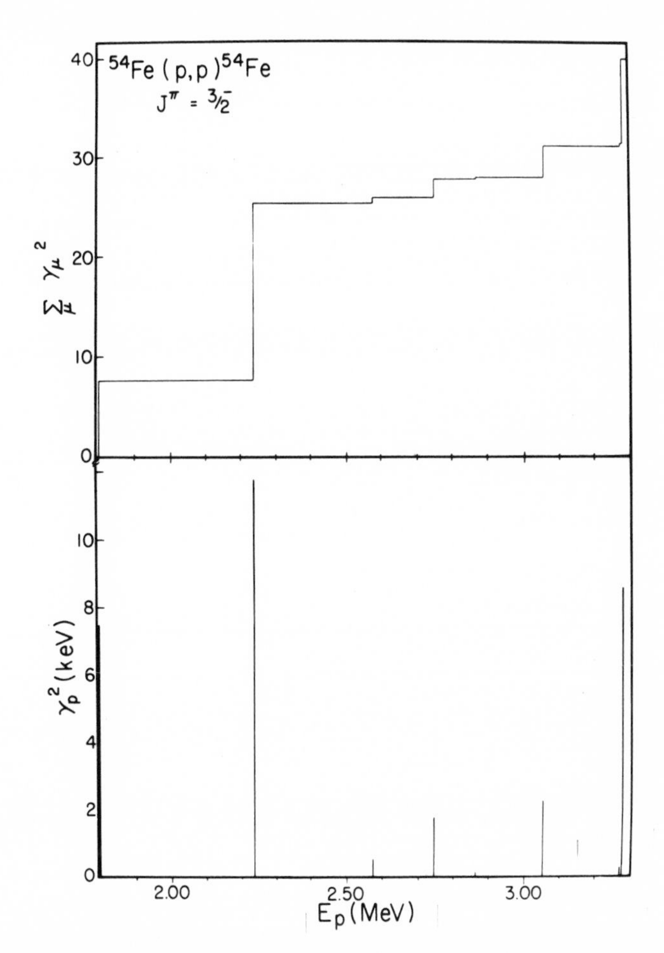

Fig. 7. Reduced widths for the $p_{3/2}$ resonances observed in Fe^{54}(p,p). The first, second, fourth and last states are analogs.

splitting) closer to the $p_{1/2}$ part of the 2p giant resonance.

In the identification of analog states one is guided on the one hand by the experimental results (most are obvious anomalies in reduced width plots), and on the other by in-

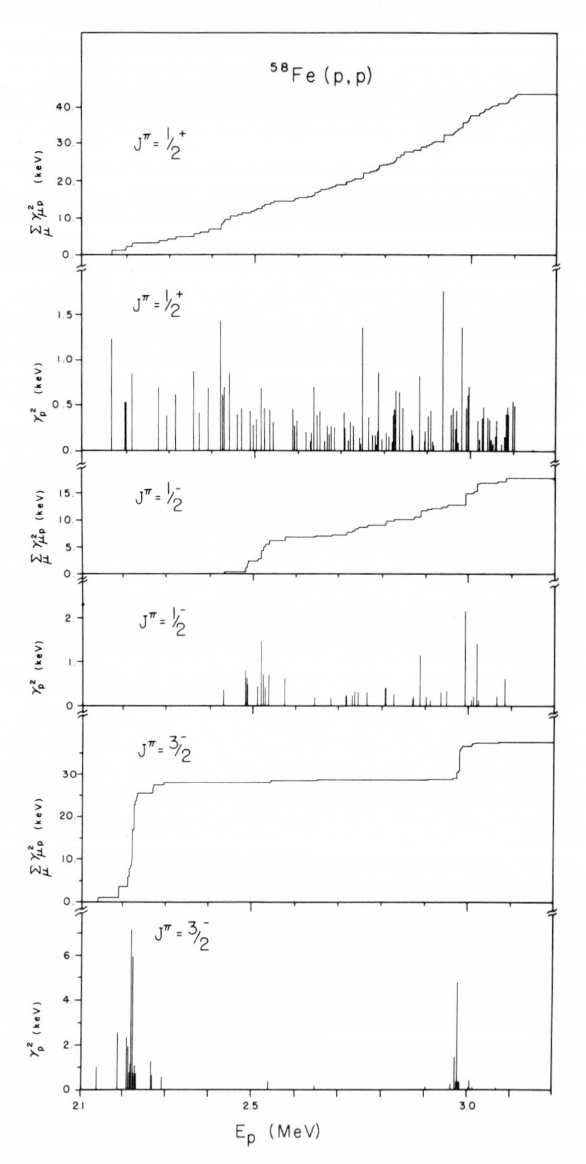

Fig. 8. Reduced widths for resonances observed in Fe^{58}(p,p). Note the smoothness of the cumulative $s_{1/2}$ strength, and the relative magnitudes of the strength functions. Fragmented $p_{3/2}$ analogs are apparent by inspection.

formation from (d,p) experiments which establish location, ℓ-value, and single particle strength of the parent states. Thus Coulomb energies and spectroscopic factors are important in the identification of analog states, as well as being interesting topics in themselves.

IV. COULOMB ENERGIES AND SPECTROSCOPIC FACTORS

The Coulomb energy difference between the parent and analog state is given by

$$\Delta E_c = B_n + E_p^{cm} - E_x$$

where B_n is the binding energy of the last neutron in the parent system, E_p^{cm} is the energy of the analog state and E_x is the excitation energy of the parent state in the parent system.

Our absolute energies are normally determined only to a few keV unless special efforts are made, although the relative energies over a narrow energy range are reproducible to a few hundred eV. Analog energies are taken to be:

(1) for analogs which occur as single levels, the observed energy,

(2) for analogs with a few fragments, the centroid of the distribution and

(3) for highly fragmented analogs, the analog energy as determined by fits to a Robson distribution (see Section V).

Typical errors on the Coulomb energies are about 5 keV. For practical purposes, such as identification of analogs, simple phenomenological formulae[19] usually predict Coulomb energies reliably at the order of 0.5%. Typical variations in ΔE_c from state to state (in a given isotope) average a few tens of keV.

The spectroscopic factor for the analog state is taken to be

$$S_{pp} = (2T_o + 1)\frac{\Gamma_p^A}{\Gamma_{sp}}$$

where T_o is the isospin of the target nucleus, Γ_p^A is the proton width of the analog state and Γ_{sp} is the single particle width. That is Γ_p^A is the sum of the proton partial widths for the fragments of the analog minus the background strength. The latter is usually rather small. To order $1/(2T_o + 1)$, S_{pp} should equal the (d,p) spectroscopic factor of the parent state. This relation can be used to estimate Γ_p^A and thus is an aid in the identification of analog states. More importantly, the analog state measurement yields an independent determination of the parent state spectroscopic factor. Uncertainties in the value of S_{pp} arise both from the calculation of Γ_{sp} and the determination of Γ_p^A.

The single particle widths were calculated using a computer program written by Harney.[20] This program calculates single-particle widths by three methods; a detailed comparison of these methods is given by Harney and Weidenmüller.[21] In calculating proton single-particle widths, the neutron potential was varied to obtain the correct neutron binding energy for the particular parent state. The corresponding proton well was then calculated by adding to the neutron well a symmetry potential of the form

$$U_{sym} = \tfrac{1}{2}\left(2T_o + 1\right)\ 125/A\ \text{MeV}.$$

The resulting potential was used to calculate the proton single-particle widths.

In addition to experimental uncertainties in Γ_p^A and to calculational uncertainties in Γ_{sp}, there are a variety of other uncertainties. For example, contributions to the analog widths from the wings of the distribution are missed if one only observes resonances in the central region of the analog. Corrections for the existence of other channels also may effect the analog strength.

In the comparison between S_{pp} and S_{dp}, the uncertainties in S_{dp} are certainly relevant. As an extreme case in point, one of our most striking disagreements involved the spectroscopic factor for a fragmented $p_{3/2}$ analog state. We obtained $S_{pp} \simeq 0.05$, while the S_{dp} value was 0.22. This appeared to be a glaring disagreement, especially since for this particular state both the analog data and the (d,p) data looked very good. However, a very recent (d,p) experiment obtained $S_{dp} = 0.045$.

Our results for about 35 p-wave analogs yield spectroscopic factors which average about 0.6–0.7 of the spectroscopic factors determined in the (d,p) experiments. This result seems to hold true for strong and weak analogs, and for fragmented and unfragmented analogs. Our results are by and large consistent with the few gross structure studies performed for the same analogs. We studied one analog several ways: by fine structure, by averaging the fine structure data and by performing a poor resolution measurement--the three results agreed.

It is probably fair to say that the analog and (d,p) results agree in a very general way, that there is some evidence that analog state spectroscopic factors are lower than (d,p) spectroscopic factors, and that the large discrepancies observed for specific states probably indicate errors in one of the two types of experiments rather than any fundamental difficulty. All of our spectroscopic factors have been obtained with the same equipment and the same analysis procedures, but we are comparing with (d,p) results obtained by many different groups at many different laboratories using different procedures. Overall the average discrepancy of 30% or so is probably not too surprising.

V. FINE STRUCTURE DISTRIBUTIONS

The characteristic pattern for a fragmented analog state is well known. The region of enhancement is roughly Lorentzian, with a typical width of a few tens of keV. The distribution is asymmetric, with stronger enhancement on the low energy side. The detailed features of the fine structure distribution are predicted by microscopic theories of analog states; they are tested experimentally by comparison with fine structure measurements. The distribution of the reduced widths is expected to be described by[13]

$$S(E) = S_o \frac{\left(E - E_\lambda - \Delta\right)^2 + \omega^2/4}{\left(E - E_\lambda\right) + \Gamma^2/4}$$

where S(E) is the local strength function ($S = \langle \gamma^2/D \rangle$), D is the level spacing, S_o is the background strength function, E_λ and Δ are the energy and displacement of the analog state, Γ is the observed width of the distribution and ω^2 is a parameter which includes the effect of other open channels; ω^2 should vanish for the one channel case. Γ is related to the spreading width W_o by

$$\frac{\Gamma^2}{4} = \frac{W_o^{\,2}}{4} + \frac{W_o D}{2\pi} .$$

Difficulties in fitting this fine structure distribution to the data arise because in most cases the analog states have too few fragments for reliable fitting, and because the background proton strength functions are often small and inaccurately measured. The procedure adopted is to average both the data and the theoretical distribution. The fitting is performed by an automatic search code which varies any or all of the parameters in order to minimize χ^2. Since some of the parameters are interdependent (empirically and theoretically), studies have been performed to understand these interdependencies and to consider the uniqueness of fit. Although "blind" fitting procedures are fairly rapid and normally lead to convergence, they may sometimes be very misleading.

The energies of the analogs are rather well determined, as are the spreading widths. The other parameters are less well determined, but approximate values can usually be obtained. Sample data and fits are shown in Figure 9. (Since the energy averaged data and the theoretical distribution are dimensionless quantities, they are actually plotted on a different scale from the original data. To simplify the drawing, these scales have been suppressed.) In practice we observe mostly p-wave analogs and usually weak mixing. That is, the analog strength is appreciably shared by at most a few fragments. The spreading widths are usually (but not always) rather small, of the order of 10 keV or less. This presumably reflects the fact that in this mass region, the analogs of low-lying parent states are rather far removed from both the antianalog and from the next p-wave giant resonance.[22]

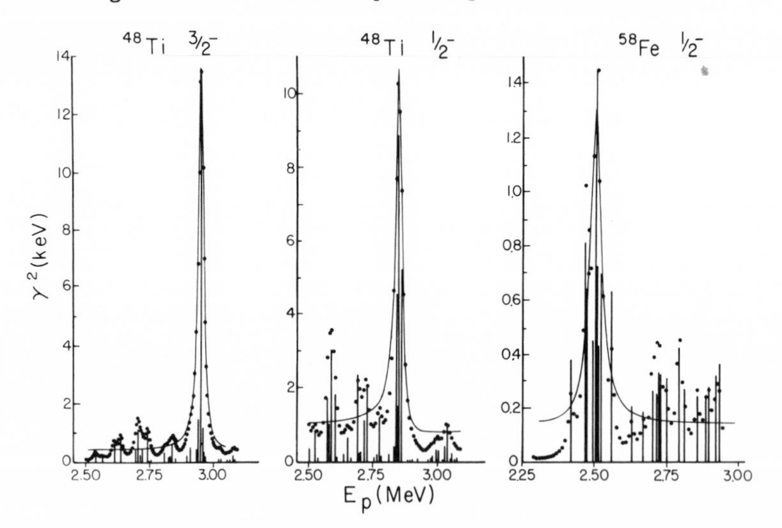

Fig. 9. Fragmented analog states with 1, 3 and several levels enhanced. The vertical lines are the reduced widths of the individual resonances, while the dots are the result of averaging these reduced widths. The solid curve is a fit--see text for discussion. The widths of the distributions are ∿1, ∿5, ∿25 keV, respectively.

VI. (p,n) REACTIONS

Neutron decay from the $T_>$ analog state $(T = T_0 + \frac{1}{2})$ to the low lying states of the residual nucleus involves an isospin change $\Delta T = 3/2$, and is forbidden if isospin is conserved. It is therefore somewhat ironic that compound nuclear analog states were first observed in the (p,n) reaction. The observed resonance in the (p,n) yield is attributed to mixing of the $T_>$ analog state with the background $T_<$ states through the Coulomb interaction. The enhancement of the entrance proton channel by the analog state is assumed sufficient to explain the observed resonance in the (p,n) yield. According to the conventional description of analog states, the neutron decay widths should not be strongly enhanced by the analog states. The simplest picture yields no enhancement, and other effects are expected to be small.

However, this question of neutron enhancement has not been thoroughly examined experimentally. Results of an early measurement of the proton and neutron partial widths of the fine structure of the analog of the ground state of Ca^{49} seem to show an enhancement of the neutron widths.[23] The evidence for neutron enhancement is based on measurement of only a few levels, and must be regarded as tentative. Work on the $Ca^{48}(p,n)Sc^{48}$ reaction is summarized by Elwyn *et al.*[24]

Measurements in our laboratory on Cr^{54} and Ni^{64} targets are relevant to this question. The (p,p) and (p,n) yields were measured[25] for the analogs of the third excited state of Cr^{55} and the first excited state of Ni^{65}. Proton and neutron yields were measured simultaneously; neutrons were detected with BF_3 counters embedded in a polyethylene matrix. Since several neutron groups were unresolved, these measurements were far from ideal experimental tests of neutron enhancement. Due to the paucity of experimental information, however, even these results are important.

If the neutron channel is strongly enhanced, it should exhibit "giant resonance" behavior similar to that displayed in the proton channel. If both the neutron and proton partial widths of the resonances near an analog state are enhanced, they should be correlated. Conversely, a statistically significant positive correlation between the neutron and proton partial widths would indicate that the neutron channel is enhanced. Figure 10 shows the proton and neutron reduced widths for the Cr^{54} data. Visually there is no obvious "giant resonance" pattern in the neutron widths nor is there any correlation between the neutron and proton widths. The linear correlation coefficient between the two sets of widths is $r \simeq 0.1$, consistent with no correlation. The experiment on Ni^{64} yielded similar results, with a correlation coefficient between the two sets of widths of $r \simeq -0.1$. Although the technique is not sufficiently sensitive to detect a weak enhancement, our results are *consistent* with no enhancement of the neutron decay.

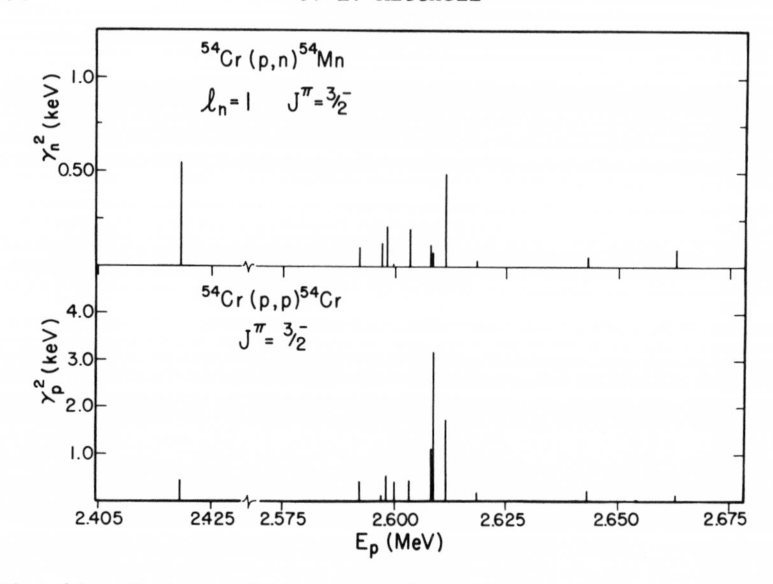

Fig. 10. Proton and neutron reduced widths for a fragmented $p_{3/2}$ analog. Widths in the two channels are not correlated.

VII. ELECTROMAGNETIC DECAY OF ANALOG STATES

The role of isospin in electromagnetic transitions is summarized by Warburton and Weneser.[26] Here we review briefly only highlights which are directly relevant to our own work, placing emphasis on correlations between the fine structure widths in various channels. Correlation phenomena in the decay of analog states appear to be neglected experimentally, although there is appreciable interest in correlations in the neutron capture reaction. The importance of correlations between partial widths in competing channels has been emphasized by Lane.[13,27]

We adopt as a working hypothesis the view that enhancement implies correlation. That is, if two or more channels are enhanced by the analog, then the fine structure widths in these channels are expected to be correlated. Conversely, if we observe a correlation between the partial widths of a given channel and the elastic widths, we conclude that the channel is enhanced by the analog. This is true even if the background partial widths in that channel cannot be measured. (The absence of such correlations between the proton and neutron widths led to the conclusion that the neutron channel is not strongly enhanced by the analog state. See Section VI.) In this section, analog to antianalog transitions, M1 transitions corresponding to Gamow-Teller beta decay of the parent state, and inelastic scattering are considered as possible "sources" of enhancement in the decay of analog states.

The antianalog state (AAS) contains the same configurations as the isobaric analog state (IAS). The AAS has isospin $T_<$. The IAS-AAS transition should be strong, if the simplest picture holds. Since the antianalog state may have appreciable proton single particle strength, this configuration should be excited by direct proton transfer, *e.g.*, a (He^3,d) reaction. The (He^3,d) spectroscopic factors may be used to operationally identify the AAS; this is the procedure usually adopted. Practical problems abound--scarcity of (He^3,d) data, lack of spectroscopic information such as spin, *etc.*

The pioneering work of Endt[28,29] established the existence of strong M1 transitions between isobaric analog states and antianalog states in the 2s-1d shell, but in the 1f-2p shell the situation is quite different, with the $p_{3/2}$ IAS to AAS transition strength greatly reduced.[30-32] Strong IAS to AAS transitions are again observed in the decay of $g_{9/2}$ analog states.[33-35] We have studied the decay of 3 fragmented $p_{3/2}$ analogs; from the systematics we expect weak IAS-AAS transitions.

Another possible origin of enhanced γ ray decay lies in the connection between the M1 strength and the strength of the Gamow-Teller beta decay from the parent state. Strong transitions from ground state analogs are sometimes observed; these enhanced transitions correspond to strong beta branches. Hanna[36] has summarized the comparison between the beta and gamma strength. In the mass region $A \simeq 30$ to 60 there is not much recent evidence to consider.[37-41] There is usually qualitative agreement between the gamma and beta transition, but not detailed agreement. We have studied the γ ray decay of one ground state analog for which the beta decay strength predicts a strong γ ray transition.

A third enhancement mechanism for which we have evidence involves inelastic decay of the analog. Suppose, as a concrete picture, that the parent state has a large admixture of excited core plus particle. The inelastic decay from the analog will be enhanced. A rather sensitive way to study this is to observe the gamma rays following inelastic scattering. Since we also obtain this information in studying the capture reaction, the inelastic data obtained in this manner are presented together with the capture data.

Figure 11 shows elastic and capture excitation curves[42] in the vicinity of the analog of the ground state of Fe^{59}. No $p_{3/2}$ resonances are observed in the elastic channel in a wide energy range above and below the analog state. The enhancement of the background states by the analog raises some of the background $p_{3/2}$ states above the level of observability and thus leads to an increase in the observed level density.

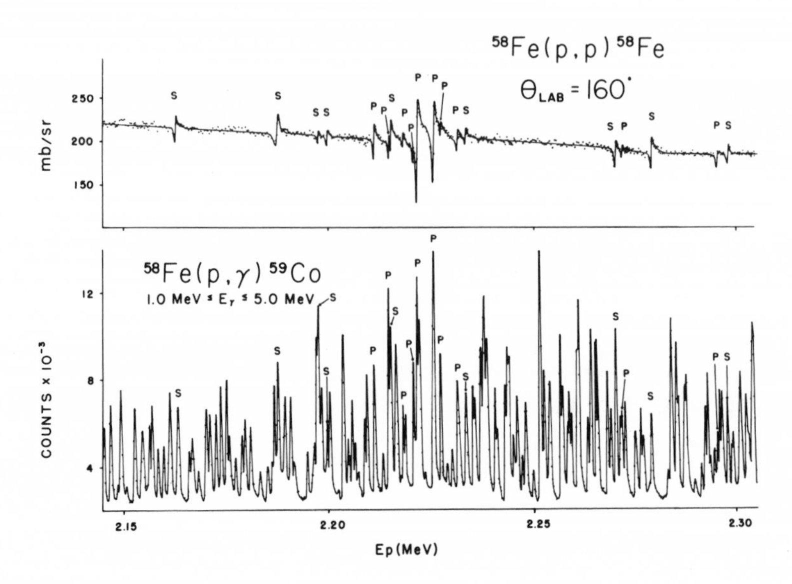

Fig. 11. Excitation functions for Fe^{58}(p,p) and (p,γ) reactions near a fragmented p-wave analog. The line through the elastic data is a fit.

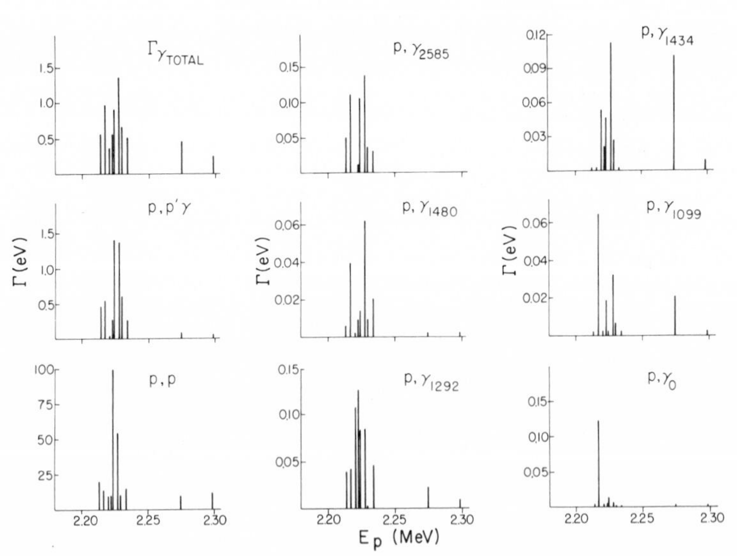

Fig. 12. Widths for decay into various channels of the analog of the ground state of Fe^{59}. Individual γ ray transitions are labeled by the final state. Note the correlations between the p, p' and Γ_γ total widths (see text).

The $p_{3/2}$ resonances at the central region of the analog were used to determine an approximate level spacing ($\bar{D} \simeq 2.6$ keV). Assuming that the level density can be written as

$$\rho \sim (2J + 1) \exp \left(- \frac{(J + \frac{1}{2})^2}{2\sigma^2} \right)$$

with the cutoff parameter $\sigma = 3$, one expects about 80 s- and p-wave resonances per 100 keV. There are about 80 resonances observed in the (p,γ) reaction in the region $E_p = 2.15$ to 2.25 MeV. Thus it appears that primarily s- and p-wave resonances are observed in the $Fe^{58}(p,\gamma)Co^{59}$ reaction at this energy.

Figure 12 shows the widths for elastic and inelastic scattering and the capture channels, as well as the total capture widths. These widths were derived from Ge(Li) spectra accumulated at the energies of the $p_{3/2}$ resonances shown in Figure 11. Decay of the $s_{1/2}$ resonances was also studied. It is clear that the p and p' widths are correlated (r = 0.88, confidence level >99%). Widths corresponding to IAS-AAS transitions are not strongly correlated. The IAS-AAS strength is 0.05 W.u. Most of the γ ray transitions are not correlated with the elastic widths. The total γ ray widths are correlated with the elastic and inelastic widths (r = 0.60 and 0.87, respectively).

Proton and γ ray partial widths[43] for the analog of the ground state of Cr^{55} are shown in Figure 13. Note the rather unusual fine structure pattern in the elastic widths. The p and p' widths are correlated (r = 0.79), but in this case the total capture widths are not strongly correlated with the elastic widths. The γ ray decay is dominated by one strong transition--to the ground state of Mn^{55}. The γ_0 widths are strongly correlated with the p widths (r = 0.73). The sum of the $\Gamma_{\gamma 0}$ widths from the eight fragments is 1.43 eV, while the prediction from the beta decay strength is 0.38 eV. The IAS-AAS transition strength is small, and the widths for IAS-AAS transitions are not correlated with the elastic widths.

The four sets of widths on the right hand side of Figure 13 correspond to transitions to final states which are tentatively assigned $J^\pi = 3/2^-$. These four sets of widths are all correlated with one another, some very strongly. However, the pattern is different from the elastic pattern. At face value this would imply another doorway state. The results are at least suggestive.

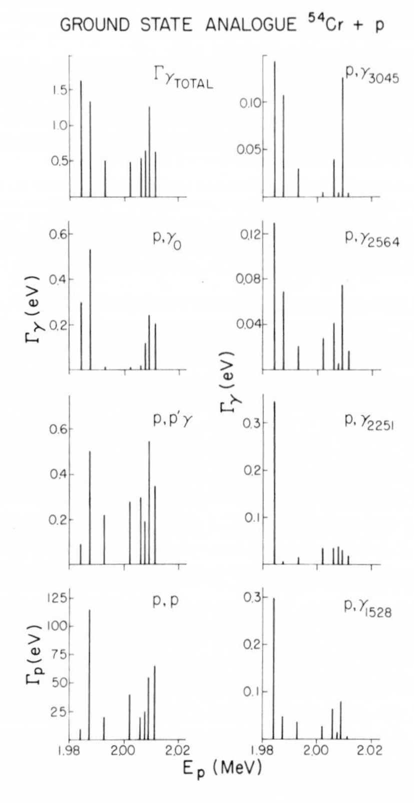

Fig. 13. Widths for decay into various channels for the analog of the ground state of Cr^{55}. Correlations are discussed in the text.

Results similar to those for Mn^{55} and Co^{59} were obtained for the analog of the first excited state of Ni^{63}. There are strong correlations between the p, p' and Γ_γ^{total} widths. The individual γ ray widths are not correlated with the analog patterns and the IAS-AAS strength is weak. The correlation between Γ_γ^{total} and Γ_p is real, since a detailed study of the decay of s-wave resonances led to a correlation between Γ_p and Γ_γ^{total} of $r = -0.06$. There is other strong evidence for correlations between elastic and total γ ray widths in the decay of fragmented analog states. Vingiani[30,31] studied the decay of analog states in Sc^{49} and Sc^{47}. From the published elastic and total capture widths one obtains a linear correlation coefficient of 0.99 for Sc^{49} and 0.98 for Sc^{47}.

On the basis of this limited information, the status of correlations between fine structure widths in various channels for analog states may be tentatively summarized:

1. There is evidence for correlations between elastic and inelastic widths for several cases (V^{49}, Mn^{55}, Co^{59}, Cu^{63}).

2. There is one case of a correlation between elastic and capture widths for an enhanced transition corresponding to the Gamow-Teller beta decay from the parent state (Mn^{55}).

3. No strong correlations have been observed between elastic and capture widths for weak IAS to AAS transitions (Mn^{55}, Co^{59}, Cu^{63}).

4. There is evidence for correlations between elastic and total capture widths (Co^{59}, Sc^{47}, Sc^{49}, Cu^{63}).

More experimental information is clearly needed. The measurement of the decay of a fragmented analog with enhanced IAS to AAS transitions would be of particular interest. Measurement of correlations should also prove fruitful in the study of intermediate structure via proton resonance reactions; this has been previously suggested in neutron capture experiments.[45]

VIII. INELASTIC SCATTERING

Inelastic scattering has been measured either by direct observation of the inelastic protons or by observation of the γ rays following inelastic scattering (see Section VII). Qualitative features of the direct measurements are described below, followed by a discussion of spectroscopic results from both types of experiments.

The elastic scattering measurements are high counting rate experiments; the inelastic protons are observed only when the cross sections are relatively large. Usually this corresponds to the upper part of our energy range (above $\sim$2.5 MeV) and to rather low-lying first excited states (below $\sim$1 MeV). In general the magnitude of the inelastic partial widths is determined by penetrability considerations. The $\ell = 2$ resonances decay in the inelastic channel by $\ell_{p'} = 0$ and thus have the largest inelastic widths. On the average, $\ell = 1$ and $\ell = 0$ resonances have inelastic widths which are progressively smaller. A second factor in determining the average inelastic widths is the strength function. For example, $p_{3/2}$ resonances usually have larger inelastic widths than $p_{1/2}$ resonances. The $p_{1/2}$ resonances can decay only with $j_{p'} = 3/2$, while the $p_{3/2}$ resonances can decay with both $j_{p'} = 1/2$ and $3/2$. Since the $p_{1/2}$ strength functions in this mass region are larger than the $p_{3/2}$ strength functions, $p_{3/2}$ resonances tend to have larger inelastic widths.

The inelastic partial widths for different channel spins are often determined. The best determination in our data is for the decay of $p_{3/2}$ resonances. Preliminary analysis indicates no striking pattern in the channel spin mixing ratios.

Inelastic scattering from analog states has been of interest primarily because this reaction yields information about the parent state. In simplest terms, just as the elastic spectroscopic factor for the analog corresponds to stripping from the target to the parent state, the inelastic spectroscopic factor corresponds to stripping from an excited state of the target to the parent state. Thus inelastic decay from analog states supplies unique nuclear information. The topic is reviewed by Morrison;[46] work in the lead region is discussed by Stein.[47] There is not much information on inelastic spectroscopic factors in the f-p shell, primarily because the most interesting analogs (of very low-lying parent states)occur at such low bombarding energies. Work by Cosman[48] on Ti^{50} and by Ramavataram[49] on Ni^{64} provide examples.

Inelastic spectroscopic factors may be determined with essentially the same procedure as for the elastic case, except that the single particle width must be calculated at a lower energy $(E_A - E_{2+})$. Since the energy of the 2^+ state is typically $\sim$1 MeV, this energy $(E_A - E_{2+})$ corresponds to an energy very low on the barrier, and the single particle widths are probably not as well determined as for the elastic case.

For the inelastic data obtained from the γ ray experiments, there was no information available on the background inelastic strength. We use the correlation results of Section VII: a strong correlation between elastic and inelastic widths is assumed to imply enhancement by the analog. The observed inelastic strength is then taken as originating primarily from analog effects. For the inelastic data obtained from charged particle measurements, there was direct information available on the background.

Inelastic spectroscopic factors were extracted for five analog states (two from charged particle experiments and three from γ ray experiments). Four of the five states had $S_{pp'}$ between 0.1 and 0.2, while one was very small. Interpreting the inelastic spectroscopic factor as a measure of the excited core plus particle admixture in the parent state, our results indicate typical amplitudes of $\sim$0.3-0.4 for the excited core configuration in low-lying states in odd A nuclei in the f-p shell.

IX. STATISTICAL PROPERTIES

It is clear that there is a large amount of statistical information in our data, and that these high resolution proton experiments may be ideally suited for some types of statistical studies. The data provide information on level densities, spacing and width distributions, strength functions, *etc*. Since these topics are not really the emphasis of this conference nor of this talk, I shall simply mention a few topics rather briefly.

Wide ranges of level densities are observed, as indicated earlier in the elastic data. As Figure 14 shows, this is displayed even more clearly in the capture data. One interesting feature of our data is that a rather large energy range is covered. So much, in fact, that one can see the energy dependence of the level spacing explicitly. Figure 15a shows the average level spacing versus proton bombarding energy for s-wave resonances in $Ti^{48}(p,p)$; the slope corresponds to a nuclear temperature of about 1.5 MeV. The spacing distribution is shown in Figure 15b with a Wigner distribution superimposed. Since the average level spacing is not constant either the analytical expression must be reformulated or the data must be "corrected." We chose to "correct" the data[50] to a constant average spacing; the results are shown in Figure 15c. A comparison of the reduced widths with the Porter-Thomas distribution is shown in Figure 15d.

Most proton strength function measurements have been based on the (p,n) reaction (*e.g.*, Johnson and Kernell[51] and references therein). There is essentially no data measuring proton strength functions via resonance reactions. We have

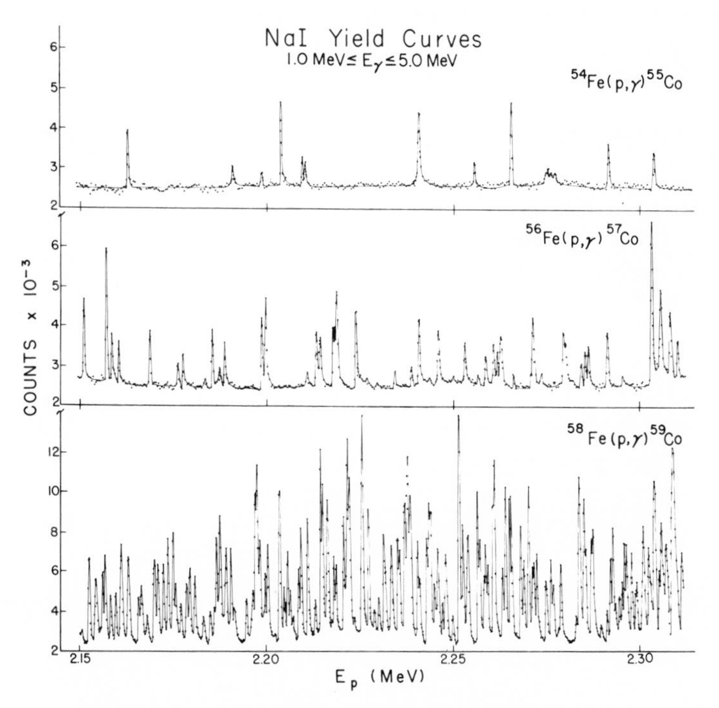

Fig. 14. Excitation functions for the $Fe^{54,56,58}(p,\gamma)Co^{55,57,59}$ reactions. The systematic change in level density is apparent.

data[52] on $s_{1/2}$, $p_{1/2}$ and $p_{3/2}$ strength functions, as well as fragmentary information on the d-wave strength functions. The most interesting results are for the $1/2^-$ strength functions, as shown in Figure 16.

As noted earlier, the $p_{1/2}$ strength function is much larger than the $p_{3/2}$ strength function, indicating the expected splitting of the 2p giant resonance. The isospin effect is obvious by inspection. The variation of the $p_{1/2}$ strength function with mass number is much stronger than predicted by conventional optical model parameters. The simplest way to account for this strong A dependence is with a surface peaked imaginary term with a smaller than usual diffuseness. Potentials of this sort are discussed by Satchler and Perey.[53] A

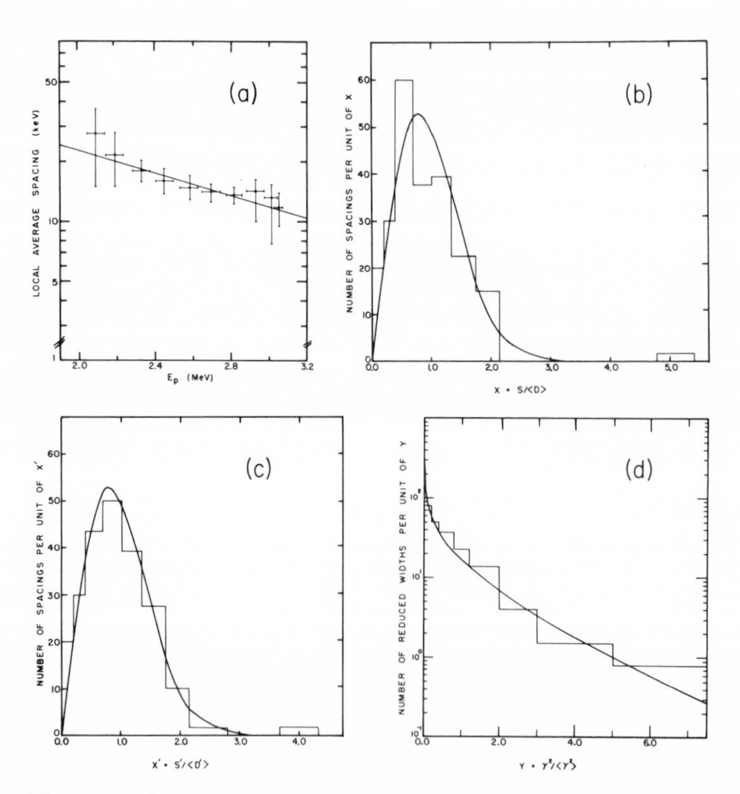

Fig. 15. Spacings and widths for s-wave resonances observed in Ti^{48}(p,p).

(a) average level spacing versus bombarding energy,
(b) uncorrected spacings with Wigner distribution superimposed,
(c) corrected spacings,
(d) reduced widths with Porter-Thomas distribution superimposed.

sharply peaked surface imaginary potential was first used by Moldauer[54] and later by Johnson and Kernell.[51]

Analysis to determine the number of missing levels and the number of misassigned levels is in progress. This involves a variety of techniques, including analysis of the width and spacing distributions and several statistical tests, such as the F-statistic of Dyson. If one can observe essentially all of the levels, then the spacings may be examined in light of recent theoretical ideas. (See Ref. 55 for a theoretical summary and Refs. 56-57 for recent experimental work applying these ideas to neutron data.) There is certainly a lot of activity and excitement in the field of statistical properties of nuclei. We feel that high resolution proton resonance studies can make significant contributions

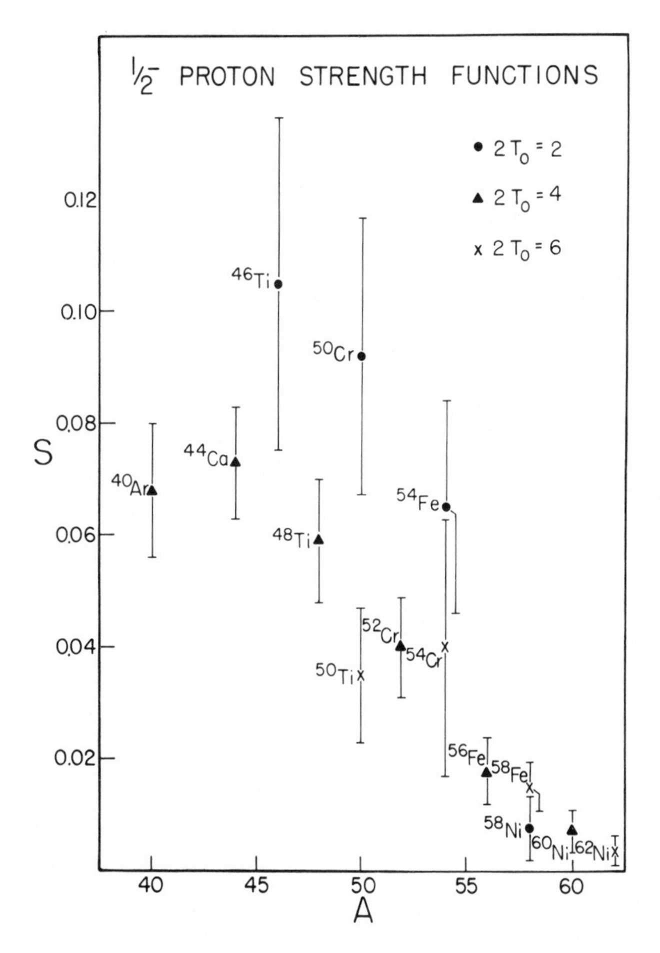

Fig. 16. The $p_{1/2}$ proton strength functions. The fractional error is assumed to be $\sqrt{2/n}$, where n is the number of levels.

in this area.

X. CONCLUSION

Rather than try to summarize this talk, I'll try to point out what we plan to do in the future. We plan (1) to measure elastic scattering on more isotopes, (2) to finish fitting the fine structure distributions, (3) to perform more capture experiments, (4) to perform high resolution experiments at higher energies, both on our 4 MV accelerator and our tandem accelerator (high resolution systems now working on both accelerators), (5) to place greater emphasis on the statistical analysis of these data and (6) to try to obtain a more complete set of data (with fewer missed levels and misassigned levels). In the longer run we hope to obtain high resolution beams of other particles (*e.g.*, alphas) and of course to try to improve the resolution even further. In summary, we find the field of high resolution proton resonance reactions to be both rich and rewarding, and anticipate that this will continue to be true in the future.

ACKNOWLEDGMENTS

The credit for this series of high resolution experiments should go to our dedicated graduate students, past and present: particularly Dr. J. C. Browne, Dr. J. D. Moses, Dr. D. P. Lindstrom, Dr. N. H. Prochnow, Dr. W. C. Peters, Mr. W. M. Wilson, Mr. J. F. Wimpey, Mr. T. Dittrich and Mr. D. Outlaw. The support and encouragement of nuclear theorists is gratefully acknowledged--special thanks go to Dr. A. M. Lane and Professor D. Robson.

REFERENCES

1. *Isobaric Spin in Nuclear Physics*, ed. J. D. Fox and D. Robson (Academic, New York, 1966).
2. *Nuclear Isospin*, ed. J. D. Anderson, S. D. Bloom, J. Cerney and W. W. True (Academic, New York, 1969).
3. *Isospin in Nuclear Physics*, ed. D. H. Wilkinson (North-Holland, Amsterdam, 1969).
4. J. D. Anderson and C. Wong, Phys. Rev. Lett. 7, 250 (1961).
5. A. M. Lane, Nucl. Phys. 35, 676 (1962).
6. J. D. Fox, C. F. Moore and D. Robson, Phys. Rev. Lett. 12, 198 (1964).
7. D. Robson, Phys. Rev. 137, B535 (1965).
8. A. M. Lane, R. G. Thomas and E. P. Wigner, Phys. Rev. 98, 693 (1955).
9. U. Fano, Phys. Rev. 124, 1866 (1961).
10. P. Richard, C. F. Moore, D. Robson and J. D. Fox, Phys. Rev. Lett. 13, 343a (1964).

11. G. A. Keyworth, G. C. Kykér, E. G. Bilpuch and H. W. Newson, Nucl. Phys. 89, 590 (1966).
12. C. Mahaux, *Nuclear Isospin*, (see Reference 2) p. 351.
13. A. M. Lane, *Isospin in Nuclear Physics*, (see Reference 3) p. 509.
14. P. B. Parks, H. W. Newson and R. M. Williamson, Rev. Sci. Instr. 29, 834 (1958).
15. D. L. Sellin, Ph.D. dissertation, Duke University (1968) unpublished.
16. D. P. Lindstrom, H. W. Newson, E. G. Bilpuch and G. E. Mitchell, Nucl. Phys. A168, 37 (1971).
17. D. P. Lindstrom, H. W. Newson, E. G. Bilpuch and G. E. Mitchell, Nucl. Phys. A187, 481 (1971).
18. N. H. Prochnow, H. W. Newson, E. G. Bilpuch and G. E. Mitchell, Nucl. Phys. A194, 353 (1972).
19. J. Jänecke, *Isospin in Nuclear Physics*, (see Reference 3) p. 297.
20. H. L. Harney (unpublished).
21. H. L. Harney and H. Weidenmüller, Nucl. Phys. A139, 241 (1969).
22. A. Z. Mekjian, Nucl. Phys. A146, 288 (1970).
23. P. Wilhjelm, G. A. Keyworth, J. C. Browne, W. P. Beres, M. Divadeenam, H. W. Newson, and E. G. Bilpuch, Phys. Rev. 177, 1553 (1972).
24. A. J. Elwyn, F. T. Kuchnir, J. E. Monahan, F. P. Mooring and J. F. Lemming (to be published).
25. J. D. Moses, J. C. Browne, H. W. Newson, E. G. Bilpuch and G. E. Mitchell, Nucl. Phys. A168, 406 (1971).
26. E. K. Warburton and J. Weneser, *Isospin in Nuclear Physics*, (see Reference 3) p. 173.
27. A. M. Lane, Ann. Phys. 63, 173 (1971).
28. P. M. Endt, *Second Symposium on the Structure of Low-Medium Mass Nuclei*, P. Goldhammer and L. W. Seagondollar, eds. (University of Kansas, Lawrence, 1966) p. 58.
29. P. M. Endt, *Third Symposium on the Structure of Low-Medium Mass Nuclei*, J. P. Davidson, ed. (University of Kansas, Lawrence, 1968) p. 73.
30. G. Vingiani, G. Chilosi and W. Bruynesteyn, Phys. Lett. 26B, 285 (1968).
31. G. B. Vingiani, G. Chilosi and C. Rossi-Alvarez, Phys. Lett. 34B, 597 (1971).
32. H. V. Klapdor, Phys. Lett. 35B, 405 (1971).
33. I. Fodor, I. Szentpétery and J. Szücs, Phys. Lett. 32B, 689 (1970).
34. I. Szentpétery and J. Szücs, Phys. Rev. Lett. 28, 378 (1972).
35. S. Maripuu, J. C. Manthuruthil and C. P. Poirier (to be published).
36. S. Hanna, *Isospin in Nuclear Physics*, (see Reference 3) p. 591.

37. E. G. Adelberger and D. P. Balamuth, Phys. Rev. Lett. 27, 23 (1971).
38. E. Gaarde, K. Kemp, Y. V. Naumor and P. R. Amundsen, Nucl. Phys. A143, 497 (1970).
39. L. G. Mann and S. Bloom, Nucl. Phys. A140, 598 (1970).
40. S. Maripuu, Phys. Lett. 31B, 181 (1970).
41. M. Sakai, R. Bertini and C. Gehringer, Nucl. Phys. A157, 113 (1970).
42. W. C. Peters, E. G. Bilpuch and G. E. Mitchell (to be published).
43. G. E. Mitchell *et al.*, *Statistical Properties of Nuclei*, J. B. Garg, ed. (Plenum, New York, 1972) p. 299.
44. J. F. Wimpey, G. E. Mitchell and E. G. Bilpuch (to be published).
45. R. Chrien, (see Reference 43) p. 233 and references therein.
46. G. C. Morrison, (see Reference 2) p. 435.
47. N. Stein, (see Reference 2) p. 481.
48. E. R. Cosman, D. C. Slater and J. E. Spencer, Phys. Rev. 182, 1131 (1969).
49. K. Ramavataram, C. S. Yang, G. F. Mercier, C. St.-Pierre, D. Sykes and S. Ramavataram, Nucl. Phys. A191, 88 (1972).
50. E. G. Bilpuch, N. H. Prochnow, R. Y. Cusson, H. W. Newson and G. E. Mitchell, Phys. Lett. 35B, 303 (1971).
51. C. H. Johnson and R. L. Kernell, Phys. Rev. C2, 639 (1970).
52. E. G. Bilpuch, J. D. Moses, H. W. Newson and G. E. Mitchell (to be published).
53. G. R. Satchler and F. Perey (to be published).
54. P. A. Moldauer, Nucl. Phys. 47, 65 (1963).
55. M. L. Mehta, (see Reference 43) p. 179.
56. H. I. Liou, H. S. Camarda, S. Wynchank, M. Slagowitz, G. Hacken, F. Rahn and J. Rainwater, Phys. Rev. C5, 974 (1972).
57. H. I. Liou, H. S. Camarda and F. Rahn, Phys. Rev. C5, 1002 (1972).

DISCUSSION

TEMMER: If you'll forgive me--I know you know this terribly well--but the audience might be slightly misled by something you said. One of your slides (Fig. 8) shows $1/2^+$, $1/2^-$, and $3/2^-$ accumulated as well as differential reduced widths. You said that there are no $3/2^-$ states between 2.1 and 3 MeV at the bottom of Fig. 8. This is not what you meant to say, I am sure; you meant to say there are no $3/2^-$ states that you can *observe*, but presumably the nucleus has just as many states per unit energy interval over the whole energy range.

MITCHELL: Yes, I tried to make that point in a later slide. The level density does not change as one goes through the analog.

FOX: How good will your resolution be on the tandem?

MITCHELL: We measured one resonance a few weeks ago and obtained 400 eV.

FOX: Do you expect to improve that?

MITCHELL: It will be difficult, since we have made all of the easy improvements.

GABBARD: If you look at these resonances in terms of a level shift, *i.e.* the deltas, all of the resonances that you showed seem to have negative deltas. Have you ever seen one that has a positive delta and if not is this understood?

MITCHELL: To answer the last question first--the negative delta is what one expects to observe according to the conventional formulation and this is basically what we do see. However, in many cases the level shift (or asymmetry parameter) is not easy to extract from the data. Most of the analogs do not show a very strong asymmetry. Occasionally one would see a fragmented analog which looks as though it might have an asymmetry to the other side, but that's probably due to statistical fluctuations. We have no clear cut cases with an asymmetry strongly to the high energy side.

MARIPUU: You mentioned that there are no strong analog to antianalog transitions in the $f_{7/2}$ shell. I wonder if it is clear to most people that this is because the single particle contribution is cancelled by the core contribution. The strong correlations you see to all the other states are due to the transitions through the core, that is, due to the $f_{7/2}$ particles.

MITCHELL: I just said that in the fp -shell the situation is different from that in the sd-shell. I am simply quoting the empirical observations that $p_{3/2}$ states have weak analog to anti-analog transitions, while for $g_{9/2}$ states the analog to antianalog strengths are strong. Since we were looking at $p_{3/2}$ analog decay, we expected the transition to the anti-analog to be weak.

IV.A. NUCLEAR LIFETIMES THROUGH THE USE OF THE CRYSTAL BLOCKING TECHNIQUE*

G. M. Temmer
Rutgers University
New Brunswick, New Jersey

I. INTRODUCTION

Considerable experience has been gained over the past decade concerning the remarkable effects of the penetration of charged particles through crystal lattices, the so-called "channeling" and "blocking" phenomena. For our purposes, we shall have reference mainly to the latter, which deal with the suppression, along crystalline axes and planes, of nuclear reaction yields of secondary charged particles from nuclei situated at lattice sites. The incident beam travels along a random direction. Figure 1 illustrates this situation for "prompt" secondary particles obtained from (elastic) Rutherford scattering, for example The detector reveals an angular distribution pattern, when normalized to reaction yield for amorphous material, of the type shown in the insert of Figure 1, labeled p_0.

It was soon realized by workers in this field[1,2] that the shape of these angular distribution patterns, or "blocking dips," could be affected by the fact that the struck nucleus might recoil some distance from the lattice site before emitting the secondary reaction particle. That is to say, the "Coulomb shadowing" of the secondary particle emission pattern from a given nucleus by its neighbors along atomic rows or planes would be less pronounced, as shown in the insert of Figure 1 by the dip labeled p_1. Clearly the significant distance here is $v_{\perp}\tau$, where $v_{\perp}$ is the component of recoil velocity perpendicular to the crystal axis, say, and τ is the mean lifetime of the recoil nucleus. The longer the lifetime, the farther the nucleus is able to fly before emitting the observed secondary charged particle, and the shallower the dip will become. Clearly, by measuring dip profiles carefully, we have here a "clock" measuring the time interval between the striking of the nucleus and the emission of the secondary particle. This, in brief, is the crystal blocking lifetime technique. We wish to emphasize that for these and most other considerations we shall completely ignore all wave aspects of the particles involved.

Let us say a few words about the nature of the blocking pattern for a prompt event. The full width at half dip is given by the simple relation at the bottom of Figure 1, involving the charges Z_1e and Z_2e of secondary particle and

*Work supported in part by the NSF and Bell Laboratories.

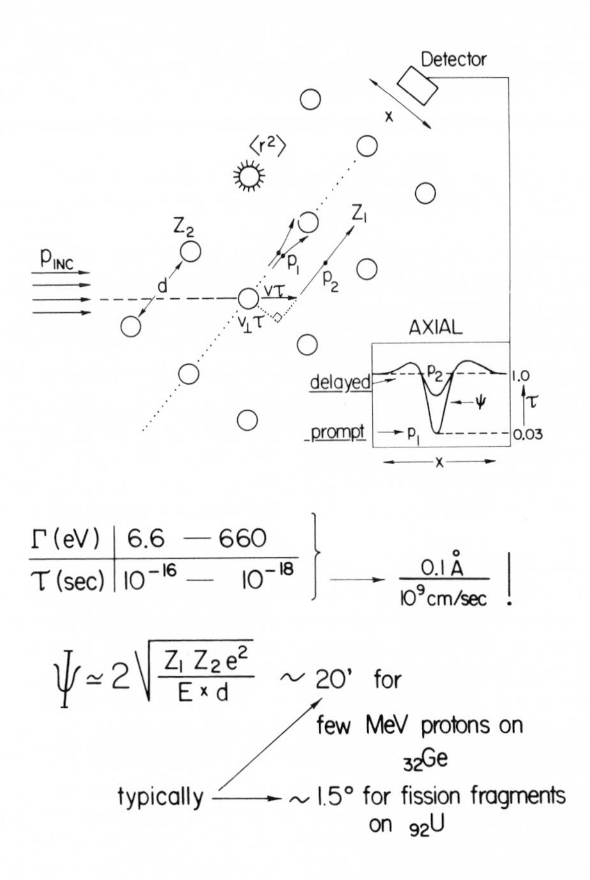

Fig. 1. Schematic description of crystal blocking lifetime method. p_o denotes outgoing charged particles of charge Z_1 without time delay, yielding angular distribution in detector along x labelled p_o in the insert, so called "prompt blocking dip", with typical minimum yield along crystal axis. Here p_1 denotes particles emitted from recoiling nucleus after flight distance $v\tau$ from original lattice site, leading to shallower angular distribution along x labelled p_1, so called "delayed blocking dip". Insert ordinate is normalized to unity yield for amorphous target, and $\langle r^2 \rangle$ symbolizes lattice vibrations. Lower part of figure lists approximate lifetime range and corresponding widths to which blocking method is sensitive, as well as expression for critical angle, *i.e.* full width at half minimum of blocking dips for two typical cases.

lattice nucleus, respectively, the kinetic energy E of the former, and the lattice spacing d. Typical values for the angular widths of blocking dips are listed. In addition to the width, an important parameter is the minimum value of the dip. This value can be as low as 0.02 of normal yield, and is influenced by such physical processes as lattice vibrations, crystal imperfections, strains, multiple scattering, *etc.*

In order to estimate the widths of the "window" on the world of lifetimes, one realizes that the flight distances are of the order of the lattice spacings, and must not be *less* than the lattice vibration amplitude ($\sim$0.1 A^o). Combining this with typical recoil velocities ($\sim 5 \times 10^7$ cm/sec for 5 MeV protons striking a nucleus of A $\sim$ 75) we obtain a range of lifetimes to which this method might be sensitive, lying between 10^{-15} and 10^{-18} second; the possibility of varying the perpendicular velocity component by varying the recoil angle relative to the crystal direction is included here.

These lifetimes are at least 100 times shorter than those accessible to the Doppler shift attenuation method. Some remarks are in order concerning the nuclear physics of energy levels in this domain, corresponding to the widths given in the bottom part of Figure 1. Of course, complementary resonance *width* measurements are possible in this region, mainly by slow neutron resonance spectroscopy. It is interesting to note that the widths of energy levels in this range are essentially always controlled by *nucleon* emission and absorption, with very few exceptions, whereas most other methods of lifetime (or transition rate) determination deal with levels governed entirely by *electromagnetic* transitions; these include, in addition to the Doppler shift method already mentioned, the delayed coincidence technique, Coulomb excitation, and resonance fluorescence. It turns out that radiative widths at excitation energies in the vicinity of nucleon binding energies are just fractions of an electron volt, so that the nucleon widths determine total widths or lifetimes. The blocking method is therefore one of the few which allows us a glimpse of truly *nuclear* lifetimes.[3]

We need to say a few words about the process of extracting lifetime values from the shapes of blocking dips. This is of necessity a complicated story, since it involves the attempt to account properly for the many factors governing the motion of charged particles through the crystal lattice field. Recently, some rather ambitious computer programs have been developed which are beginning to yield meaningful and consistent results.[7,8,9] Figure 2 illustrates the effect of finite lifetimes on the minimum blocking yields for protons on thin germanium crystals.[8] If these results check out, they allow us to relate the decrease in blocking dip depth to the

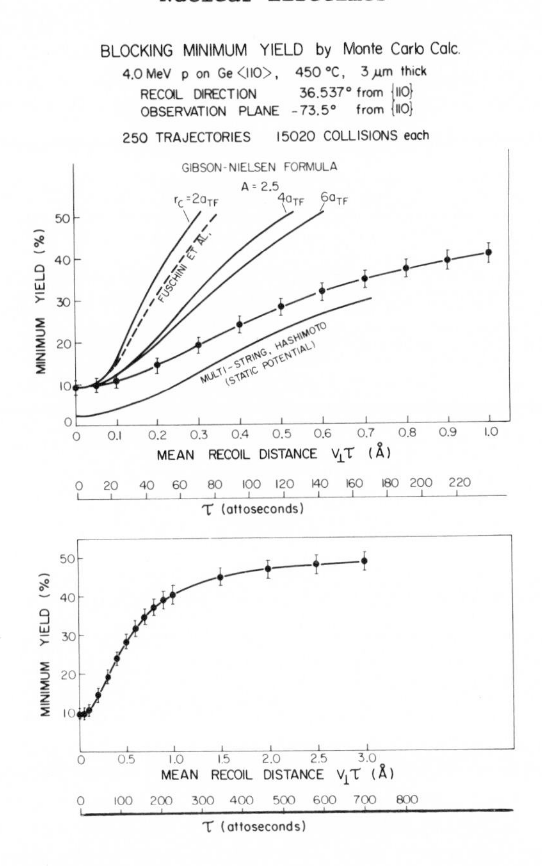

Fig. 2. Blocking minimum yield in per cent of normal yield, vs. mean perpendicular recoil distance in angstroms as calculated by various models. (Refs. 7, 8, and 9.) Bottom curve extends to large recoil distances; upper curves show enlarged portion for less than one angstrom of recoil distance. Here a_{TF} refers to Thomas-Fermi screening radius, and A = 2.5 to a certain choice in the analysis used in Reference 7. Calculations correspond to conditions of the experiments of Reference 21. Approximate times of flight in attoseconds (10^{-18} second) for germanium recoil nucleus are also given along the abscissae.

mean lifetime value. It is beyond the scope of this article to go into the details of these very elaborate calculations, requiring as much as 150 hours on a medium size computer to generate a blocking dip.[8] Note, however, the large discrepancy between the recent, flatter curves given in Figure 2, and the earlier, crude approximations,[10] which neglected all but one string of nuclei, and used a sharp cut-off r_c.

These blocking dips contain considerably more information than is represented by the value of the minimum yield; the half-widths, as well as higher moments of these shapes can be shown to have different sensitivities to physical properties, such as long lifetime components, prompt components, and others. Only careful auxiliary experimentation in conjunction with improved calculations will permit us to gain confidence in these procedures. We wish to emphasize, however, that the time ordering of events within the sensitivity window of this method does not depend on these detailed considerations, but only on the minimum yield ratios.

II. BLOCKING LIFETIME EXPERIMENTS

A. First Generation

The first demonstrations of a delay in the appearance of nuclear reaction products by the shallowing of blocking dips occurred in 1968 simultaneously and independently at the Rutgers-Bell Nuclear Physics Laboratory in the United States, and at the Japan Atomic Energy Research Institute in Japan, using very different reactions and techniques. We shall briefly recall these results.

(1) *Delay in Fission*[10]

The Rutgers-Bell experimenters used thin crystals of ordinary uranium oxide and incident protons of 10 MeV from our tandem accelerator. The detector consisted of a sheet of cellulose acetate in which fission fragment tracks could only be made visible by appropriate etching.[11] Both axial and planar blocking (star patterns) could be observed clearly. One problem encountered in this type of experiment is the establishment of a known "prompt" dip. Without entering into the many ramifications of the complicated fission process--these workers believe the fissions to occur *after* the emission of a neutron, so-called "second-chance" fission--a *prompt* dip could be approximated by first sending the incident proton beam almost parallel to the crystal axis, but well outside the critical angle discussed in the introduction. Even though the recoil nuclei travel some distance along the atomic row, the emitted fission fragments will still be blocked by the neighboring atoms along the direction of observation, *i.e.* along the crystal axis. One then looks for a shallowing of the dip when bombarding along a direction nearly perpendicular to the crystal axis. Using an early, crude method of extraction,[10] a lifetime of 1.4×10^{-16} second was estimated.

The fission lifetime problem is currently being pursued in several laboratories; using heavy ions, in the Soviet Union,[12-15] and at Brookhaven National Laboratory;[16] induced by 2-4 MeV neutrons in Studsvik,[17] and by 14 MeV neutrons in Basel.[18] Delays were observed in all cases, as well as variations in mean lifetime with excitation energy.

(2) *Delay in Inelastic Proton Scattering*

A very different approach was taken by Maruyama *et al.*[19] These workers used natural germanium crystals about 8 μm thick, *i.e.* ∿ 200 keV to the incident protons of around 5 MeV. Since this energy is well below the Coulomb barrier (∿8.5 MeV), the elastic scattering is essentially Rutherford scattering, and provides a convenient, built-in monitor for a prompt dip in the sample under bombardment, integrating over all shallowing effects existing in such crystal samples, with the exception of the possible reaction delays of interest. Simultaneously, these workers examined the *planar* blocking dips associated with inelastic proton groups, leading to the lowest excited 2^+ states in the more abundant even-even isotopes Ge^{70} and Ge^{72}. For this type of experiment, plastic detectors are clearly inappropriate, and semiconductor position-sensitive detectors had to be used. The main results from their pioneering investigations are:

(a) a clear demonstration of a lifetime effect for inelastic protons in Ge^{70} and Ge^{72} below 5 MeV proton energy;

(b) in the case of Ge^{72}, a disappearance of the delay above about 5.2 MeV, *i.e.* the appearance of a prompt blocking dip for the inelastic group in question.

The latter effect can be simply understood in terms of the (p,n) threshold in Ge^{72} which occurs at 5.204 MeV (but at 7.104 MeV in Ge^{70}). The opening of the neutron channel entails a drastic widening of the compound resonances contributing to the inelastic scattering, say to widths in excess of one keV, so that the blocking method no longer yields a measurable lifetime. A more detailed investigation of the threshold effect is planned at Rutgers.

These inelastic scattering results were confirmed by *axial* blocking experiments at the tandem accelerators at Rutgers University,[20,21] and at Harwell,[22] the former employing crystals almost an order of magnitude thinner than previous workers (∿30 keV to incident protons). An experimental setup for this type of experiment is shown in Figure 3.

A few words of explanation are in order concerning these inelastic scattering experiments. At about 5-MeV proton bombarding energy, the compound nuclei whose narrow levels are responsible for the observed delays are excited to energies between 10 and 12 MeV. From reasonable theoretical estimates of level densities at these excitations,[23] involving merely a moderate extrapolation through a few MeV, from empirical densities obtained by counting $J = 1/2^+$ s-wave resonances at the neutron binding energy (∿7.5 MeV), we expect thousands of

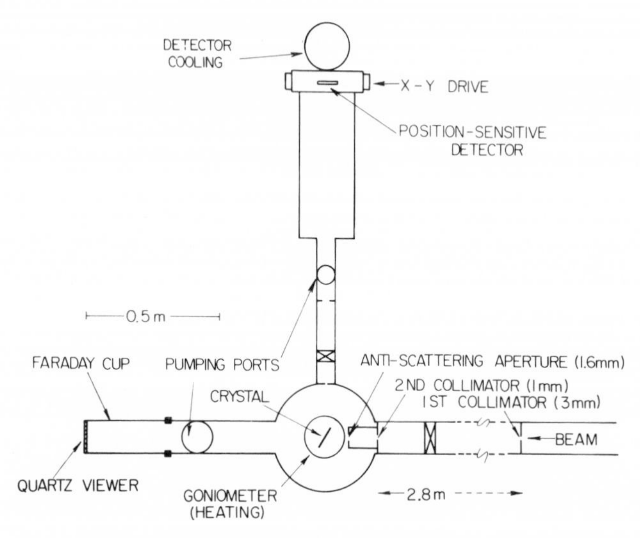

Fig. 3. Experimental set-up used at the Rutgers-Bell tandem accelerator.

states to contribute to the scattering from the thicker crystals used in Japan[19] and Harwell,[22] and hundreds from those used in our Laboratory.[21] These states, in the odd-even compound nuclei of arsenic, have half-integral spins up to a value determined by the penetrabilities; each sub-population levels of given spin and parity will have a variety of widths distributed according to some law, such as the Porter-Thomas distribution.[24] The values of lifetimes deduced from these experiments, even if properly extracted (*cf.* discussion in the Introduction), therefore represent *grand* averages over many parameters, especially if, as has been true hithertofore, only the blocking yield at the dip minimum is used. Moreover, if there were an appreciable prompt component present in the inelastic scattering yield--which is considered unlikely here--it would merely reduce the average observed lifetime. This is just one of many subtle effects which must be realized and treated with care when interpreting blocking experiments.

(3) *Reaction Delay from a Single Compound-Nucleus Reaction.*

Recently, two ingenious groups in Japan[25] and Denmark[26] independently succeeded in determining the shallowing of the blocking dips when observing resonant alpha particles from the (p,α) reactions on Al^{27} in an aluminum crystal, and in both Al^{27} and P^{31} in a GaP crystal, respectively. These nuclei are, in fact, in the s-d shell, and serve to legitimize this talk! Once again, plastic sheet detectors insensitive to protons could be used. No energy discrimination was possible or needed for the outgoing alpha particles, and essentially *one resonance* only could be examined at one time, even though using very low energy protons (<1 MeV), and crystals thicker than the proton range. This could be achieved because in this energy region the compound levels lie far apart (∿100 keV) and there is no observable alpha particle yield between resonances. The promptness calibration was carried out both by observing a resonance so wide as to yield no measurable shallowing of the blocking dip, as well as by using the trick described in Section A(1). Figure 4 shows the blocking dips for emitted alpha particles obtained at two different proton bombarding energies, and three different angles, the lower one exciting only a narrow resonance ($\Gamma \sim 4$ eV) with measurable blocking lifetime, the higher one leading to a resonance so wide ($\Gamma \sim 1$ keV) as to yield a prompt blocking dip.[26] The deduced lifetimes agreed with the widths known from resonant reaction yields, within the rather wide margins of error of both the lifetime and width determinations (∿factor of five overall).

Such measurements, while not yielding essentially new nuclear information, are nevertheless extremely important in an auxiliary way to investigate the many crystallographic properties which enter into the determination of blocking lifetimes.

B. Second Generation

By second generation blocking experiments we mean the observation of differences in lifetimes occasioned by variations of nuclear parameters such as specific proton bombarding or emission energies, residual state spin, or possible structural differences among the compound nuclear states which decay to specific residual states in nuclear reactions.

(1) *Enhancement of Compound Nucleus Fine Structure Near Analog Resonances*

Since all the fine structure compound states possessing the spin and parity of an isobaric analog state located in their midst will mix with the latter via the Coulomb interaction, they will be broadened over and above their "normal" widths Γ_i, their new enhanced widths Γ_λ being given by an expression of the form[27]

$$\Gamma_\lambda = \frac{\left(E_\lambda - E_a - \langle i|V|a\rangle \left(\frac{\Gamma_a\uparrow}{\Gamma_i}\right)^{\frac{1}{2}}\right)^2}{\left(E_\lambda - E_a\right)^2 + \left(\frac{\Gamma_a\downarrow}{2}\right)^2}\,\Gamma_i\ ,$$

the factor in brackets being referred to as the *enhancement factor*. The symbols occurring in this expression are: E_λ, fine structure resonance energy; E_a, analog resonance energy; $\langle i|V|a\rangle$, Coulomb mixing matrix element; $\Gamma_a\uparrow$, $\Gamma_a\downarrow$, escape and spreading widths of the analog resonance, respectively. Near the center of the analog resonance ($E_\lambda \simeq E_a$), the expression simplifies to the following form:

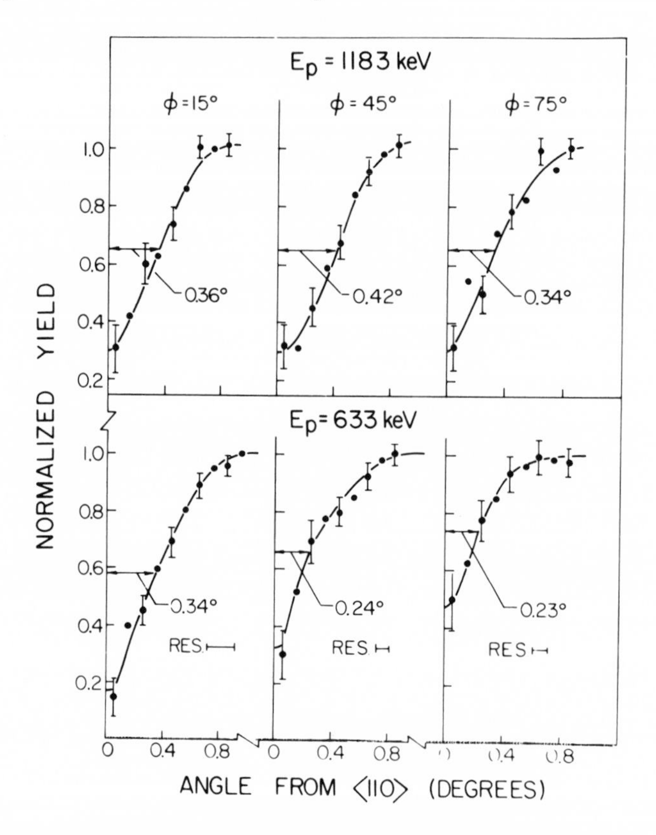

Fig. 4. Blocking dips observed along three <110> directions and at two proton bombarding energies, for ∿2 MeV alpha particles from $Al^{27}(p,\alpha_o)Mg^{24}$ reaction in aluminum crystal. Three upper curves: proton energy just exceeds compound state at 12.724 MeV in Si^{28}, known to have large width (∿1 keV). "Prompt" dip observed, independent of angle. Three lower curves: lower proton energy just exceeds resonance at E_p = 633 keV, corresponding to a state in Si^{28} at 12.192 MeV. Dip minimum depends on angle of observation relative to recoil direction. Deepest dip at $\phi = 15^o$ simulates prompt dip; shallow dip at $\phi = 75^o$ indicates measurable lifetime of ∿1.5 x 10^{-16} sec. From Reference 26.

$$\Gamma_\lambda \simeq \frac{2}{\pi} \frac{\Gamma_a\uparrow}{\Gamma_a\downarrow} \bar{D} ,$$

where $\bar{D}$ stands for the mean level separation. This, in fact, allows one to determine the level density of a given spin and parity at the analog resonance.

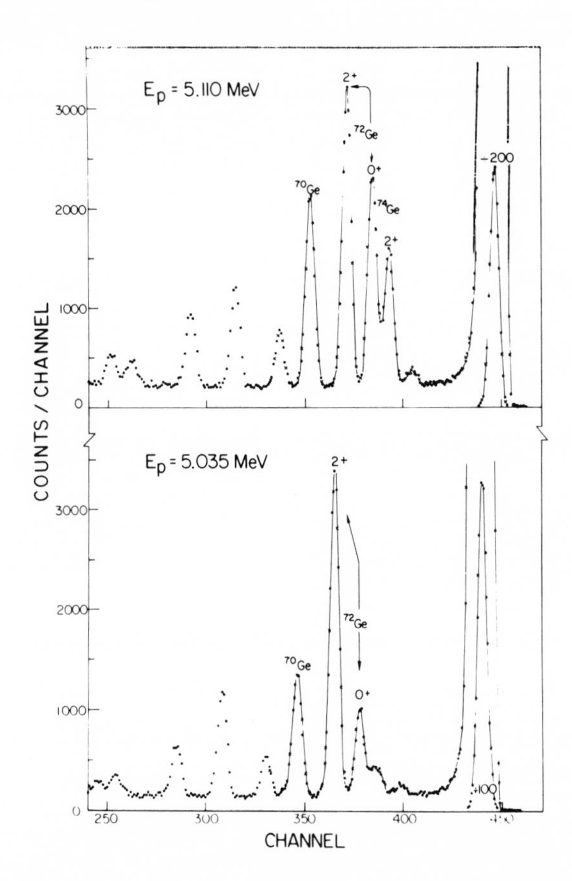

Fig. 5. Inelastic proton spectra from natural germanium crystal at two bombarding energies, observed in position sensitive silicon detector, identifying the even-even nuclei giving rise to various peaks. Note reduction factor for elastic peak. Peak at channel ~310 corresponds mainly to 2^+ excited state at 1.465 MeV in Ge^{72}. Practically no impurities appear owing to heating of crystal to 450°C during bombardment. From Ref. 21.

As was explained at the end of Section A(2), there are usually many compound resonances with various spins and parities contributing to nuclear reactions in experiments of the type described there. When we set our proton bombarding energy to correspond to an isobaric analog resonance with $J = 5/2^+$, say (which might be of the order of 50 keV wide), all nearby $5/2^+$ fine structure states will undergo an enhancement, *i.e.* a shortening of their lifetimes, while all other states of $J \neq 5/2^+$ will retain their average, off resonance behavior. The observed *net* enhancement is therefore considerably diluted. Nevertheless, we succeeded in clearly demonstrating this effect for a $J = 5/2^+$ and a $J = 1/2^+$ analog resonance in As^{73}.

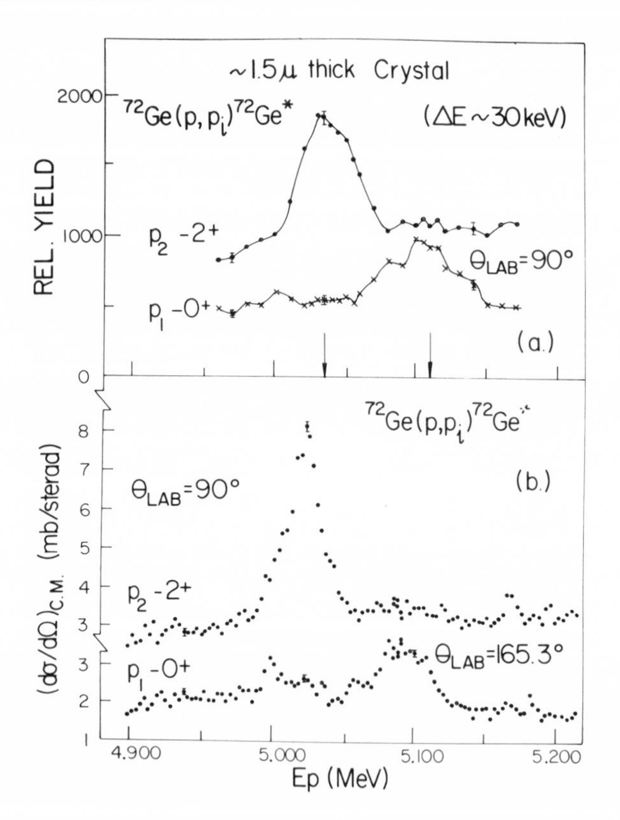

Fig. 6. (a) Excitation curve for 90° inelastic proton scattering on Ge^{72} to first excited 0^+, and second excited 2^+ states for ~30 keV thick crystal. Arrows denote *on* and *off* resonance conditions described in text. (b) Thin target (3 keV) excitation curves for the same states as in (a), $J = 5/2^+$ isobaric analog resonance (IAR) in As^{73} at 5.022 MeV; $J = 1/2^+$ IAR at 5.094 MeV. Reference 21.

Figure 5 shows the inelastic proton spectra observed from thin Ge crystals at two bombarding energies, with identification of peaks with even-even isotopes of germanium. Figure 6 shows the excitation curves for inelastic proton scattering leading to the first excited 0^+ state, as well as the second excited 2^+ state of Ge^{72}, for both a 3 keV thick target, and the actual 30 keV thick crystal target used in the blocking experiments. Figure 7 shows the blocking dips both *on* and *off* the $5/2^+$ analog resonance, as well as the appropriate, prompt elastic scattering calibration dip measured at the energy of the *outgoing* inelastic protons. The quantity R refers to the ratio of dip minimum to average shoulder value; W denotes the full angular width at half dip. The dip *on* resonance is significantly shallower than *off* resonance, corresponding to the speeding up, on the average, of the decays of the J = $5/2^+$ fine structure states; that is to say, their broadening (enhancement)[21] by Coulomb mixing with the analog of the $5/2^+$ state at 1.64 MeV in Ge^{73}. A similar but less pronounced effect for the J = $1/2^+$ state at 1.75 MeV was also observed.[21] Figure 8 contains all the

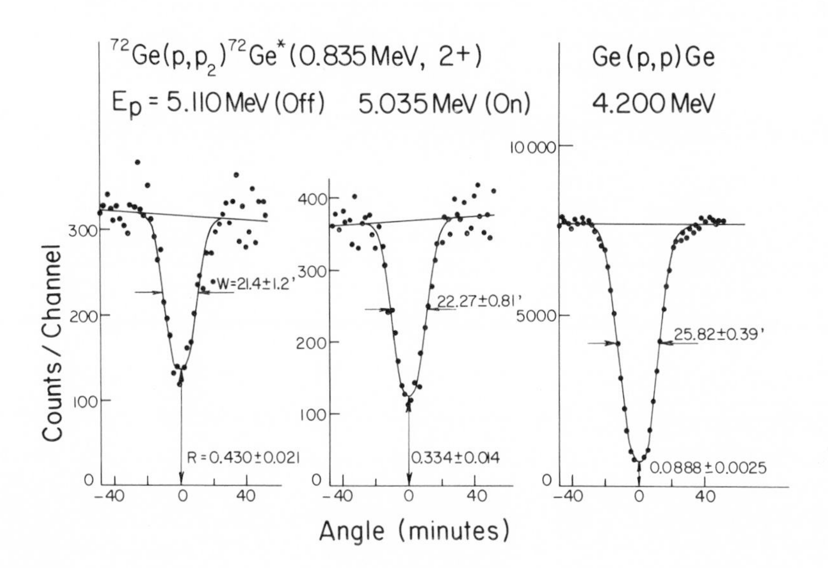

Fig. 7. Comparison of blocking dips for inelastic protons to 0.835 MeV, 2^+ state in Ge , *on* and *off* J = $5/2^+$ analog resonance, and "prompt" elastic calibration dip at the appropriate outgoing proton energy (4.200 MeV). Note deeper minimum *on* resonance, reflecting the enhancement of J = $5/2^+$ fine structure. Here R = ratio of minimum to average shoulder value; W = full angular width at half minimum. From Reference 21.

blocking patterns obtained in two runs at appropriately chosen energies, together with the prompt reference dips. Table I summarizes the numerical results, and gives lifetime estimates, using both the earlier approach of extraction,[10] and the more recent approach of Hashimoto.[8] Note the appreciably longer values (up to a factor of three) when using the latter approach.

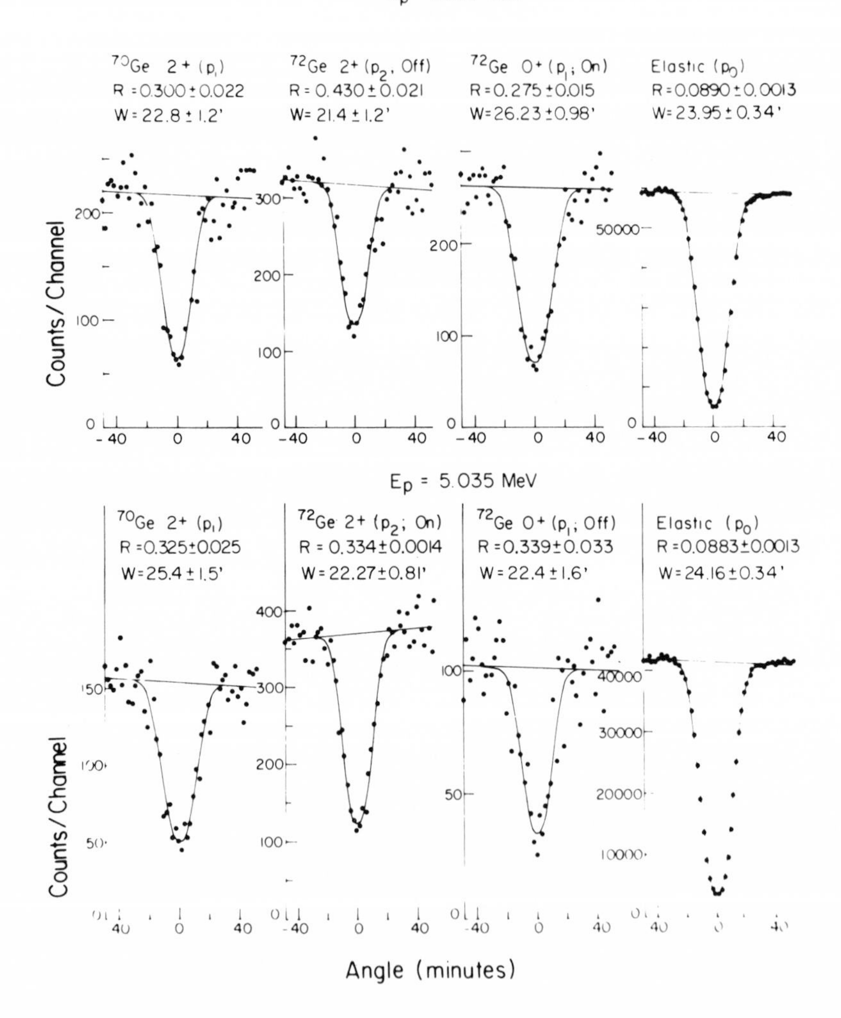

Fig. 8. Blocking dips for most inelastic proton scattering peaks shown in Figure 5, along with comparison "prompt" elastic dips. For meaning of R and W, see caption of Figure 7.

In the course of these runs, we observed two additional effects which can be discerned from Figure 8 and Table I: a difference in R-values for *off* resonance dips, at the same bombarding energy, for the targets Ge^{70} and Ge^{72}. This can be qualitatively understood in terms of the larger level density in Ge^{72} due to the higher excitation energy, by about 2 MeV, caused by larger proton binding energy in the latter nucleus. This effect, contributing to a narrowing of the fine structure levels on the average, is offset by increased penetrability of the inelastic protons leading to the lower lying 2^+ state in Ge^{72} (*higher* outgoing proton energy), and leads to a net decrease of level widths or increase of lifetimes in this nucleus relative to Ge^{70}.

Another effect will be discussed in the next paragraph.

(2) *Residual State Spin Effect on Average Compound Nucleus Lifetime*[21]

Figure 8 also contains a comparison of two blocking dips for the same compound nucleus As^{73}, at nearly the same (off analog resonance) excitation energy, for inelastic protons leading to two different final states in Ge^{72}, one having spin 0^+ at 690 keV, and one having spin 2^+ at 835 keV. They are compared with the appropriate calibration dip in Figure 9. The dip for the 2^+ state is considerably shallower, indicating a longer average life. This can be understood on the basis of the higher compound nuclear level spins J required ($3/2^+$ and $5/2^+$) to minimize the angular momentum needed for the rather low energy inelastic protons to reach $I' = 2^+$, compared to the $J = 1/2^+$ compound states allowing $\ell = 0$ for both in and outgoing protons when exciting the $I' = 0^+$ state. The latter compound states are wider, on the average, and hence lead to shorter lifetimes and deeper blocking dips. This difference is slightly increased by the increase in penetrability for the slightly lower lying 0^+ state.

(3) *Residual State Structure Effect on Average Compound Nuclear Lifetime*[28]

Very recently, we repeated our experiment with the germanium crystal, extending the inelastic excitation spectrum under observation to include the three additional inelastic peaks seen immediately to the left of the Ge^{70} 2^+ peak in Figure 5. The center one of the three consists mainly (>90%) of protons leading to the second 2^+ state in Ge^{72} at 1.465 MeV. The blocking dips for the two 2^+ states in Ge^{72} are compared in Figure 10 with the appropriate elastic calibration dips obtained at the correct outgoing proton energies. We see a significant difference in the inelastic dips. A fraction of this difference is ascribable to the energy difference of 600 keV between the proton groups, but the rest

TABLE I

Blocking lifetime results for Ge+p along <110> axis.

		R or R'	R-R'
E_p = 4.200 MeV			
$Ge(p,p_o)$		0.0888 ± 0.0025 = R_1'	
E_p = 5.035 MeV			$R-R_1'$
$Ge(p,p_o)$		0.0883 ± 0.0013	
$Ge^{72}(0^+)p_1$	(off)[d]	0.339 ± 0.033	0.250 ± 0.033
$Ge^{72}(2+)p_2$	(on)[d]	0.334 ± 0.014	0.245 ± 0.014
$Ge^{70}(2+)p_1$		0.325 ± 0.025	0.236 ± 0.025
E_p = 5.110 MeV			
$Ge(p,p_o)$		0.0890 ± 0.0013	
$Ge^{74}(2+)p_1$		0.161 ± 0.017	0.072 ± 0.017
$Ge^{72}(0^+)p_1$	(on)[d]	0.275 ± 0.015	0.168 ± 0.015
$Ge^{72}(2+)p_2$	(off)[d]	0.430 ± 0.021	0.341 ± 0.021
$Ge^{70}(2^+)p_1$		0.300 ± 0.022	0.211 ± 0.022
E_p = 3.708 MeV			
$Ge(p,p_o)$		0.1105 ± 0.0020 = R_2'	
E_p = 4.358 MeV			
$Ge(p,p_o)$		0.1096 ± 0.0018 = R_3'	
E_p = 5.210 MeV			
$Ge(p,p_o)$		0.1057 ± 0.0014	
$Ge^{72}(2+)p_2$	(off)[d]	0.446 ± 0.012	0.3355 ± 0.012
$Ge^{70}(2^+)p_1$		0.278 ± 0.013	0.1675 ± 0.013
$Ge^{70}(0+)p_2$		0.168 ± 0.024	0.0584 ± 0.024
$Ge^{72}(2+)p_3$	(off)[d]	0.511 ± 0.022	0.3854 ± 0.021

[a]Obtained with formula of Reference 10 and $r_c = 3a_{TF} = 0.422$Å.

[b]From Monte Carlo calculation of Reference 8.

[c]1 as = 10^{-18} second.

[d]Refers to $5/2^+$ analog resonance.

TABLE I (CONTINUED)

Single String[a]		Monte Carlo[b]	
$V_{\perp}\tau$ (Å)	τ (as)	$V_{\perp}\tau$ (Å)	τ (as)[c]
0.269 ± 0.027	63.1 ± 6.3	0.705 ± 0.130	165 ± 30
0.265 ± 0.011	62.1 ± 2.6	0.670 ± 0.053	156.5 ± 12.4
0.258 ± 0.020	58.8 ± 4.6	0.636 ± 0.087	144 ± 20
0.132 ± 0.014	31.6 ± 3.3	0.244 ± 0.037	58.3 ± 8.8
0.219 ± 0.011	51.0 ± 2.6	0.498 ± 0.037	115.5 ± 8.6
0.349 ± 0.021	81.2 ± 4.9	$1.29^{+0.33}_{-0.22}$	300^{+77}_{-52}
0.238 ± 0.017	53.9 ± 3.8	0.562 ± 0.61	128 ± 14
0.344 ± 0.012	79.4 ± 2.8	$1.24^{+0.16}_{-0.18}$	285^{+37}_{-30}
0.205 ± 0.009	46.0 ± 2.1	0.455 ± 0.031	102 ± 7
$0.122^{+0.021}_{-0.024}$	$27.4^{+4.7}_{-5.4}$	0.214 ± 0.053	48.1 ± 11.9
0.413 ± 0.030	95.2 ± 6.9	>2.1	>485

can only be explained in terms of differences in *structure* between the compound states which decay to one or the other residual state of identical spin. In other words, the parentages of the two 2^+ states of Ge^{72}, which are quite likely to be different, seem to reflect on the subpopulations of compound resonances decaying to one or the other state. This means that there must be other parameters, in addition to angular momentum, which label compound resonances; a not too surprising fact, but one not allowed for in our usual statistical description of the complicated fine structure states in the nuclear continuum. It is important to collect additional cases of this type to discern possible systematic trends.

C. Third Generation

Here we shall discuss some experiments which seem within the realm of experimental possibility in the not too distant future. They mainly require the availability of thin, single monoisotopic crystals.

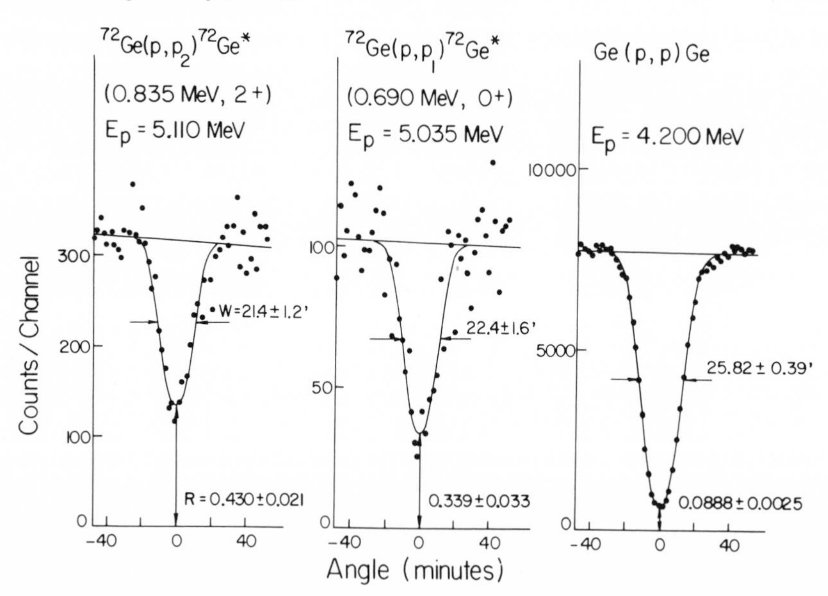

Fig. 9. Comparison of inelastic proton blocking dips corresponding to spin 2^+ (0.835 MeV) and 0^+ (0.690 MeV) excited states of Ge^{72}, along with comparison "prompt" elastic dip. The 2^+ dip is shallower, revealing longer mean lifetime or narrower widths of contributing compound states in As^{73}.

For meaning of R and W, see caption of Figure 7.

(1) *Distinction of Compound and Shape Elastic Scattering by Time Delay*

We have been assuming until now that the blocking patterns associated with elastic scattering, the latter being mainly Rutherford scattering, represented a prompt event, *i.e.* times of the order of nuclear traversal times, or about 10^{-20} seconds or less. This assumption is borne out by the fact that the R-values for the prompt blocking dips were identical within the rather small uncertainties, independently of the specific bombarding energy; that is to say, it seemed not to matter whether one formed isobaric analog resonances or not. Figures 7 through 10 contain a number of these elastic dips at different bombarding energies. Now one must realize that the elastic peaks in the spectra of Figure 5 are composed of the contributions from all isotopes of germanium, no two of which have their analog resonances at exactly the same bombarding energy. It is therefore essential to use isotopically enriched single crystals. This has now become technically possible.[29] Moreover, as can also be discerned from Figure 5, the *inelastic* peaks represent less than one percent of the elastic peak intensity; it is not unreasonable to expect the compound elastic component to represent about the same fraction of the elastic peak--a frac-

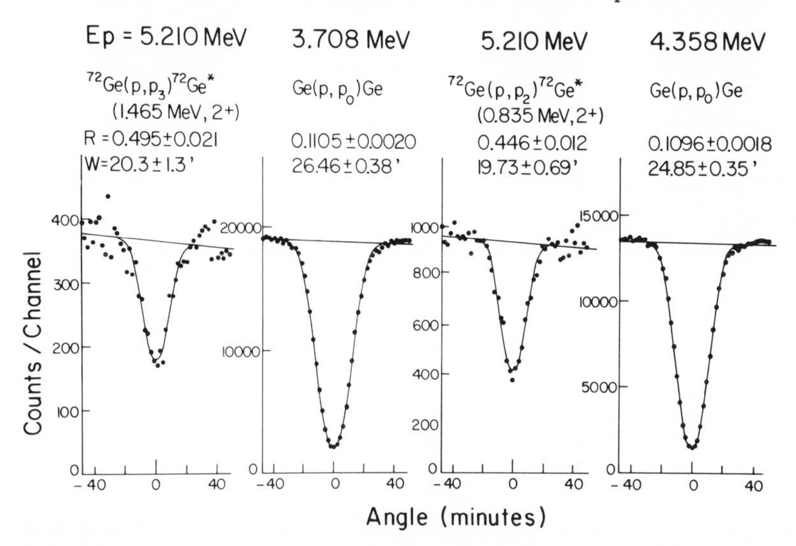

Fig. 10. Comparison of inelastic proton blocking dips corresponding to the first two excited 2^+ states in Ge^{72} at 0.835 and 1.465 MeV, respectively, along with comparison "prompt" elastic dips. Upper 2^+ dip is shallower, revealing structural differences in the compound continuum. For meaning of R and W, see caption of Figure 7. From Reference 28.

tion clearly insufficient to be revealed in the blocking pattern. Conditions must be found to maximize the compound elastic fraction, which means that we must approach the single channel case as closely as possible by being below neutron threshold, minimizing inelastic scattering, *etc*. There are basically two approaches to detect possible compound elastic contributions: one involves the comparison of *on* and *off* analog resonance elastic scattering dips, following the lead of some experiments using polarized and unpolarized protons revealing enhanced compound elastic components[30,31,36] near analog resonances. Another would involve the comparison of elastic blocking dips along different crystallographic directions, at the same bombarding energy, in the spirit of the fission experiments discussed in Section A(1).

(2) *Lifetimes in* $(p,n\bar{p})$ *and* $(d,n\bar{p})$ *Reactions*

The emission of protons from isobaric analog states following the (p,n) or (d,n) reactions has been thoroughly studied.[32,33] The $\bar{p}$ protons emitted from the residual nuclei in the first (perhaps direct) reaction have often been referred to as "delayed", presumably relative to the neutrons emitted first. In fact, the protons from the analog state having $T = T_z + 1$, are "speeded up" relative to other nearby proton groups in the observed spectrum, following the (p,n) or (d,n) reaction to unbound states with lower isospin $T = T_z$. This speed up is exactly the enhancement which we discussed in Section B(1). All other protons in the spectrum, coming from the (p,p') and (d,p) reaction, respectively, are likely to be prompt because the latter are direct, or because they go through highly excited compound nuclear regions where many nucleon channels are open and the mean level widths are large. This proposed experiment is equivalent to those described in Section B(1), except that

(a) both on-resonance and off-resonance conditions can be achieved simultaneously in the same outgoing proton spectrum, and

(b) analog states far below the Coulomb barrier for emerging $\bar{p}$ protons can be formed in the first step reaction.

Once again, isotopically separated crystals are required here, since usually only certain even-odd isotopes are favorable targets.[32] Having germanium crystals at hand from our other experiments, we made an attempt to observe the $Ge^{73}(p,n)As^{73*}(\bar{p}_o)Ge^{72}$ reaction via the ground state analog of Ge^{73}, an experiment which equivalently would have to be carried out at about 3.3 MeV proton energy on Ge^{72} to form this $9/2^+$ state with $\ell = 4$ protons, clearly an impossibility. Unfortunately, no $\overline{p}$ peak could be observed with about 13 MeV protons incident, for reasons which are discussed in Reference 32.

III. CONCLUSION AND OUTLOOK

The future of crystal blocking lifetime experiments is dependent to a large extent on technical developments. Every new target nucleus is a major undertaking in terms of crystal growing, especially when separated isotopes are required. Blocking and channeling properties have to be investigated before one is ready to undertake a lifetime measurement. Nevertheless, in principle, any nucleus is amenable to such determinations.

An essential role is played by the large computer calculations of the trajectories of charged particles through single crystals; it is the only avenue to the correct extraction of absolute lifetime values from observed blocking profiles. We wish to emphasize that thus far, essentially all information could be obtained merely from a relative ordering in time, of various nuclear events, for which no detailed calculations were necessary. The future undoubtedly holds out much promise concerning our more complete understanding of the blocking profiles, involving not only the values at minimum yield, but their widths, slopes at half minimum, and other as yet unspecified parameters describing the detailed shape of blocking dips. Each of these will be sensitive to different segments of the time evolution spectrum of nuclear reactions, and is likely to lead to new insights not previously accessible by any other method.[34]

Finally, looking far into the future, we realize that the blocking phenomenon exists at all energies and for all outgoing charged particles. In Figure 11 we present an example of a possible lifetime measurement of an "elementary" particle, the η^o. It could be produced by the "Primakoff effect"[35] with incident high energy photons, and its (relativistic) flight distance $\beta\gamma c\tau_0$ could be determined from the decreased blocking of its charged decay products. The relativistic critical angle for blocking is given at the bottom of the figure in milliradians, m_o being the rest mass of the outgoing charged particle. These angles, although much smaller than those found in low energy nuclear physics, are of the order of those commonly encountered with high energy accelerators and correspond to the resolution of associated detectors, such as wire chambers; it is therefore quite feasible to observe blocking dips with half widths of a few milliradians. It will be left to the ingenuity of the individual experimenters to devise methods of promptness calibration. For highly relativistic particles, the flight distance becomes proportional to the energy, so that the window of lifetime sensitivity will move toward shorter times with increasing energy. It is conceivable that one might someday measure the times of flight of "resonances" having widths of several MeV!

LIFETIMES OF ELEMENTARY PARTICLES OR RESONANCES BY CRYSTAL BLOCKING?

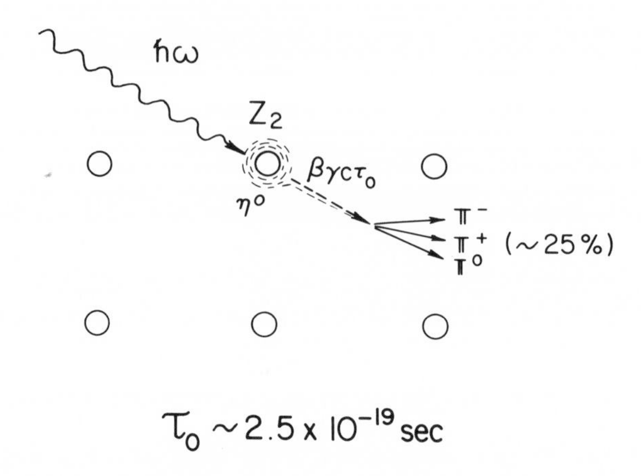

$$\psi_c^{rel} = 2\sqrt{\frac{Z_1 Z_2 e^2}{p\beta c d}} \sim \frac{4.5}{\sqrt{m_o}}\sqrt{\frac{\gamma}{\gamma^2-1}}\ \text{mrad}$$

Fig. 11. Speculation on the possibility of measuring elementary particle (or resonance) lifetimes at high energies. Example: η^o photo production, and decay mode into three pions; π^+ and π^- would be blocked less than if emitted from lattice site. Expression for relativistic critical blocking angle in milliradians is given at bottom; m_o = rest mass of blocked charged particle, and γ = total energy in units of the rest mass.

ACKNOWLEDGMENTS

I wish to acknowledge my colleagues, especially M. Maruyama, W. M. Gibson, Y. Hashimoto, and S. Yoshida for what I have learned from my association with them over the past few years.

REFERENCES

1. A. F. Tulinov, Soviet Physics Doklady 10, 463 (1965).
2. D. S. Gemmell and R. E. Holland, Phys. Rev. Lett. 14, 945 (1965).

3. The so-called rescattering effect[4,5] has in a few cases been used to infer nuclear lifetimes, although the situation is confused at present. Another method, not yet successfully applied, is the soft bremsstrahlung technique.[6]
4. C. Kacser and I. J. R. Aitchison, Rev. Mod. Phys. 37, 350 (1965).
5. J. Lang *et al.*, Nucl. Phys. 88, 576 (1966).
6. R. M. Eisberg, D. R. Yennie, and D. H. Wilkinson, Nucl. Phys. 18, 338 (1960).
7. J. H. Barrett, Phys. Rev. B3, 1527 (1971).
8. Y. Hashimoto, Bull. Am. Phys. Soc. 17, 560 (1972), and private communication.
9. E. Fuschini *et al.*, Nuovo Cimento 10A, 177 (1972).
10. W. M. Gibson and K. O. Nielsen, Phys. Rev. Lett. 24, 114 (1970).
11. R. L. Fleischer, P. B. Price, and R. M. Walker, Ann. Rev. Nucl. Sci. 15, 1 (1965).
12. S. A. Karamyan *et al.*, Sov. J. Nucl. Phys. 13, 543 (1971).
13. V. V. Kamanin *et al.*, preprint P7-6302, JINR Dubna, 1972.
14. S. A. Karamyan, F. Oganesyan, and F. Normuratov, Sov. J. Nucl. Phys. 14, 279 (1972).
15. V. V. Kamanin *et al.*, preprint P7-6291, JINR Dubna, 1972.
16. J. U. Andersen, W. M. Gibson, E. Laegsgaard, and K. O. Nielsen, private communication.
17. J. U. Andersen, K. O. Nielsen, and J. Skak-Nielsen, preprint, Aarhus, Denmark, 1972.
18. U. Noelpp *et al.*, Helv. Phys. Acta 45, 55 (1972).
19. M. Maruyama *et al.*, Nucl. Phys. A145, 581 (1970).
20. W. M. Gibson *et al.*, Bull. Amer. Phys. Soc. 16, 557 (1971); G. M. Temmer, Bull. Amer. Phys. Soc. 17, 577 (1972).
21. W. M. Gibson *et al.*, Phys. Rev. Lett. 29, 74 (1972).
22. G. J. Clark *et al.*, Nucl. Phys. A173, 73 (1971).
23. J. R. Huizenga, private communication.
24. C. E. Porter and R. G. Thomas, Phys. Rev. 104, 483 (1956).
25. K. Komaki *et al.*, Phys. Letters 38B, 218 (1972).
26. R. P. Sharma, J. U. Andersen, and K. O. Nielsen, to be published.
27. See *e.g.* W. M. MacDonald and A. Z. Mekjian, Phys. Rev. 160, 730 (1967).
28. G. M. Temmer *et al.*, Bull. Amer. Phys. Soc. 18, 119 (1973).
29. W. M. Gibson, private communication.
30. W. Kretschmer and G. Graw, Phys. Rev. Lett. 27, 1294 (1971).
31. H. L. Scott, C. P. Swann, and F. Rauch, Nucl. Phys. A134, 667 (1969).
32. P. S. Miller and G. T. Garvey, Nucl. Phys. A163, 65 (1971).

33. C. F. Moore *et al.*, Phys. Rev. Lett. 17, 926 (1966); N. Cue and P. Richard, Phys. Rev. 173, 1108 (1968).
34. For a review of crystal blocking lifetime measurements with emphasis on the solid state aspects, see W. M. Gibson and M. Maruyama, in *Particle Channeling*, D. V. Morgan, ed. (John Wiley and Sons, New York, in press).
35. C. Bemporad *et al.*, Phys. Lett. 25B, 380 (1967).
36. E. P. Kanter *et al.*, Bull. Amer. Phys. Soc. 18, 119 (1973).

DISCUSSION

ROBERTSON: Two questions: first, was the temperature you ran your crystal at for annealing?

TEMMER: Right, that is presumably why.

ROBERTSON: It's just to keep things going?

TEMMER: All I can say is that we searched for a year at what temperature we should set. People said we should set at liquid nitrogen temperature, and that particular crystal lasted 5 seconds! And it took a month to grow! 450°C is the magic number, 400° is too cold and 500° is too hot. That's all I can tell you.

ROBERTSON: The second question is: do you have some idea of how many levels you would have been averaging over in those two states where you saw a difference in the compound lifetime?

TEMMER: You mean, how many levels of the compound nucleus are contributing within our target thickness of 30 keV?

ROBERTSON: Yes.

TEMMER: About 1000 levels of various spins and parities.

TEMMER: I might just say that every nucleus which can be formed as a crystal can be done, but just give the man who grows the crystal a year's notice. At the moment we are trying to grow isotopically enriched lead crystals, epitaxially on NaCl--this looks promising. We are also trying to do molybdenum. We're trying to do the compound *versus* shape elastic scattering case and that can only be done with separated isotopes.

IV.B. DOPPLER SHIFTS WITH GAS BACKING

D. J. Donahue
Department of Physics, University of Arizona
Tucson, Arizona

The Doppler shift attenuation (DSA) method has proved to be a powerful and widely used tool for the measurements of mean lives of nuclear states. As generally used, with nuclei in excited states recoiling into solid backings, mean lives from a few picoseconds down to nearly 10^{-14} seconds can be measured. In the experiments which I will discuss, nuclei recoil in a gas, and thus depending on the gas pressure used, mean lives from a few picoseconds up to nanoseconds can be measured.

The recoil distance, or plunger method, is generally used for mean lives in this range. However, if recoil ions have speeds less than about 1% the speed of light, which is typical of speeds that result from reactions produced by particles from a 5MV accelerator, and if the uncertainty in the distance from the target to the plunger is 5 microns, then for mean lives less than about 30 picoseconds uncertainties in plunger measurements are greater than about 10%. This is comparable with uncertainties in the DSAM resulting from stopping power uncertainties. At any rate, what I am saying is that there are some instances in which DSA measurements are at least as good as the plunger measurements, and are technically easier to make. Having said this I can proceed to the rest of the talk.

There are several laboratories that use gas-backing methods, in particular at Cal Tech and Padova. The experiments which I will describe were all done at Arizona. They were started by Bob Hershberger, who is now at Munich, building a plunger.

A schematic outline of our measurements is shown in Figure 1.

i) The beam enters from the left and

ii) passes through an annular particle detector (1).

iii) Part of the beam strikes a target at (2). This target is several hundred micrograms/cm^2 thick, evaporated onto a thick tantalum backing. The target is made in two halves whose separation can be varied to adjust the amount of beam it intercepts.

iv) γ rays produced in reactions in this target are detected in a Ge(Li) detector (6), and coincident γ ray and particle pulses from detector (1) are stored together on magnetic tape.

For the mean lives in which we are interested, the nuclei emitting these γ rays are stopped in the tantalum before emission, so the γ rays have energy E_o.

v) The remainder of the beam not stopped by the upstream target passes through a second annular particle detector (3) through a pressure-retaining foil (either 0.1 mil platinum or 0.2 mil tantalum) and strikes a thin ($\leq$100 μgm/cm^2) target evaporated onto the foil.

vi) Excited nuclei produced in this target recoil into the gas cell (5) which contains krypton gas at a pressure which can be varied from vacuum up to 20 atmospheres. Gamma rays from these nuclei in coincidence with particles in detector (3) are stored with those protons on magnetic tape. After a run is completed, a Nova 1200 computer is used to transfer the spectra of γ rays of interest from the magnetic tape to the memory of an 8192 channel pulse height analyzer.

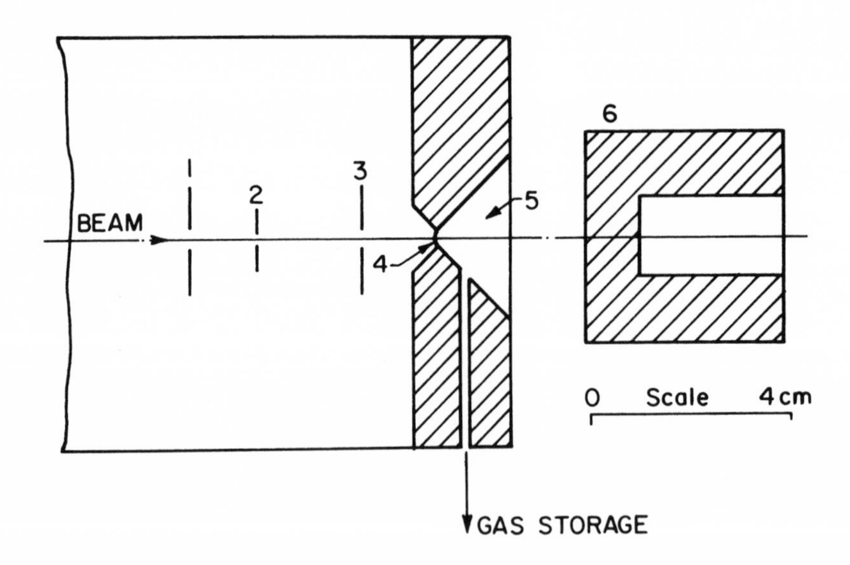

Fig. 1. Experimental arrangement. The numbers refer to: (1) upstream particle detector; (2) thick target; (3) downstream particle detector; (4) thin target evaporated on downstream side of pressure-containing foil; (5) pressure cell; and (6) Ge(Li) detector.

The difference in energy of γ ray peaks from the two targets is the numerator of the equation

$$F = \frac{\left\langle E_\gamma(v)\right\rangle_{0^o} - E_o}{\left\langle E_\gamma(v_i)\right\rangle_{0^o} - E_o} \,.$$

The brackets indicate an average over time and over the finite geometries of the particle and γ ray detectors. The denominator of the equation can be either measured by running the experiment with the gas cell evacuated or calculated. The coincidence requirement fixes the direction of the recoiling nuclei so that the calculation can be done accurately. The coincidence requirement also allows us to know from which target a particular γ ray comes, and greatly increases the signal to background ratio. The coincidence requirement also results in long running times, from eight to 24 hours or more, but the fact that both energies in the numerator or the denominator are measured simultaneously eliminates problems associated with the change of gain of electronics.

Figure 2 shows a series of curves of F *vs.* τ for Fe^{55} nuclei recoiling in different pressures of krypton. Foils of 0.2 mil tantalum will hold about 20 atmospheres, so as can be seen, mean lives as short as a few picoseconds can be measured.

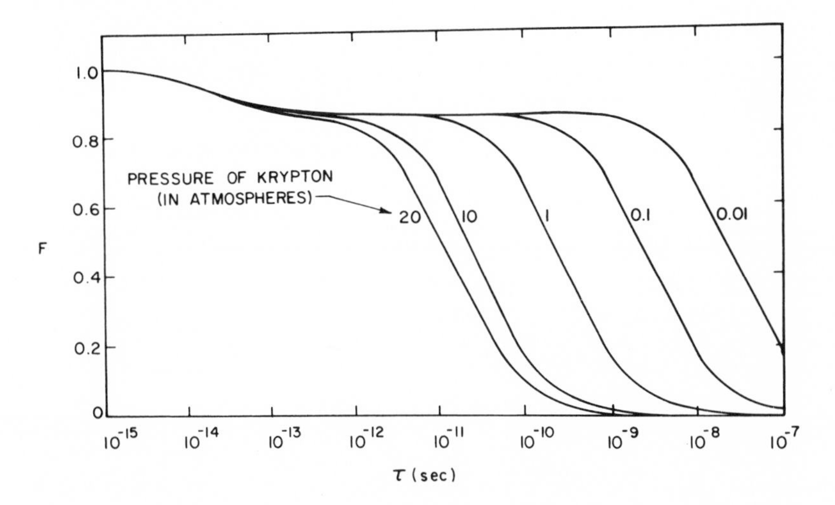

Fig. 2. Doppler shift attenuation factor F, *vs.* mean life, τ, for Fe^{55} nuclei recoiling in krypton gas at various pressures. Curves were calculated for 3.5 MeV deuterons incident on a Fe^{54} target 50 $\mu g/cm^2$ thick.

There are at least three things that cause concern in the analysis of Doppler-shift measurements in gas. The first of these is the condition of the targets. In our experiments, the upstream target serves only to establish the unshifted γ ray energy. The only requirement of the target is that there be no large voids between the evaporated target and the tantalum backing. The downstream target is made very thin ($\lesssim$100 $\mu g/cm^2$) so that not much slowing down of the recoil ions will occur in the target material. Thus uncertainties in target composition, thickness, or slowing down power will not much affect the results of the measurement.

The second point of concern is the possibility that heating of the gas by the beam will change the density of the gas in just that volume which is seen by the recoil ions. Figure 3 shows measurements of F for the reaction $O^{16}(d,p)O^{17}$ (0.871 keV, τ = 2.6 x 10^{-10} sec) as a function of pressure in the gas cell, for currents of 2 namps and 20 namps. As can be seen there is no difference between the two measurements. From these data we conclude that if the current is kept below 20 namps, density changes of the gas by the beam are not important. For most measurements we have tried, the current must be kept below 20 namps in order to keep the counting rate in

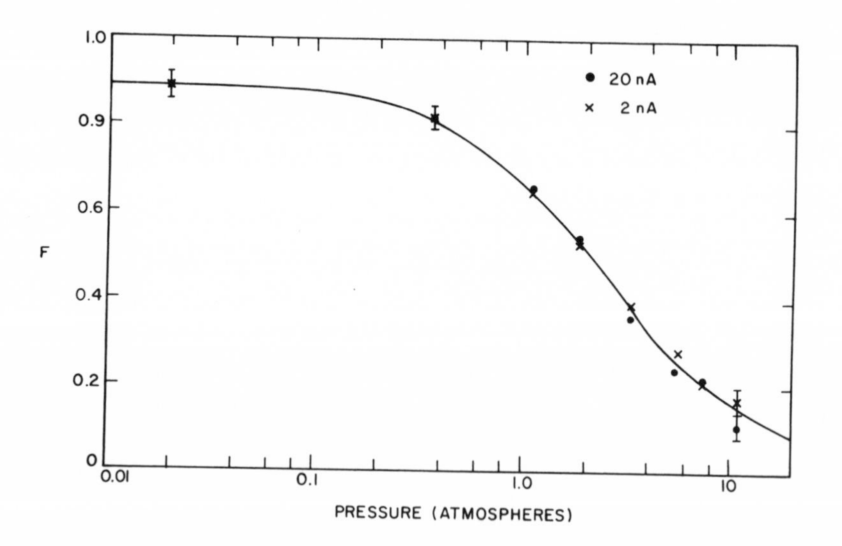

Fig. 3. Doppler shift attentuation factor, F, for 0.87 MeV γ rays from O^{17} recoiling through krypton gas, as a function of the pressure of the gas. The dots (⊙) and crosses (x) were measured using 20 and 2 nA, respectively, of 2 MeV deuterons. The solid curve was calculated for a mean life of the O^{17} state of 2.587 x 10^{-10} sec.

the γ ray detector at an acceptable level. The solid curve was calculated using the known mean life of the 0.871 MeV state in O^{17}, and the slowing down properties of the thin SiO_2 target and krypton calculated from the theory of Lindhard, Scharff, and Schitt.[1]

The other source of concern with gas backings, indeed with all DSA measurements, is the slowing down properties of the media in which the recoil nuclei move. We have been, and are, working on this problem and I would like to show some results. We are doing experiments in which Doppler shifts are measured for γ rays from states whose mean lives are well known, and using the results to gain some information about stopping powers.

Consider

$$F = F\left(\tau, \frac{dE}{dx}\right)$$

$$\frac{dE}{dx} = f_e\left(\frac{dE}{dx}\right)_{elec} + f_n\left(\frac{dE}{dx}\right)_{atomic} .$$

We chose experiments with $v_o \simeq 10^{-2}c$ and $\bar{v}_f \simeq 3$ to $5 \times 10^{-3}c$. For this range $(dE/dx)_{atomic}$ is small, the second term in dE/dx does not contribute much, and uncertainties in f_n are not crucial. We then choose $f_n = 1.0$, measure F, know τ, calculate $(dE/dx)_{elec}$ from the theory of Lindhard and his coworkers[1] and solve for f_e.

Figure 4 shows some preliminary results. We have plotted f_e *vs.* Z_1, the charge of the recoiling nuclei, for $Z_2 = 36$ (krypton). In most instances the errors are from uncertainties in the measurement of F. In fact, the errors are so large because, at $F = 1/2$, the per cent error in f_e is twice the per cent error in the measurement of F. The results give some confidence that in the region of Z_1 studied here, stopping powers calculated by Lindhard's theory are not wildly wrong, and in fact errors from uncertainties in stopping powers are comparable to the other errors in the experiment. These results might seem to contradict some of the recent slowing-down-power measurements. There are certainly shell effects in the slowing down process, and the calculated stopping powers that we used do not include those effects. However, those effects vary according to the ions involved in the slowing down and to the speed of the moving ions. There are no measurements for ions of the initial velocities we use slowing down in krypton.

Finally, on this subject I would like to show some re-

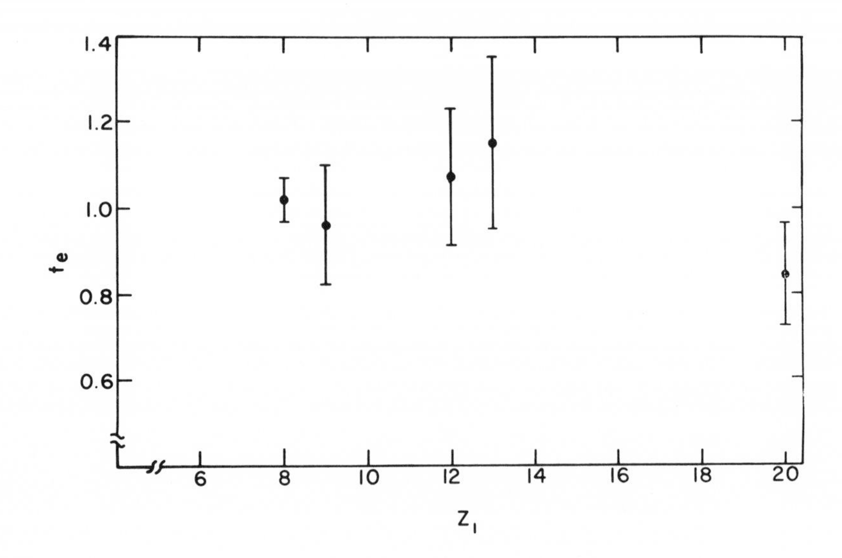

Fig. 4. Measured values of f_e for various ions recoiling in krypton.

sults obtained by Bini, Bizzeti, Bizzeti-Sona, and Ricci at Padova, and published in the last Physical Review.[2] They are illustrated in Figure 5. It shows a comparison between measured and calculated attenuation factors for γ rays from the 0.5 MeV-state in F^{17} recoiling in N^{14}. The mean life of this state is well known, so that from their results they can deduce f_e for F^{17} in nitrogen. They conclude that with $f_e = 1.1$, as suggested by the measurements of Ormrod, they got agreement with their experimental results, but in fact, $f_e = 1.0$ gives a better fit to their data.

The best way to be reasonably confident of the stopping powers used in a given experiment is to find a close by state whose mean life has been measured electronically (or with a plunger) and determine f_e in the manner I have indicated above. For example, John McCullen in our laboratory is interested in measuring the mean lives of states in Ca^{44}. This is a nucleus in which the best guess for the mean life of seven of the eight lowest states is between 3 and 30 picoseconds. Thus, it is an ideal case for the gas backing method, and we need to know the stopping power of krypton gas for calcium ions. There is a state in Ca^{42}, the second 0^+ state at 1.85 MeV, whose mean life is known to 6%. It is one of the points on the previous slide, and we can use this state to find, to about 12%, the f_e for calcium ions moving at a particular initial and final speed through krypton. As long as one can find such a state, the gas backing method should be a reliable

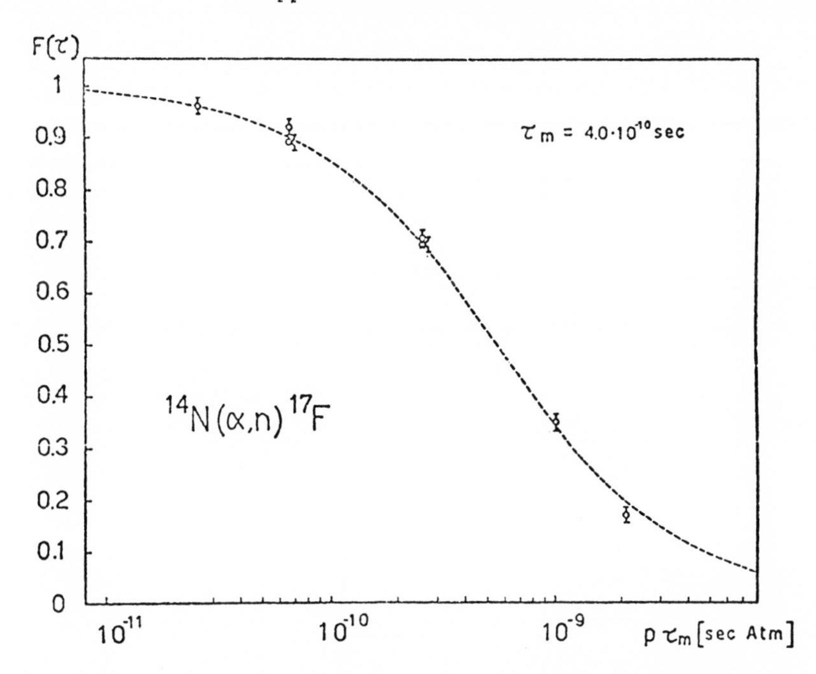

Fig. 5. F *vs.* pτ for F^{17} ions in the 0.495 MeV state recoiling in nitrogen. The dotted curve is calculated using the known mean life of this state. This curve is from the annual report of the Laboratori Nazionali De Legarno (Padova). See also Reference 2.

and relatively simple way to determine mean lives that are longer than a few picoseconds.

Finally, I will show some preliminary results on Ca^{44}. As I mentioned, this is an experiment being done by John McCullen and myself in our laboratory. Figure 6 shows a spectrum of protons produced in the $K^{41}(\alpha,p)Ca^{44}$ reaction with 8 MeV α particles. These are protons in coincidence with γ rays. The figure illustrates two problems encountered in these experiments. The first is the poor resolution. The alpha particles must traverse 0.1 or 0.2 mils of material, and because of straggling, they have an energy spread of ≈125 keV when they get to the target. In addition, there is kinematic broadening, so we wind up with resolutions of about 150-200 keV. The second problem is illustrated by the large peaks designated as P^{31}. They result from the $Si^{28}(\alpha,p)P^{31}$ reaction in silicon pumping oil which always seems to be present on our pressure foils.

Nevertheless, the protons to many states in Ca^{44} can be seen. The γ rays in coincidence with these protons are all stored on magnetic tapes, and the next time we meet, we will tell you what the mean lives for all these states are.

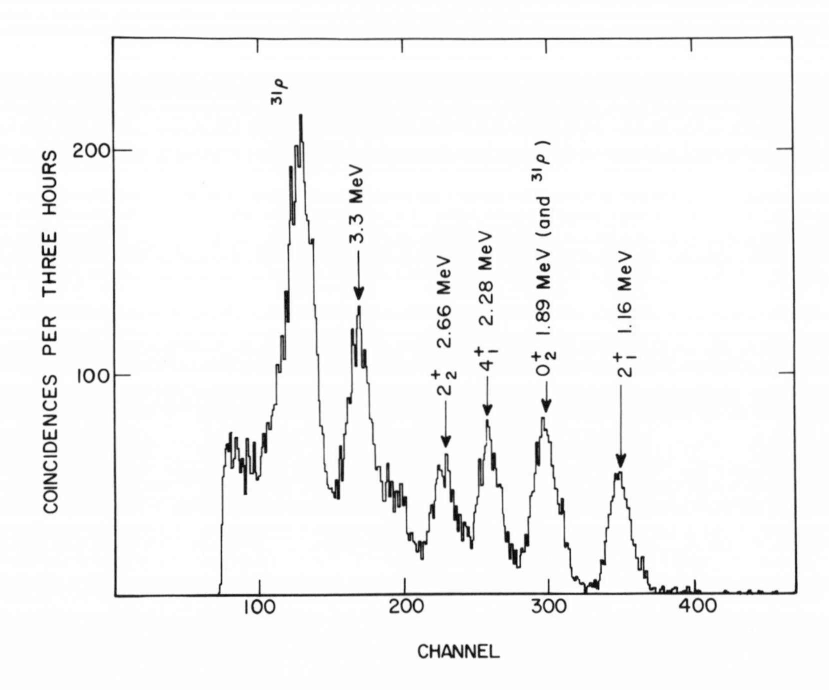

Fig. 6. Spectrum of protons from the reaction $K^{41}(\alpha;p)Ca^{44}$ in coincidence with gamma rays. The peaks marked P^{31} are from the $Si^{28}(\alpha,p)P^{31}$ reaction in impurities in the target.

REFERENCES

1. J. Lindhard, M. Scharff, and H. E. Schitt, Kgl. Danske Videnskab. Selskab Mat. Fys. Medd. 33, No. 14 (1963).
2. M. Bini *et al.*, Phys. Rev. C6, 784 (1972).

DISCUSSION

DE VOIGT: You find that the Lindhard theory applies well for the description of this slowing down process. This is nice, but if I understand well you performed tests for nitrogen and krypton only. It is well known that there are many problems in this respect. I would like to ask you opinion about the paper of Broude *et al.* who measured a certain lifetime using many different backings and observed large discrepancies, even for very light slowing down materials like carbon. Do you have any comment on this?

DONAHUE: Well, there are obviously shell effects in the slowing down process and Lindhard's theory does not include these, so I would say no amount of information would make you completely confident that you were doing the right thing, unless you measure the slowing down for the experiment that you happen to be doing right then. Now if there were really a disagreement outside of a factor of two, I'd be inclined to suspect that it's due to a target condition more than to the stopping power.

SESSION CHAIRMAN (WARBURTON): I have a comment on that. For years people who have been doing Doppler shifts have said that you can trust the electronic stopping of Lindhard's theory (theory in quotes because he never claimed that it was accurate), but with a certain coefficient put in front that he said was a good guess. You could trust it to about 15%. If you take the mean and standard deviation on Broude's results you get 15%.

DONAHUE: That was my impression also.

WARBURTON: There's one very bad point which is carbon and I'm sure that that is due to a target problem.

SHARPEY-SCHAFER: I would just like to confirm that. Broude feels that the two points that are really off on his results are due to oxidation of the targets. This is what occurred to most people when they saw the results, I think. Incidentally, the gamma line was in Na^{22}.

DONAHUE: I would like to say that the one curve showed, among other things, that gas stopping powers are not off by factors of two, and as was pointed out by the Italians, at the pressures of gas that are used in these experiments, the gases are more like solids than gases. That is, the time between collisions is less than the de-excitation time of states in the atom.

DE VOIGT: Well, of course, a lot of problems are connected with lifetime measurements but it seems to me that if you scan the whole scale of available lifetimes you will find a terrible discrepancy in many cases of similar Doppler shift lifetime measurements, probably due to the interpretation of the slowing down theory. So I think the best thing you can do is to test your particular slowing down material on a well known lifetime determined by other methods.

DONAHUE: It's not as easy to do for a solid backing. There aren't as many well known lifetimes in the range in which you use solid backings, but if you can, you certainly should.

DE VOIGT: On the other hand for calculating the electronic stopping power it is quite possible to have experimentally determined values instead of this Lindhart theory. Then you can include the velocity dependent constant C_a which is a real constant in the Lindhard theory. But well, I'm wondering why you get this nice agreement. I don't understand it, but it might apply in your case.

DONAHUE: I think there were six cases including the one from Italy and they are all in very good agreement, which does sound improbable.

ROBERTSON: I would like to give what I think is a possible answer to the question of dE/dx and this applies particularly for low velocity. You are in the velocity region where you can "see" physical target structure, so it is not necessarily the dE/dx curve but also the assumption of continuity in the slowing down process which can be under question, not just the dE/dx process itself. This is my feeling at any rate.

IV.a. PULSED BEAM DIRECT TIMING LIFETIME MEASUREMENTS USING Ge(Li) DETECTORS

Barry C. Robertson
Queen's University
Kingston, Ontario, Canada

The Ge(Li) detector has received extensive use because of its excellent energy resolution, but relatively little work has been done with it as a timing device due to the complexity of its pulse shape. However the development of the constant fraction trigger[1] and the amplitude and rise-time compensation (ARC) technique[2] has changed this and made it possible to increase the range of lifetimes that can be measured by the direct timing technique using Ge(Li) detectors.

In the direct timing technique both the generation and decay time of a nuclear level are measured; the lifetime of a long-lived state can be obtained from the shift of the time interval centroid from the prompt value. The main problem is making sure that the times measured depend only on the real event time. This can be done for the start signal by using the arrival of short (<1 ns) beam bursts at the target. For the stop time determination, the complexity of the γ decay spectrum normally encountered makes it necessary to use a Ge(Li) detector, so that spurious time dependences are introduced in this branch. Nevertheless it is possible both to reduce these unwanted effects using the ARC technique and to directly measure the remaining dependence by using the pulsed beam arrangement.

The output pulse from a Ge(Li) detector ideally consists of two risetime components. The changeover point from one component to the other is determined by the signal generation site in the detector. The ARC timing technique accounts for pulse amplitude variations by employing an effective constant fraction trigger (Figure 1) and risetime variations, by using the lowest practicable trigger fraction so that most of the signals are timed before they change risetimes. Using this technique the residual time walk can be reduced to approximately 1 ns/MeV, which must then be directly measured.

The procedure for measuring this walk under running conditions, where several prompt γ rays are present, is indicated in Figure 2. For each event both time and energy information are taken so that the time spectrum for each γ ray can be identified. In this way a prompt time centroid *vs.* photopeak energy curve can be built up. When doing this it is important to correct the centroids for the contribution from background under the energy peak. This is done by using windows placed next to the appropriate photopeaks. At the same time γ rays from levels whose lifetimes are to be meas-

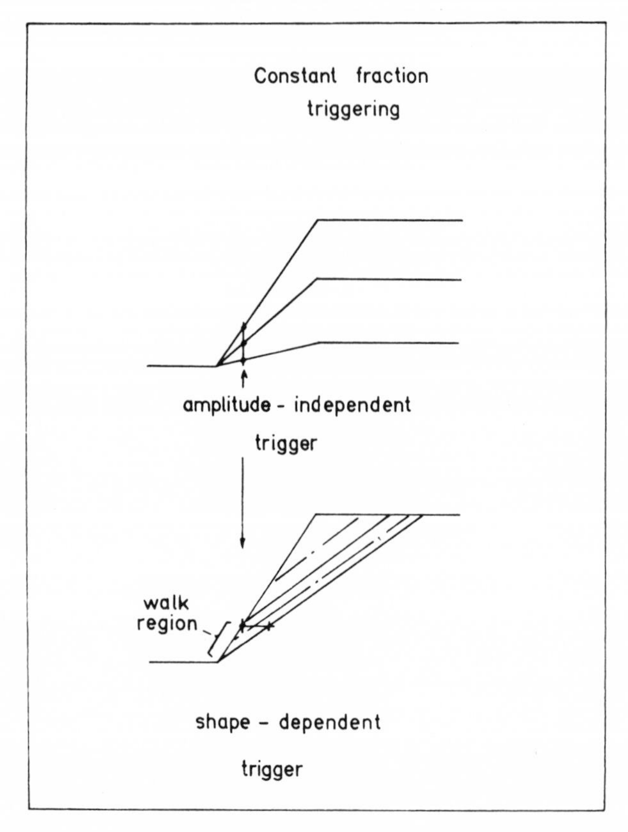

Fig. 1. Operation of constant fraction trigger for Ge(Li) pulses.

ured are processed, and their time centroid can be compared directly with the prompt time curve.

A typical time spectrum under experimental conditions is shown in Figure 3. This is the 803 keV γ ray of Fe^{55}, produced by the $Mn^{55}(p,n\gamma)Fe^{55}$ reaction at 4 MeV with 0.5 ns beam pulses. The detector was a 23 cm^3 coaxial Ge(Li) crystal with a resolution of $\sim$ 2 ns. Neutron-induced events in the detector are well separated in time from the target γ rays, and the background correction completely removes them. The prompt time curve for several Fe^{55} γ rays which is shown is Figure 4 exhibits the expected walk of $\sim$ 1 ns/MeV and is quite linear. Also shown is the centroid of the 1409 keV γ ray, which was expected to be delayed and in fact a lifetime of τ = 49 ± 10 ps was measured.[3] This is in good agreement with Donahue's measurement[4] of $\tau = 40 \pm ^{40}_{20}$ ps, using the gas backing Doppler shift technique. However there is a slight disagreement between Donahue's value for the lifetime of the 1317 keV level of τ = 14 ± 3 ps and the upper limit of 10 ps

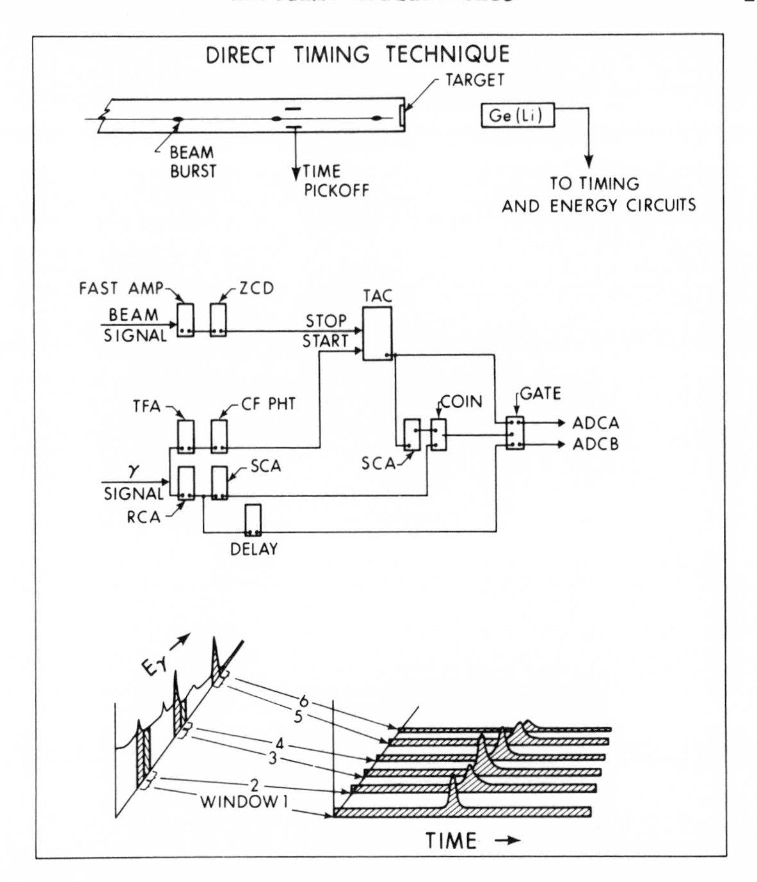

Fig. 2. Schematic diagram of the direct timing experimental arrangement. The zero-crossing discriminatory (ZCD) was used to provide a stop signal for the TAC. The single channel analyzers (SCA) were used to provide timing information for the coincidence requirement; the γ ray energy windows were digitally set after the coincidence information was accepted by the computer.

from the direct timing measurement. There are also a number of solid backing Doppler shift measurements for this level[3,5,6] which give a lifetime of approximately 0.8 ps.

The existence of this discrepancy then leads to the question of whether it could be resolved using the direct timing technique, or rather what is the limiting accuracy of the technique. There are two main effects which must be considered in determining the timing accuracy of a Ge(Li) detector. The most important is the correction of the photopeak spectrum for the underlying background. This background generally occurs about 20 ps later. The time of a prompt comp-

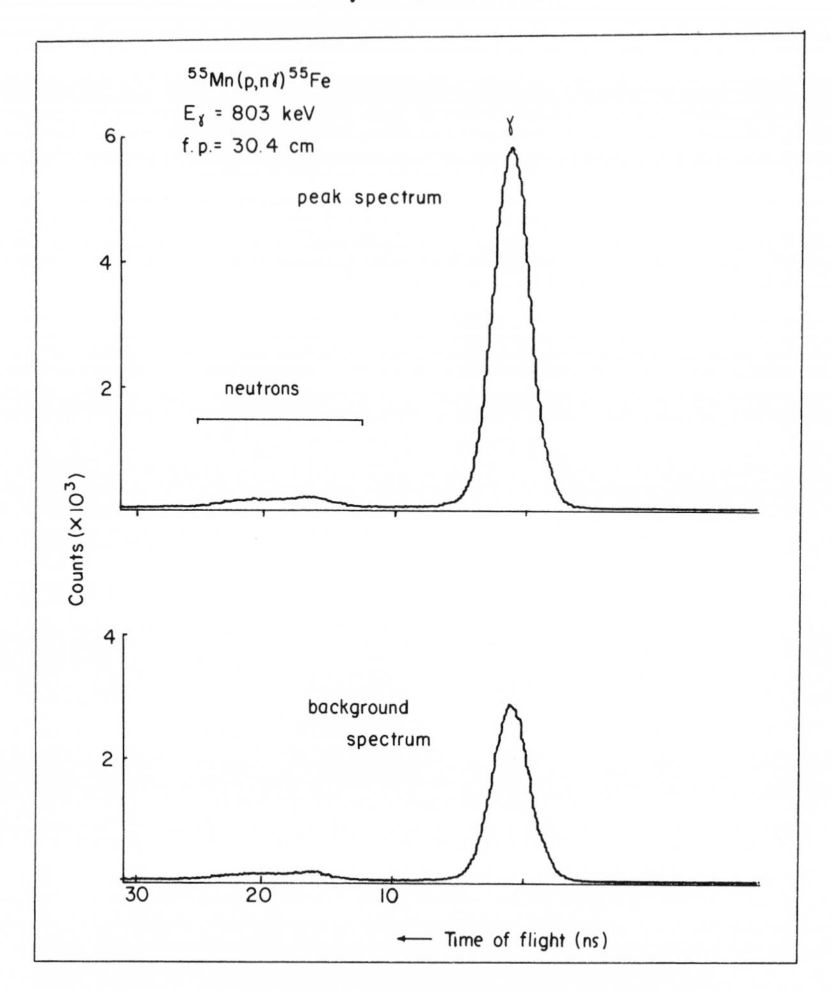

Fig. 3. Time spectrum for 803 keV γ rays from Fe^{55} observed for a flight path of 30.4 cm.

ton scattered γ ray can vary by as much as 50 ps, depending on where in the detector it scatters, and neutron-induced events can occur up to nanoseconds late. However, with sufficient care, this correction should be good to better than 1 ps.

Due to the geometry of coaxial detectors a time centroid variation which depends on the γ ray distribution over the face of the detector can also be expected. This is caused by the correlation of pulse shapes with γ interaction position and makes it important to use the lowest trigger fraction possible. Typically a pulse which changes before reaching the trigger fraction (which was 0.04 in this exper-

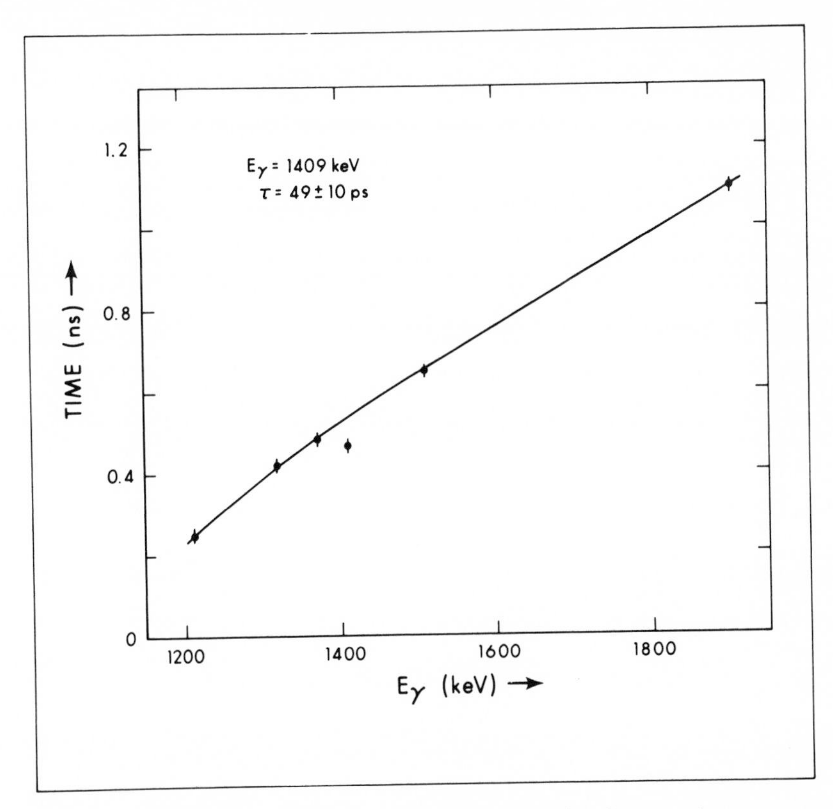

Fig. 4. Centroid position of time signals for various γ rays belonging to Fe^{55} as a function of γ ray energy. The solid line is a least squares fit to the prompt γ rays. The centroid positions have been corrected for background contribution to the photopeak.

iment) can give a signal that is late by up to $\sim$ 2 ns, and it is the fractional change in the number of these pulses which produces the centroid shift. Assuming that the sensitive region lies on the outer edge of the detector, a centroid shift of about 5 ps would be produced by an a_4 term of 0.3 in the intensity distribution if the detector were placed at 0° to the beam axis 5 cm from the target. Moving the detector out to 30 cm reduces this shift to $\sim$ 0.3 ps, and moving it away from 0° will provide a cancellation. Consequently this effect should introduce an error of less than 1 ps *provided* a low trigger fraction is used, and the detector is reasonably positioned.

In addition to these effects, a real time zero shift can occur if the mean production yield in the target for different levels occur at different energies and therefore different times. However for targets up to about 1 mg$\cdot$cm^{-2} thick the beam crossing time is on the order of 50 fs or less, so that this effect will normally be negligible.

The conclusion to be drawn from this is that by using a Ge(Li) detector to obtain good energy background suppression and a pulsed beam for good time background suppression and by using the pulsed beam to measure on-line the time behavior of both photopeak and background in the Ge(Li) detector, it should be possible to obtain reliable lifetime measurements in the ps range using the direct timing technique.

REFERENCES

1. D. A. Gedke and W. J. McDonald, Nucl. Instr. and Meth. 58, 253 (1968).
2. R. L. Chase, Rev. Sci. Instr. 49, 1318 (1968).
3. B. C. Robertson, G. C. Neilson and W. J. McDonald, Nucl. Phys. A189, 439 (1972).
4. D. J. Donahue and R. L. Hershberger, Phys. Rev. C4, 1693 (1971).
5. B. C. Robertson *et al.*, Nucl. Phys. A160, 137 (1971).
6. R. L. Hershberger, Ph.D. Dissertation, University of Arizona, 1970 (unpublished).

DISCUSSION

MCELLISTREM: That looks very nice, Barry. I wonder if you are presently limiting yourself to only those cases where you know that you have a series of prompt peaks. Do you have to restrict yourself to having a target which has a series of prompt lines so you can have a calibration curve?

ROBERTSON: I must know what time zero is, and in fact I've got a favorable condition here. I have high energy gamma rays; I've got a lot of them, and I've got only one that I think is delayed. But what you can do once you can do twice.

INTRODUCTORY REMARKS

Louis Brown
Carnegie Institution
Washington, D.C.

Previous Symposia have not included sessions on polarization. This was not an oversight on the part of the organizers but reflected the small contribution that this branch has made to nuclear spectroscopy during the first score years of its life. Indeed at the 1970 International Symposium on Polarization in Madison, G. R. Plattner surveyed what had been accomplished in the study of compound nuclear states in light nuclei, concluding that it was almost zero, if the number of level parameters determined using the technique was a measure. I was quite flattered by his talk because he used measurements and analyses made by some colleagues and myself on which we had expended no small amount of effort, as prime examples of this futility. If one looks for the stated objective in a large fraction of all polarization publications, one will find that it was to calibrate an analyzer for possible future use in nuclear spectroscopy, leading a colleague to quip: "You fellows sure spend a lot of time calibrating your beams". Or as Georges Temmer remarked to me when Cor van der Leun showed his first slide, depicting the progress of nuclear spectroscopy: "I don't see any bumps on those curves showing when polarized ion sources came on". Although things have progressed in the last few years, it is perhaps worth noting that the most recent article on polarization in Physical Review Letters[1] is entitled *New Method for Determining Polarization Standards in Nuclear Reactions.*

On the other hand polarization's birth was auspicious. Heusinkveld and Freier[2] showed in 1952 in the first polarization experiment with $He^4(p,p)He^4$ that the spin orbit force was large and the splitting of the p-levels inverted, as required by the shell model. I think it is worth mentioning, especially at a conference where the shell model is so important, that twelve years before the first polarization experiment and nine years before Mayer and Jensen's proposal of a strong spin-orbit force Wheeler and Barschall[3] analyzed $He^4(n,n)He^4$ and concluded from the phase shifts that there must be such a force. This eveidence seems to have been forgotten. This initial direction, *i.e.*, observing the spin-orbit force in elastic scattering, led to important developments in the optical model[4] but not to very much in spectroscopy.

In 1966 Moore and Terrell,[5] who were not in the polarization club and who had probably never even calibrated a beam, introduced polarization techniques for determing spin assignments of isobaric analog resonances that have been widely used. A year later Yule and Haeberli,[6] who were members of the polarization club and who had calibrated many beams, demonstrated how data from stripping reactions initiated with vector polarized deuterons could be interpreted for determining the j of the transferred nucleon as easily as its ℓ had been previously. Recently even the phase shifters have determined a few level parameters. With these techniques polarization has finally earned its way into nuclear spectroscopy.

REFERENCES

1. P. W. Keaton *et al.*, Phys. Rev. Lett. 29, 880 (1972).
2. M. Heusinkveld and G. Freier, Phys. Rev. 85, 80 (1952).
3. J. A. Wheeler and H. H. Barschall, Phys. Rev. 58, 682 (1940).
4. L. Rosen *et al.*, Ann. Phys. 34, 96 (1965).
5. C. F. Moore and C. E. Terrell, Phys. Rev. Lett. 16, 804 (1966).
6. T. J. Yule and W. Haeberli, Phys. Rev. Lett. 19, 756 (1967).

V.A. POLARIZED BEAMS IN NUCLEAR SPECTROSCOPY*

H. F. Glavish
Department of Physics, Stanford University
Stanford, California

I. INTRODUCTION

There are two specific applications of polarized beams in nuclear spectroscopy which I shall discuss. One of these is the study of the giant dipole states of nuclei observed with the polarized proton capture reaction $A(\vec{p},\gamma)B$. The other is the study of particle-gamma correlations following a nuclear reaction that is initiated by a polarized incident beam. An example of this latter application is the study of the angular correlation of gamma rays detected in coincidence with inelastically scattered protons produced by a polarized incident proton beam in a $(\vec{p},p'\gamma)$ reaction.

In the case of the giant dipole states of nuclei recent experiments at Stanford (see Section II) show that there are large polarization effects in the proton capture reaction. By combining the polarized and unpolarized data much more specific information about the reaction matrix elements can be obtained. Large polarization effects have also been observed in the neutron channel by Bertozzi *et al.*[1] and more recently by Cole and Firk.[2] They carried out the very difficult experiment of measuring the polarization of the photo neutrons produced in the reaction $O^{16}(\gamma,n)O^{15}$. The value of these polarization measurements in relation to our understanding of the structure of the giant dipole states of nuclei is discussed in detail in Section II.

The other part of this talk, polarized particle-gamma correlations, is considered in Section III where we first discuss how, in a nuclear reaction, polarization can be transferred from the incident particle to the residual nucleus. Two particular applications are then considered. In one of these, multipole mixing ratios are determined by detecting the emerging particle at 0° or 180°. A measurement of this type has recently been made at Stanford and is discussed in Section III.B. The reaction investigated was $Mg^{24}(\vec{p},p'_{180}\gamma)Mg^{24}$, $3^+(5.22\text{ MeV}) \rightarrow 2^+(1.368\text{ MeV})$ using incident polarized protons. An accurate value for the M1/E2 ratio in the $3^+ \rightarrow 2^+$ transition was obtained. A feature of this type of polarization correlation is the precision, and the absence of ambiguity, in the value found for the mixing ratio. As is well known, this is not always the case for unpolarized particle-gamma correlations. The other application

---

*Supported in part by the National Science Foundation

which is discussed in Section III.C utilizes the spin-flip geometry. Here the gamma ray is detected perpendicular to the reaction plane. Measurements of this type have been made at Stanford and Rutgers in a collaborative experiment, where analog resonances of states in Sr^{89} were studied using the reaction $Sr^{88}(p,p'\gamma)Sr^{88}$, $2^+(1.836\text{ MeV}) \rightarrow 0^+(\text{g.s.})$. In Section III we also discuss the possibility of measuring g-factors with a polarized beam.

Before I continue I should like to mention that there are two important applications of polarized beams in nuclear spectroscopy which I have omitted from this talk. One of these is the study of bound states using a transfer reaction, such as $A(\vec{d},p)B$, with a polarized incident beam. As discussed by Yule and Haeberli,[3,4] and Haeberli,[5] the j-value of the captured neutron can often be found from this type of measurement. The other application is the study of isobaric analog resonances by elastic and inelastic scattering. I have omitted these two topics from the present talk partly because I am less familiar with them than the topics I have chosen, and partly because they have been discussed in detail quite recently at the polarization conference at Madison.[5,6] There are also other applications which may be found in the proceedings of that conference.[7]

II. THE GIANT DIPOLE STATES OF NUCLEI

A. Particle-Hole Model

The giant dipole states in nuclei have been one of the most extensively studied phenomena in nuclear physics. Experimentally these states are observed as gross resonances in the cross sections of the photo nuclear reactions (γ,p) and (γ,n) and the inverse capture reactions. On the theoretical side the giant dipole states are interpreted in terms of nuclear shell theory as single particle excitations out of a closed shell to a higher lying unfilled shell. Such states can be coupled strongly to the closed shell configuration by E1 radiation, and at the same time they have sufficient energy to decay by nucleon emission. This rather elementary picture is the essence of Wilkinson's single particle model of photo nuclear reactions proposed in 1956.[8] The simplicity of this model and the implication of applying shell theory to excited states which are particle unbound, has undoubtedly been responsible for the great wealth of experimental information that now exists on photo nuclear reactions.

While explaining in a satisfactory way the magnitude of cross sections, the width of the giant resonance maxima and the angular distribution of the reaction products, the simple single particle model had serious difficulties if used to interpret the energy position of a giant resonance, the relation of (γ,p) and (γ,n) cross sections, and the gross structure of the giant resonance observed in lighter nuclei. It turned out that the missing ingredient in the model originally proposed by Wilkinson was the interaction between the excited particle and the corresponding hole left in the lower shell. Elliot and Flowers[9] took account of this interaction in their work on the O^{16} giant dipole resonance (GDR) and a physical picture describing the effects of this interaction has been elucidated by Brown and collaborators.[10,11] It has been found that the particle-hole interaction mixes the unperturbed single particle levels in such a way as to push one (or few) of the levels a long way from the unperturbed level and endow it with most of the strength for E1 decay.

As an example to illustrate what happens when the residual particle-hole interaction is taken into account, we consider the sd-shell particle-hole excitations in O^{16}. In Table I we list the unperturbed particle-hole configurations and corresponding energies, taken from Wang and Shakin.[12] In Table II we list the T = 1 energy eigenvalues obtained when a zero range particle-hole interaction is included. (There are also some low lying T = 0 states which are not included since these cannot decay via E1 radiation to the T = 0 ground state of a self-conjugate nucleus such as O^{16}.) In the same table the configuration mixing amplitudes are given. These are just the expansion coefficients for the energy eigenstates $\phi_{J^{\pi}T}$:

$$|\phi_{J^{\pi}T}> = \sum_{j_p\ell_p j_h\ell_h} C^{J^{\pi}T}_{j_p\ell_p j_h\ell_h} |j_p\ell_p, j_h^{-1}\ell_h^{-1}> \qquad (1)$$

The point to notice is that two of the eigenstates corresponding to a predominantly $(d_{5/2}p_{3/2}{}^{-1})$ and $(d_{3/2}p_{3/2}{}^{-1})$ configuration respectively, are pushed up in energy to 22.31 and 24.45 MeV. Both states have $J^{\pi} = 1^-$ and can therefore decay to the 0^+ ground state of O^{16}. Entirely consistent with the picture given by Brown,[10] the 22.31 MeV state is found to carry 65% of the E1 strength, the 24.45 MeV carries 28%, and all the remaining states only 7% between them.

The 1^- levels of O^{16} determined from the $N^{15}(p,\gamma_o)O^{16}$ reaction by Tanner, Thomas and Earle[13] are shown in Figure 1. (Note: γ_o refers to γ decay to the ground state of O^{16}, γ_1 to the first excited state, *etc.*) Comparing these experimental results with the theoretical predictions of Table II, we see that the collective particle-hole excitations at eigen

TABLE I

Unperturbed configurations in O^{16} (Reference 12).

Configuration		Unperturbed Energy
$p_{1/2}^{-1}$	$d_{5/2}$	11.53 MeV
$p_{1/2}^{-1}$	$s_{1/2}$	12.40 MeV
$p_{1/2}^{-1}$	$d_{3/2}$	16.64 MeV
$p_{3/2}^{-1}$	$d_{5/2}$	17.70 MeV
$p_{3/2}^{-1}$	$s_{1/2}$	18.57 MeV
$p_{3/2}^{-1}$	$d_{3/2}$	22.77 MeV

energies of 22.31 and 24.45 MeV correspond with the two main peaks in the O^{16} GDR.

B. Structure in the Giant Dipole Resonance

As Figure 1 shows, a total of five peaks can definitely be identified when the O^{16} GDR is observed via the $N^{15}(p,\gamma_o)O^{16}$ reaction. The same peaks can also be identified[14] in the related inverse photo neutron reaction $O^{16}(\gamma,n_o)O^{15}$. Collective single particle hole excitations account for two of these five peaks. Including higher order excitations in theoretical calculations, such as 2p-2h, 3p-3h, *etc.*, can produce additional peaks. For example, Wang and Shakin[12] have found it possible to account for the remaining three peaks in the O^{16} GDR by selecting appropriate 3p-3h excitations. The positions and dipole strengths predicted for these three intermediate peaks are in moderate agreement with experiment.

In the GDR's of nuclei with A > 40 the additional structure observed is usually explained by building into the 1p-1h calculation many-particle many-hole configurations. The most widely recognized model is the dynamic collective model[15] (DCM) which in essence selects the many-particle many-hole configurations contained in low energy vibrational states. Thus the physical picture involved is the coupling of low energy surface vibrations (such as the first 2^+ state) to the 1p-1h excitations.

TABLE II

One-particle one-hole energy eigenstates in O^{16} when particle-hole interaction is included. The mixing amplitudes $C^{J^{\pi}T}_{j_p \ell_p, j_h \ell_h}$ are listed (Reference 12).

(J T)	Energy (MeV)	$(s_{1/2}p_{3/2}^{-1})$	$(d_{5/2}p_{3/2}^{-1})$	$(d_{3/2}p_{3/2}^{-1})$	$(d_{3/2}p_{1/2}^{-1})$	$(s_{1/2}p_{1/2}^{-1})$	$(d_{5/2}p_{1/2}^{-1})$
(1^-1)	24.45	-0.0607	-0.0740	0.9714	-0.2173	0.0021	
(1^-1)	22.31	0.1225	0.9345	0.1452	0.2973	0.0483	
(1^-1)	17.43	-0.0539	-0.3160	0.1806	0.9297	-0.0162	
(1^-1)	13.48	-0.0507	-0.0440	-0.0084	0.0011	0.9977	
(2^-1)	12.66	0.0243	0.2020	0.0530	0.0164		0.9775
(3^-1)	13.18		-0.2210	0.0064			0.9753

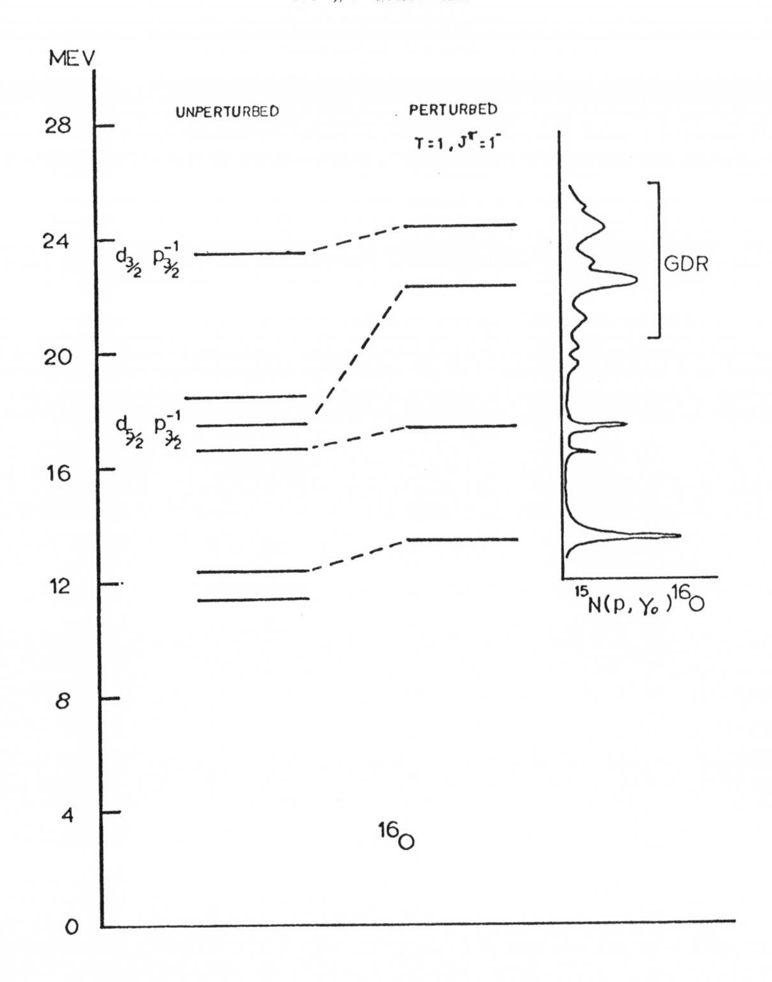

Fig. 1. The dipole levels (1^-) in O^{16} observed with the $N^{15}(p,\gamma_o)O^{16}$ reaction.[13] The theoretical unperturbed levels taken from Tables I and II are also shown.

For nuclei in the range $20 \lesssim A \lesssim 40$ a great deal of structure is observed, as illustrated in Figures 2 and 3 for the $F^{19}(p,\gamma_o)Ne^{20}$ and $Al^{27}(p,\gamma_o)Si^{28}$ reactions.[16,17] In the case $Al^{27}(p,\gamma_o)Si^{28}$ the four peaks that persist when the energy averaging interval is 600 keV (see Figure 3) can be identified with ($T = 1$, $J^{\pi} = 1^-$) single particle-hole excitations.[18] As discussed by Meyer-Schutzmeister *et al.*[19] the finer structure is most simply explained in terms of the $T = 0$ impurity in the $T = 1$ giant resonance. This impurity arises through the weak interaction with many overlapping $T = 0$ compound nucleus states and enables strong Ericson fluctuations to be

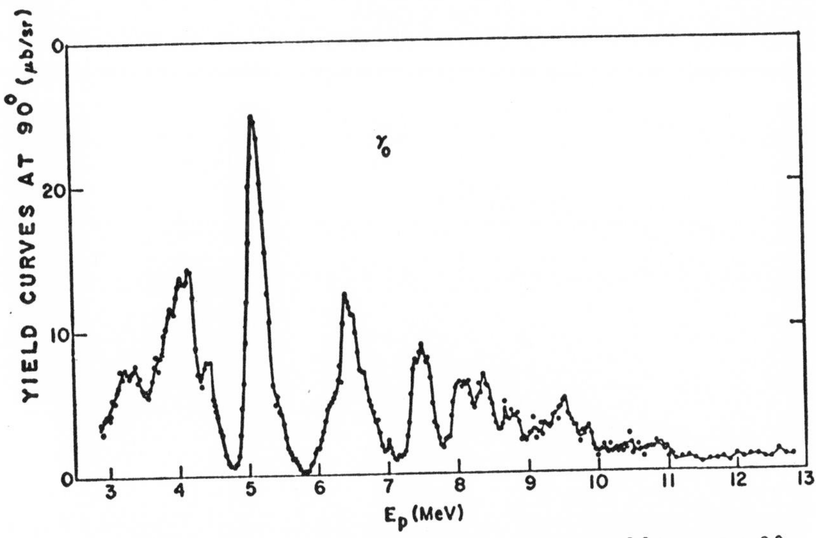

Fig. 2. Yield curve at 90° for the reaction $F^{19}(p,\gamma_0)Ne^{20}$.

observed in the GDR. The problem with the giant dipole states in nuclei with $20 \lesssim A \lesssim 40$ is that $\langle\Gamma\rangle/\langle D\rangle$ for the compound states at these excitations, ranges in value from less than unity to greater than ten. Thus, it may be difficult to distinguish between fluctuations and true resonances.

C. Polarized Angular Distributions

Most of the experimental support for the nuclear shell theory of the giant dipole states of nuclei is derived from measurements of total cross-section, or yield at a given angle, as a function of energy. This serves to locate the energy positions of the main peaks, the resonance widths and dipole strengths. By and large, as we shall see below, angular distribution measurements, most common in the case of (p,γ) capture reactions, have only provided additional qualitative support for the theoretically predicted single particle-hole configurations. This is because there are generally two or more particles' channels which can lead to E1 radiation (and at the same time satisfy parity and angular momentum conservation laws). Thus, the reaction amplitudes or T matrix elements corresponding to the allowed particle channels, cannot be delineated by angular distribution and total cross-section measurements.

For instance, in the case of the reaction $N^{15}(p,\gamma_0)O^{16}$, the T matrix elements may be specified in a representation where the proton orbital angular momentum $\vec{\ell}$ and spin $\vec{s}$ are coupled to form an angular momentum $\vec{j} = \vec{\ell} + \vec{s}$. By angular

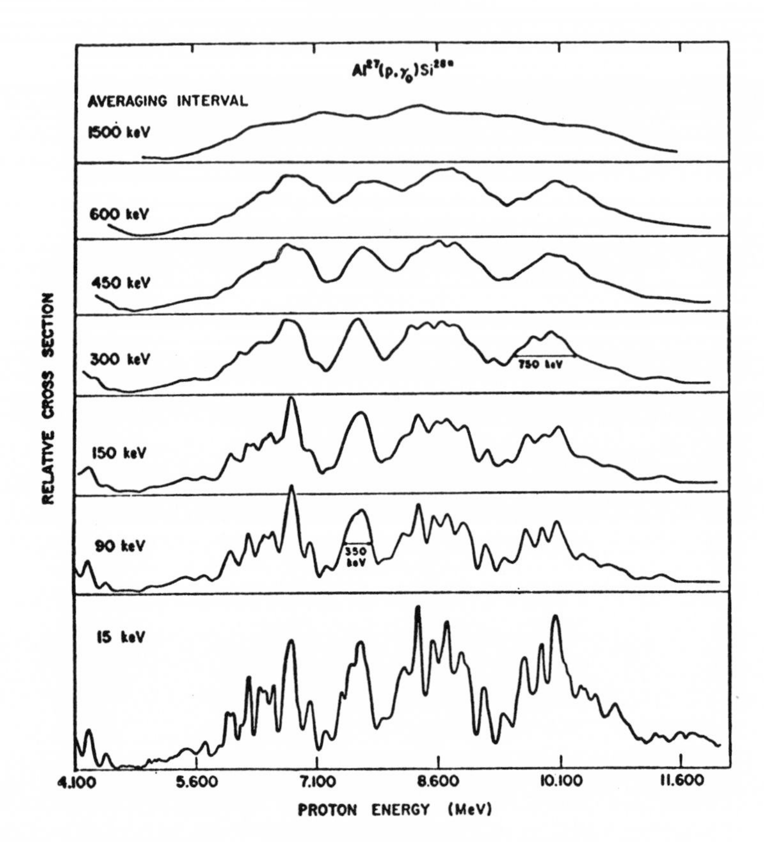

Fig. 3. Yield curve for the reaction $Al^{27}(p,\gamma_o)Si^{28}$ for various energy averaging intervals.

momentum and parity conservation, only the incident proton waves $s_{1/2}$ and $d_{3/2}$ can couple with the $1/2^-$ ground state of N^{15} to form a 1^- state in O^{16}. The T matrix elements for these two channels are generally complex and may be written as $s_{1/2}e^{i\phi_s}$ and $d_{3/2}e^{i\phi_d}$ where $s_{1/2}$ and $d_{3/2}$ refer to the real amplitudes and ϕ_s and ϕ_d are the real phases. Neglecting all radiations except E1, we find that the angular distribution must have the form

$$\sigma_u(\theta) = A_o\left(1 + a_2P_2(\cos\theta)\right) \tag{2}$$

where

$$a_2 = -0.5d_{3/2}^2 + 1.414s_{1/2}d_{3/2}\cos(\phi_d - \phi_s) \tag{3}$$

and the normalization is

$$s^2_{1/2} + d^2_{3/2} = 1. \quad (4)$$

The experimentally measured[20] value of a_2, as a function of energy is shown in Figure 4. Now one might argue that the average value measured for a_2 is close to -0.5 and therefore Eq. (3) suggests that the capture is almost purely $d_{3/2}$ (*i. e.* $d_{3/2} \simeq 1$ and $s_{1/2} \simeq 0$). However, it will be realized that Eq. (3) allows a range of other solutions, depending on the value of the phase difference $\phi_d - \phi_s$.

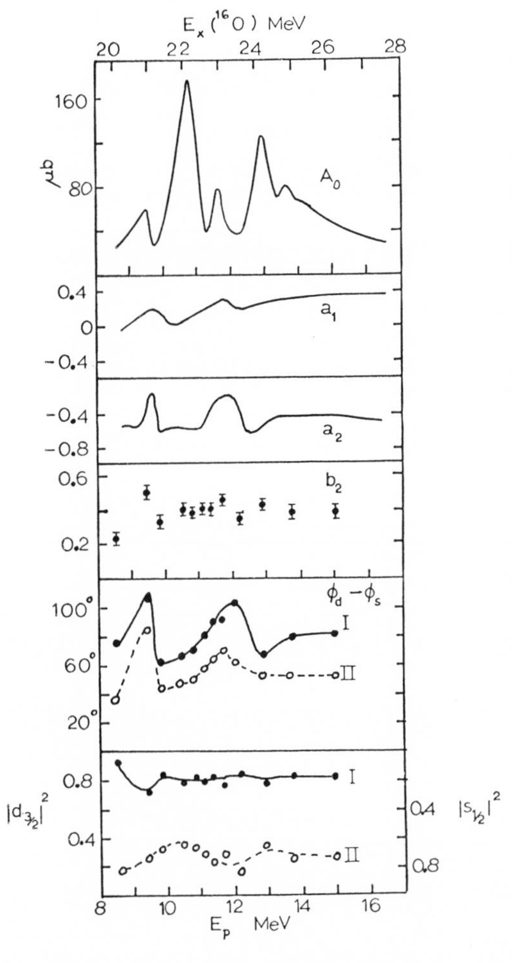

Fig. 4. Summary of the existing information on the reaction $N^{15}(p,\gamma_o)O^{16}$. The curves for the total yield A_o and the unpolarized angular distribution coefficients a_1 and a_2 are from Ref. 20. The values for b_2 were obtained from the data in Figure 5 by use of Eq. (8). The two solutions for $|s_{1/2}|^2$ (right hand scale), $|d_{3/2}|^2$ and $\phi_d - \phi_s$ were obtained by fitting the theoretical expressions (3), (4), and (6) to the experimental values of a_2 and b_2 at each energy.

If the incident proton beam has a polarization P then the angular distribution is

$$\sigma_p(\theta) = \sigma_u(\theta)\left(1 + \vec{P}\cdot\vec{n}A(\theta)\right) \quad (5)$$

where $\vec{n}$ is the normal to the reaction plane and the analyzing power $A(\theta)$ is given by

$$\sigma_u(\theta)A(\theta) = \left(-1.06 s_{1/2} d_{3/2} \sin(\phi_d - \phi_s)\right)\sin 2\theta. \quad (6)$$

Thus the value of measuring $A(\theta)$ using a polarized incident beam is immediately apparent, since combining Eqs. (3), (4) and (6) enables unique solutions to be obtained for $s_{1/2}$, $d_{3/2}$ and the phase difference $\phi_d - \phi_s$. We note that Eq. 6 illustrates the well known fact that polarization effects arise from interference, in this case it is interference between the $s_{1/2}$ and $d_{3/2}$ proton channels.

The procedure for measuring $A(\theta)$ is the usual one in vector polarization experiments. The spin of the protons are oriented normal to the reaction plane and for a given counter angle θ, the yields N_u and N_d are measured, corresponding to proton spin up and proton spin down respectively. The analyzing power $A(\theta)$ is then given by

$$A(\theta) = P^{-1}(N_u - N_d)/(N_u + N_d).$$

The above considerations apply to the case when there is only E1 radiation. In general, if there are no restrictions on the multipolarity of the γ radiation, $\sigma_u(\theta)$ and $A(\theta)$ have the general form

$$\sigma_u(\theta) = A_o\left(1 + \sum_{k=1} a_k P_k(\cos\theta)\right) \quad (7)$$

and

$$A(\theta) = \left(\sigma_u(\theta)\right)^{-1} \sum_{k=1} c_k P_k^1(\theta) = \left(\sigma_u(\theta)\right)^{-1} \sum_{k=1} b_k \sin k\theta. \quad (8)$$

The complexity of $\sigma_u(\theta)$ and $A(\theta)$ depends on how many of the coefficients a_k and b_k are non-vanishing. The rules governing this are the same for a_k and b_k:

(a) For radiation of multipolarity L the value of k is less than or equal to 2L, and k is even;

(b) For interfering radiations of multipolarity L and L' the value of k must be less than or equal to L + L' and k is even, or odd, when the radiations have the same, or opposite parity.

Unpolarized angular distribution measurements in the region of a GDR usually reveal the presence of radiations other than E1. The coefficient a_1 is often found to depart significantly from zero implying (E1-E2) or (E1-M1) interference. The added information obtained from polarization measurements should prove useful for identifying impurity radiations of this type, and with accurate measurements, quantitative estimates of the T matrix elements for the impurity radiations should be possible.

Finally, we mention that because of time reversal symmetry, the equivalent of measuring an analyzing power $A(\theta)$ in a capture reaction, is to measure the polarization of the outgoing nucleon in the inverse photo nuclear reaction. Although difficult, such measurements have been made[1,2] for the reaction $O^{16}(\gamma,n)O^{15}$, as mentioned in the introduction.

D. Reaction Models

In the previous section we explained how polarization measurements can enable unique, quantitative values to be found for the T matrix elements of a capture or photo nuclear reaction. The T matrix elements can be expressed in terms of the nuclear structure parameters of a giant dipole state by employing a reaction model. It is in this respect that giant dipole resonance studies are in their infancy. This is partly because of the paucity of angular distribution measurements and the almost total lack of polarization measurements, and partly because of the tremendous difficulty of transforming the well formulated results of bound state shell model calculations, into a form which takes into account the fact that the giant dipole states are in the continuum and can decay by particle emission. This is in contrast with, say, transfer reactions such as (d,p) or (p,d) where quite elaborate reaction models have been developed and proven. In fact, the primary motivation for making polarization measurements in particle reactions, such as elastic scattering, inelastic scattering and transfer reactions has seemed to be to study the reaction mechanism and establish a useful reaction model.

Polarization measurements can be expected to play a similar role in giant dipole studies, as well as providing structure information directly as discussed in the previous section.

The T matrix element for a capture reaction in the region of a GDR can be written as

$$T_{fi} = \langle \phi_f | D_\gamma | \psi_i^{(+)} \rangle \qquad (9)$$

where $\psi_i^{(+)}$ is the nucleon target scattering wave function, ϕ_f is the final nuclear state after γ decay, and D_γ is the electric dipole operator. Since the giant dipole states should be 1p-1h excitations (neglecting intermediate structure), one has the opportunity with this simplifying feature to solve the Schrödinger equation directly to obtain $\psi_i^{(+)}$. This is the basis of the so called coupled channel calculations, applied to the O^{16} GDR by Buck and Hill[21] and Raynal *et al.*,[22] and to the C^{12}, Si^{28} and Ca^{40} GDR's by Marangoni *et al.*[23,24] The theory of these calculations is based on just a few fundamental assumptions:

(i) $|\psi_i^{(+)}\rangle$ is expressible as a superposition of 1p-1h eigenstates of an independent particle Hamiltonian H_o with the holes being restricted to states below the Fermi surface.

(ii) $|\psi_i^{(+)}\rangle$ is an eigenstate of the Hamiltonian

$$H = \sum_{i=1}^{A} T_i + \sum_{i,j}^{A} V_{ij} = T + V \qquad (10)$$

where T_i is the K.E. operator of the ith nucleon and V_{ij} is the usual two-nucleon interaction. The fits obtained with these calculations are moderate, being somewhat better for C^{12} and O^{16} than for Si^{28} and Ca^{40}. In all instances, an absorptive potential must be included in the independent particle Hamiltonian H_o to obtain satisfactory magnitudes for the total cross-sections. The intermediate structure in the total cross-section is of course not reproduced. If, say 2p-2h excitations coupled to the 1p-1h states could be included, this might reproduce the intermediate structure, but it is not expected[23,24] to change the angular distributions significantly. Including such higher order excitations would greatly increase the complexity of the calculations.

An alternative approach for finding the T matrix elements is to use the concept of doorway states. This has been applied to O^{16} by Wang and Shakin.[12] The use of the doorway hypothesis greatly simplifies the structure of the T matrix. The doorway states ϕ_d are chosen as the normal collective 1p-1h states which are generated in bound state calculations. These doorway states are coupled strongly to the nucleon channel via the residual nucleon-nucleon interaction V_{ij} (Eq. (10)) and to the γ channel via the dipole operator D_γ. The coupling to the continuum produces an energy shift Δ_d and width Γ_d for each doorway state. The T matrix element is expressed in terms of the doorway states as follows (see Reference 12)

$$T_{fi} = <0|D_\gamma|\chi_i^{(+)}> + \sum_d \frac{<0|D_\gamma|\phi_d><\phi_d|V|\chi_i^{(+)}>}{E - E_d - \Delta_d - \Delta_x + i(\Gamma_d + \Gamma_x)/2} \tag{11}$$

where Δ_x and Γ_x are weakly energy dependent parameters describing the coupling of the doorway states to other compound states. To account for intermediate structure, secondary doorways can be considered, such as 1p-1h coupled to 2p-2h states. Then Γ_d and Δ_d contain weakly energy dependent parameters describing the coupling of the primary doorways to the secondary doorways, but otherwise are completely calculable. The final state is taken as just the vacuum state $|0>$ thus representing a closed shell. The initial state $|\chi_i^{(+)}>$ describes the nucleon scattering state and can be taken as a solution of a single particle Schrödinger equation with a real optical potential.[12] The first term in Eq. (11) describes the direct contribution to T_{fi} and depends on energy only weakly through $\chi_i^{(+)}$.

The doorway state approach seems to have several advantages over coupled channel calculations, especially as there is still a great deal of uncertainty in reaction models for the GDR and therefore a great need to make comparisons with experimental data. One of the great advantages of the doorway model is the explicit dependence of T_{fi} on the 1p-1h states ϕ_d which come from simple bound state calculations. Since the T_{fi} can be explicitly determined through polarization experiments, the simple explicit dependence of T_{fi} on ϕ_d provides a tremendous bridge between experiment and elementary theory. The more complicated interactions, although not of interest, are taken into account and not neglected. These appear in a simple parameterized fashion through the weakly energy dependent factors such as Δ_x and Γ_x. Furthermore, these parameters may be varied to facilitate comparison with experimental data.

With the aid of polarization measurements, the phases of T_{fi} are determined. Since the radial wave function for $\chi_i^{(+)}$ in Eq. (11) can be written as $e^{i\delta_{\ell j}}$ x [real radial function], the theoretically predicted phases for T_{fi} as a function of energy are easily found. Thus from the direct term in Eq. (11) we have a phase $\delta_{\ell j}$ and from each resonance term a phase $\delta_{\ell j} + \theta_d$ where

$$\theta_d = -\tan^{-1}\left\{\frac{\Gamma_d + \Gamma_x}{2(E - E_d - \Delta_d - \Delta_x)}\right\}. \tag{12}$$

Polarization and angular distribution measurements are sensitive to the phase difference $\phi_{\ell j} - \phi_{\ell' j'}$ between the two nucleon channels (ℓj) and $(\ell' j')$. The doorway state model gives the following expression for $(\phi_{\ell j} - \phi_{\ell' j'})$:

$$e^{i(\phi_{\ell j} - \phi_{\ell' j'})}$$

$$= e^{i(\delta_{\ell j} - \delta_{\ell' j'})}\left[\frac{A_{\ell j} + \sum_d B^d_{\ell j}\, e^{i\theta_d}}{A_{\ell' j'} + \sum_d B^d_{\ell' j'} e^{i\theta_d}}\right]\frac{|T_{\ell' j'}|}{|T_{\ell j}|}. \tag{13}$$

where $A_{\ell j}$ and $B^d_{\ell j}$ are the real amplitudes of the direct and resonant parts of the T-matrix element. At the peak of an isolated doorway state only one $B^d_{\ell j}$ is important and $\theta_d = \pi/2$. Therefore

$$|\phi_{\ell j} - \phi_{\ell' j'}| = |\delta_{\ell j} - \delta_{\ell' j'}| \tag{14}$$

assumming $A_{\ell j} << B^d_{\ell j}$ at a resonance.

Finally, I would like to mention that it would be of tremendous value if authours would publish the values they obtain in calculations for T matrix elements, in addition to publishing fits to existing experimental data. It would then be possible to compare the predictions of the calculations with any new experimental data.

E. Examples

O^{16}: A summary of the existing experimental information on the capture reaction $N^{15}(p,\gamma_o)O^{16}$ is given in Figure 4. The total yield A_o and the angular distribution coefficients a_1 and a_2 are taken from the unpublished work of O'Connell.[20] The coefficients b_2 in the analyzing power $A(\theta)$ of the reaction $N^{15}(\vec{p},\gamma_o)O^{16}$ (see Eq. (8)) are also included in Figure 4 and are derived[25] from the polarization data shown in Figure 5. It will be noted that analyzing powers at 45° and 135° are approximately equal in magnitude but have opposite signs, while the analyzing power at 90° is much smaller and nearly zero. Thus the analyzing power is almost pure $\sin 2\theta$ implying that impurity radiations in interference with E1 are small.

Using equations (3), (4) and (6) solutions are obtained for $|s_{1/2}|^2$, $|d_{3/2}|^2$ and the phase difference $\phi_d - \phi_s$ as shown in Figure 4. (The T matrix elements are defined as $s_{1/2}e^{i\phi_s}$ and $d_{3/2}d^{i\phi_d}$.) There are two solutions: solution I (solid curve) is characterized by a dominant $d_{3/2}$ capture with an approximately constant value of 0.80 for $|d_{3/2}|^2$. Solution II (dashed curve) corresponds to mainly $s_{1/2}$ capture with a nearly constant value of 0.70 for $|s_{1/2}|^2$.

The intermediate structure peaks at excitations of 21 and 23 MeV are both characterized by large fluctuations in the coefficient a_2 which seems to be correlated with substantial variations in the phase-difference $\phi_d - \phi_s$. There is a possibility of interpreting this in terms of the doorway state model discussed in Section D. Evidently on passing through the region of the intermediate states at 21 and 23 MeV the individual amplitudes $A_{\ell j}$ and $B^d_{\ell j}$ and phases θ_d must vary in such a way as to maintain the amplitude ratio $|T_{\ell' j'}|/|T_{\ell j}|$ constant. But this does not necessarily mean that the phase of the term $[A_{\ell j} + \sum_d B^d_{\ell j} e^{i\theta_d}]/[A_{\ell' j'} + \sum_d B^d_{\ell' j'} e^{i\theta_d}]$ in Eq. (13) must also remain constant. In other words the fluctuations in the phase difference $\phi_d - \phi_s$ and hence the coefficient a_2 might arise purely from a reaction phenomena associated with the variation of the resonance phases θ_d.

Of the two solutions I and II given in Figure 4, presumably only one of them is the correct physical solution, the other being just a mathematical solution. Wang and Shakin give form factors for the matrix elements $\langle\phi_d|V|\chi_i^{(+)}\rangle$ (Eq. (11)) for the reaction $O^{16}(\gamma,n_o)O^{15}$ at the 1p-1h doorways at

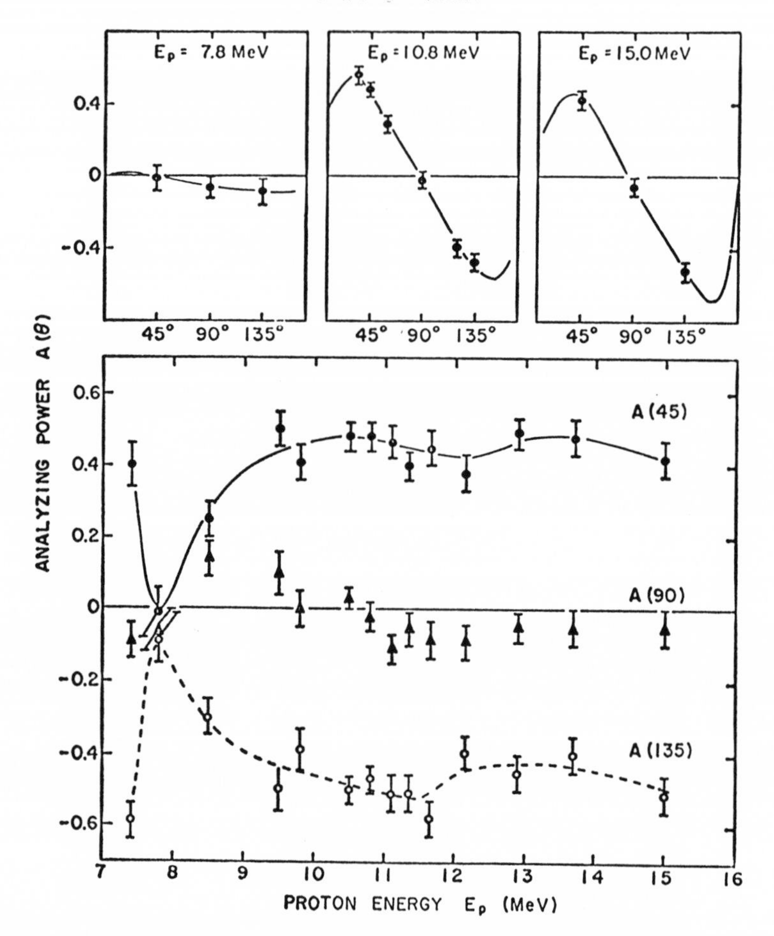

Fig. 5. Measured analyzing powers for the reaction $N^{15}(\vec{p},\gamma_0)O^{16}$. Top: data plotted as a function of angle at three selected energies. Bottom: all the data at the angles 45°, 90°, and 135° presented as a function of energy.

24.5 and 22.3 MeV. For the doorway at 24.5 MeV the form factor for $d_{3/2}$ neutron capture is about 4 times larger than the form factor for $s_{1/2}$ neutron capture. At this energy the same should be true for proton capture, since the neutron and and proton penetrabilities for s and d waves are quite comparable. This would imply that solution I is correct which, of course, is entirely consistent with the simple, single particle, bound state model, which predicts only a small $s_{1/2}$ amplitude. However, for the lower energy doorway at 22.3 MeV, Wang and Shakin[12] find the s wave form factor in the reaction $O^{16}(\gamma,n_0)O^{15}$ is dominant over the d wave form factor. Even allowing for the difference in neutron and proton penetrabilities this prediction is not entirely consistent with Solution I.

Bertozzi *et al.*[1] have measured the neutron polarization at an angle of 45°, for the reaction $O^{16}(\gamma,n_o)O^{15}$. The data is shown in Figure 6. Although these neutron polarization measurements are not as precise as the analyzing power measurements shown in Figure 5 for the reaction $N^{15}(\vec{p},\gamma_o)O^{16}$, there is general agreement with the two sets of data. In particular, both show the anomalous dip to zero at 19 MeV excitation, which corresponds to a resonance at $E_p = 7.8$ MeV (see Figure 1) in the reaction $N^{15}(p,\gamma_o)O^{16}$. In their coupled channel calculations, Buck and Hill[21] produce a theoretical fit to the neutron polarization data and this is included in Figure 6. The fit is moderate over the main part of the GDR and zero polarization is predicted near 19 MeV excitation. However the theoretical curve continues to become negative at lower energies which is inconsistent with the data in Figure 5. Furthermore, the a_2 coefficient for the reaction $N^{15}(p,\gamma_o)O^{16}$ is not very well reproduced in the Buck and Hill calculations especially near 19 MeV. A possible explanation[25] is that there is a 1^+ state in the region of 19 MeV.

In Figure 7 we show the total cross-section fit for (γ,n_o) obtained by Buck and Hill[21] and in Figure 8 the corresponding fit obtained by Wang and Shakin.[12] It will be noted that the intermediate structure is quite well reproduced in the latter doorway state calculation.

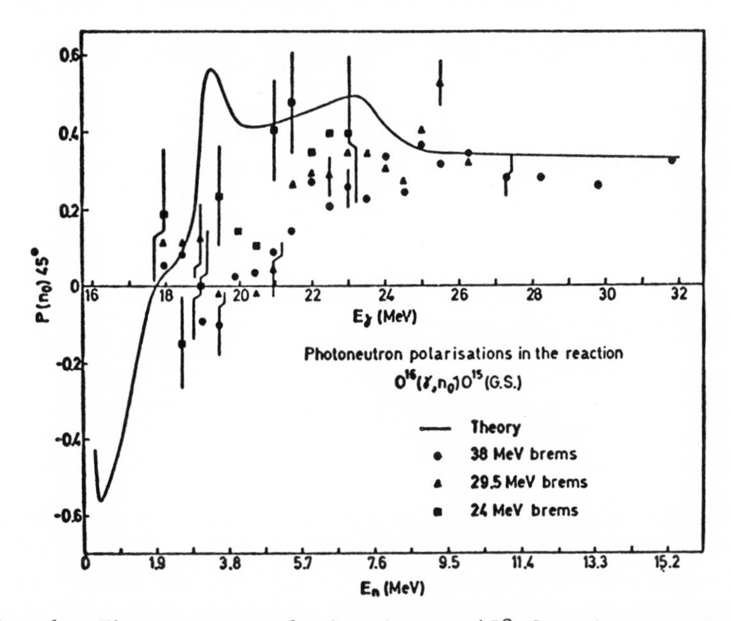

Fig. 6. The neutron polarization at 45° for the reaction $O^{16}(\gamma,n_o)O^{15}$. The data is from Bertozzi *et al.*[1] The solid curve is the prediction of the Buck and Hill coupled channel calculations.[21]

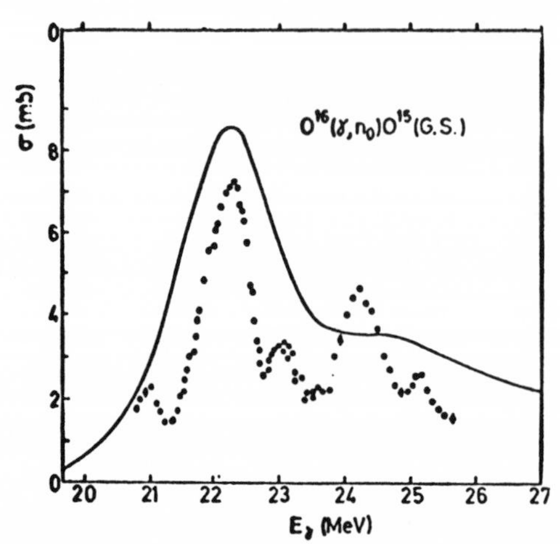

Fig. 7. Fit obtained by Buck and Hill[21] for the total cross-section of $O^{16}(\gamma,n_o)O^{15}$. The experimental data is from Caldwell *et al.*[14]

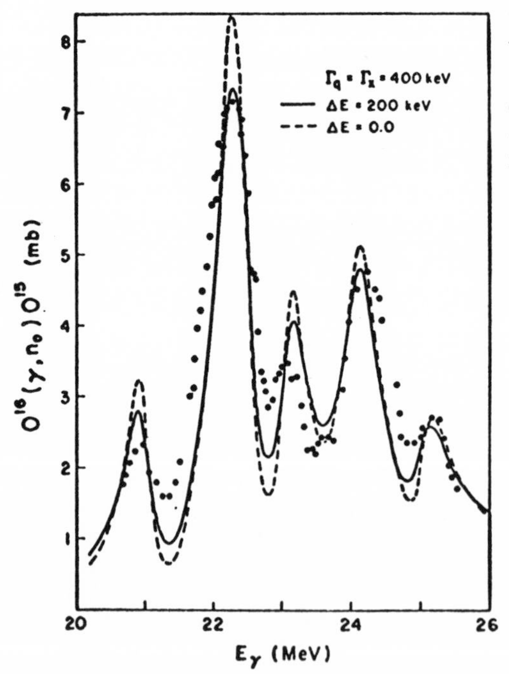

Fig. 8. Fit obtained by Wang and Shakin[12] for the total cross-section of $O^{16}(\gamma,n_o)O^{15}$. The experimental data is from Caldwell *et al.*[14]

Ne^{20}: The giant dipole states of Ne^{20} have been extensively studied using the proton capture reaction $F^{19}(p,\gamma)O^{16}$. The total yield and angular distributions of γ_o and γ_1 radiations measured by Segel *et al.*[16] are shown in Figures 9 and 10. The analyzing power for the polarized capture reactions $F^{19}(\vec{p},\gamma_o)O^{16}$ and $F^{19}(\vec{p},\gamma_1)O^{16}$ have been measured recently at Stanford using a similar technique to that described[25] for the

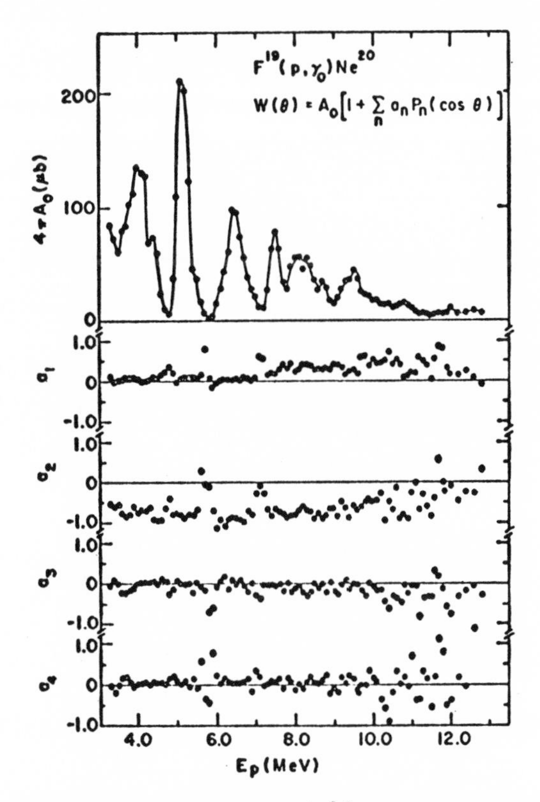

Fig. 9. The data of Segel *et al.*[16] for the total yield and angular distribution coefficients for the reaction $F^{19}(p,\gamma_0)Ne^{20}$.

reaction $N^{15}(\vec{p},\gamma_0)O^{16}$. The data for the γ_0 radiation is shown in Figure 11. The analyzing power for the γ_1 radiation was found to be essentially zero within statistics (± 0.06) at angles in the range of 40° to 140°.

The energies for the polarization measurements were selected to coincide with the most prominent peaks in the total yield curve for the γ_0 radiation. It may be noted from Figure 11 that the analyzing power A(θ) for the γ_0 radiation is remarkably constant throughout the GDR and has a predominantly sin2θ dependence, implying that impurity radiations in interference with E1 are small. For instance the curve drawn through the data at 6.5 MeV corresponds to the values $b_1 = 0.01$, $b_2 = 0.23$ and $b_3 = -0.05$ (for the definition of these coefficients see Eq. (8)). The values found for b_2 using Eqs. (7) and (8) and the unpolarized angular distribution data of Segel *et al.*[16] are shown in Figure 12.

Since the ground state spins and parities of F^{19} and Ne^{20} are $1/2^+$ and 0^+ respectively, only $p_{1/2}$ and $p_{3/2}$ inci-

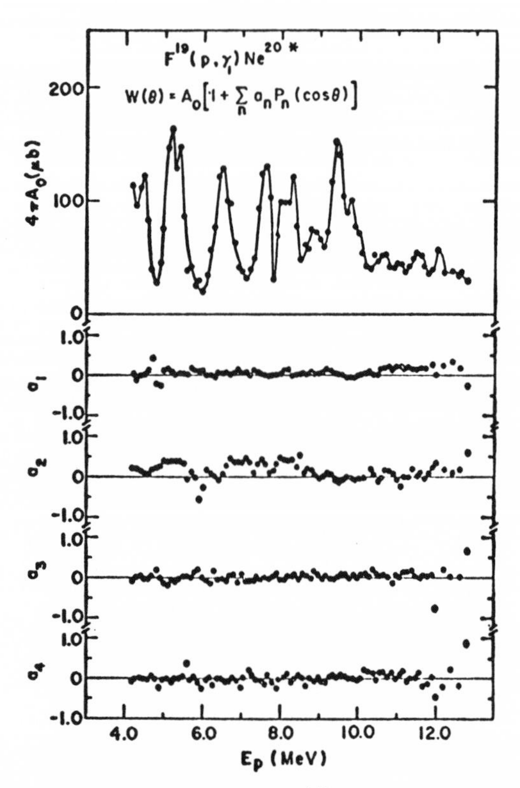

Fig. 10. The data of Segel *etal.*[16] for the total yield and angular distribution coefficients for the reaction $F^{19}(p,\gamma_1)Ne^{20*}$.

dent protons can produce E1 radiation. Neglecting all other radiations, the amplitudes and relative phases of the corresponding T matrix elements have been found by the procedure described in Section C, and are included in Figure 12. As for the case of O^{16} two solutions are possible, although in the Ne^{20} case both solutions have comparable values for the $p_{1/2}$ and $p_{3/2}$ amplitudes. Even the phase differences for the two solutions are approximately equal. It is difficult to eliminate either one of these solutions on theoretical grounds.

An alternative approach is to specify the T matrix elements in a channel spin representation. There are again two T matrix elements $R_o e^{i\phi_o}$ and $R_1 e^{i\phi_1}$, corresponding to channel spin 0 and 1 respectively. In terms of these quantities, if all radiations except E1 are neglected, we have

$$a_2 = 0.5R_1^2 - R_o^2$$

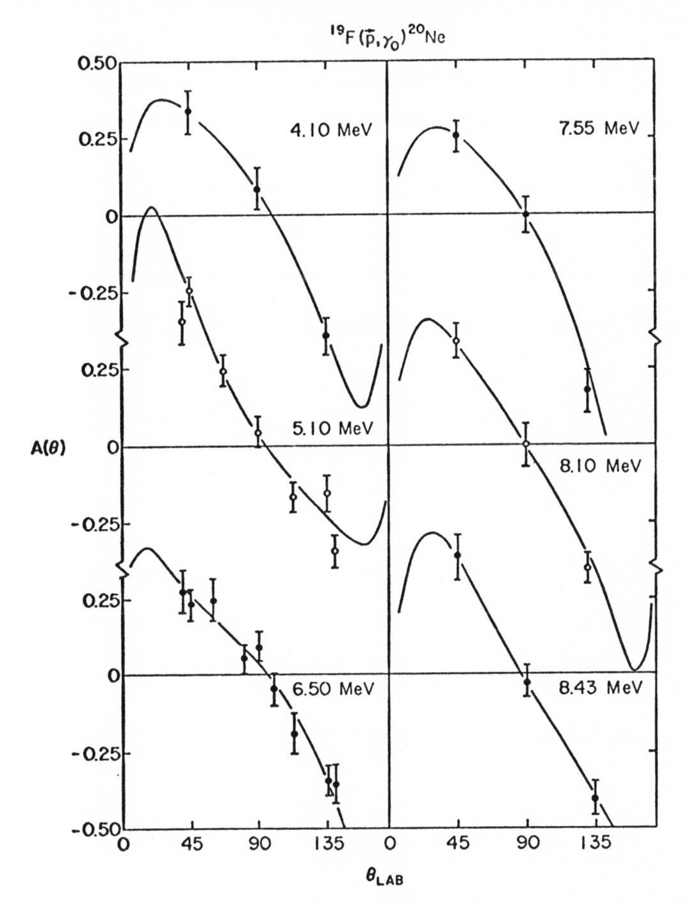

Fig. 11. The analyzing power A(θ) measured at Stanford for the reaction $F^{19}(\vec{p},\gamma_0)Ne^{20}$.

$$b_2 = -1.06R_0R_1\sin(\phi_0 - \phi_1) \tag{15}$$

leading to the average values over the GDR of $R_0 \simeq 0.91$, $R_1 \simeq 0.41$ and $\sin(\phi_0 - \phi_1) \simeq 0.76$. The preponderance of channel spin zero is a consequence of the fact that the dipole operator does not flip the nucleon spin. It should be noted that these results are very similar to the case[26] $H^3(p,\gamma)He^4$ where as in F^{19} the ground state of H^3 is $1/2^+$. This would seem to suggest that the giant dipole states of Ne^{20} would be more naturally described in a scheme where the four valence nucleons are LS coupled rather than jj-coupled. Presumably, the discrete peaks observed in the Ne^{20} GDR,

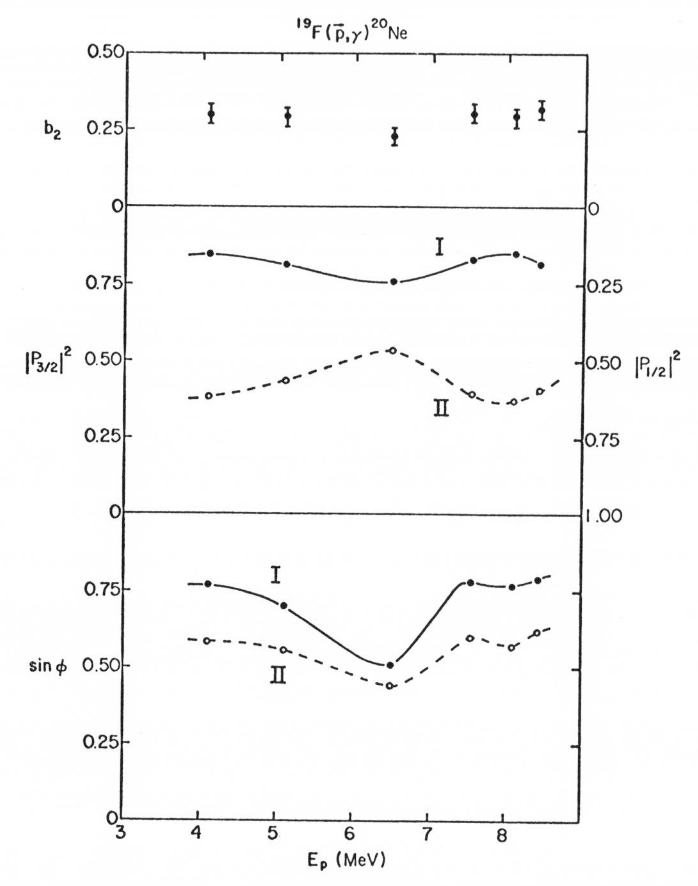

Fig. 12. The values of b_2 obtained from the data in Figure 11 by use of Eq. (8). The quantities $|p_{1/2}|$ and $|p_{3/2}|$ are the amplitudes of the T matrix elements for $p_{1/2}$ and $p_{3/2}$ proton capture. The angle ϕ is the relative phase of the T matrix elements.

which is unique in this mass region, could be accounted for by supposing the valence nucleons are subject to a non-spherical potential.

Si^{28}: Similar measurements to those above for Ne^{20} have been made for Si^{28} using the reaction $Al^{27}(p,\gamma_0)Si^{28}$. The total yield measurements of Singh *et al.*[17] are shown in Figure 3 and their angular distribution measurements in Figure 13. The unusual feature of this reaction is the value of a_2 which is essentially zero throughout the GDR. The polarization measurements taken at Stanford are shown in Figure 14. Thus although the unpolarized angular distribution is approximately isotropic, the polarization effects are substantial

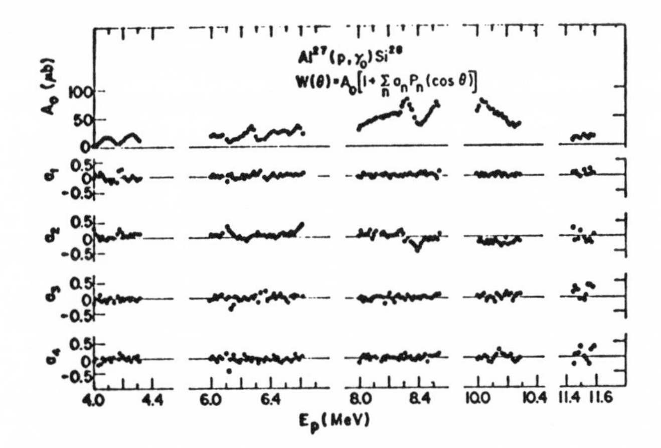

Fig. 13. The angular distribution coefficients for the $Al^{27}(p,\gamma_o)Si^{28}$ reaction obtained by Singh *et al.*[17]

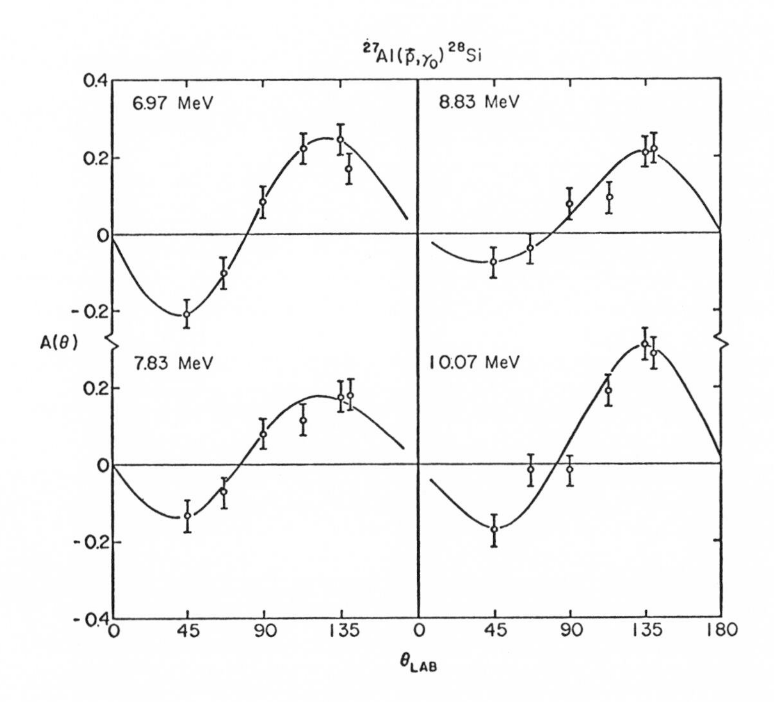

Fig. 14. The analyzing power A(θ) measured at Stanford for the reaction $Al^{27}(\vec{p},\gamma_o)Si^{28}$.

and, as usual, constant throughout the GDR. The energies chosen for the polarization measurements coincide with the four gross peaks in the total yield that persist at energy averaging intervals of 300 keV (see Figure 3). The extracted values for b_2 are shown in Figure 15. It may be observed that the analyzing power at 90° shows a significant departure from zero of about the same amount at each energy. Furthermore, if $A(\theta)$ is fitted using Eq. (8) we find that $b_3 \simeq -1/2b \simeq -0.03$ which implies that a substantial amount of E2 radiation is present.

Since Al^{27} has a $5/2^+$ ground state, $p_{3/2}$, $f_{5/2}$ and $f_{7/2}$ incident protons can produce E1 radiation. If the T matrix elements for these channels are denoted by α, β, γ respectively, and if all other radiations are neglected, then

$$a_2 = -0.1|\alpha|^2 + 0.46|\beta|^2 - 0.36|\gamma|^2$$

$$-0.32\mathrm{Re}(\alpha\beta^*) + 0.38\mathrm{Re}(\beta\gamma^*) + 1.436\mathrm{Re}(\alpha\gamma^*) \qquad (16)$$

$$b = -0.40\mathrm{Im}(\alpha\beta^*) - 0.66\mathrm{Im}(\beta\gamma^*) - 0.67\mathrm{Im}(\alpha\gamma^*)$$

with $|\alpha|^2 + |\beta|^2 + |\gamma|^2 = 1$. Since the giant dipole states are in general mixed configurations there is not a strong

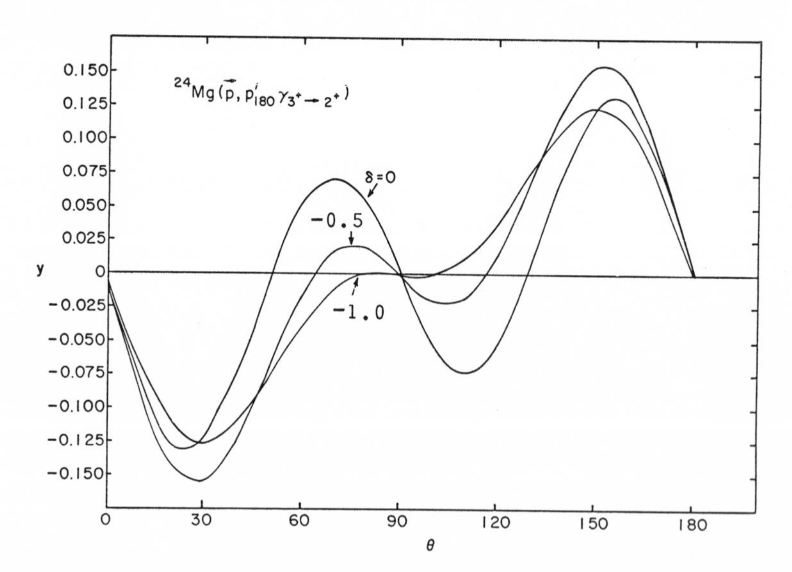

Fig. 15. Curves for the polarized angular correlation $Y(\theta_\gamma)$ for a reaction $(0^+)(\vec{p},p'_{180}\gamma)(3^+ \to 2^+)$ as a function of the mixing ratio $\delta = M1/E2$.

argument for eliminating β on the basis that the p-h state $f_{5/2}d_{5/2}^{-1}$ has a weak dipole strength. If however we make the assumption that $\beta \simeq 0$ regardless, then two solutions are found for α and γ:

Solution I: $|\alpha|^2 \simeq 0.10,\ |\gamma|^2 \simeq 0.90,\ |\phi_\alpha - \phi_\gamma| = 90^o$

Solution II: $|\alpha|^2 \simeq 0.90,\ |\gamma|^2 \simeq 0.10,\ |\phi_\alpha - \phi_\gamma| = 90^o$.

At this point we mention how valuable it would be to have a reaction model even if this only predicted phase differences reliably. If we were in a position to predict the phases then unique solutions could be found for $|\alpha|$, $|\beta|$, and $|\gamma|$ using Eq. (16).

III. PARTICLE-GAMMA ANGULAR CORRELATIONS FOLLOWING NUCLEAR REACTIONS

It was long ago realized that the angular correlation between an emerging particle following a nuclear reaction and an ensuing de-excitation gamma ray provided useful additional spectroscopic information. These ideas can be extended by initiating the reaction with a polarized beam. In general, the results of a correlation measurement made with a polarized incident beam depend upon combinations of T matrix elements which are different from those appearing in the unpolarized formula.

For instance if the emerging particle is detected at 0° or 180° relative to the incident polarized beam (this is the Litherland-Ferguson II geometry[27]), the particle-gamma correlation factorizes into two parts: one part depends on spectroscopic factors such as the level spin, parity and the multipole mixing ratio of the final excited level. The other part depends on the detailed reaction mechanism which leads to the formation of the final level. In particular, studies of correlations of this sort enable E2/M1 mixing ratios to be found accurately and quite unambiguously which is not always the case when the incident beam is unpolarized.

Another application utilizes the spin-flip geometry.[28] Here the gamma ray is detected in a direction perpendicular to the reaction plane. Correlations of the type $(\vec{p},p'\gamma)$ where the gamma decay is a $2^+ \rightarrow 0^+$ transition have been used to obtain spectroscopic information at analog resonances,[29] which otherwise cannot be obtained except by the very difficult experiment of measuring the polarization of the inelastically scattered protons when the analog resonance is excited by an unpolarized incident beam.[30]

A. Polarization Transfer

In a reaction $A(\vec{a},b)B$ initiated by a polarized incident beam the residual nucleus B is in general left in a polarized state. The magnitude and nature of the polarization of B depends on the polarization state of the incident beam and the direction of the emerging particle. There are some general results concerning the polarization of B which arise from angular momentum and parity conservation, and time reversal symmetry. This is quite well known and the most recent account and reformulation of the theory has been given by Debenham and Satchler.[31]

It turns out that the polarization state of B can be related to the polarization state of A by transfer coefficients A^{kq}_{KQ}:

$$\rho_{KQ}(J_B) = \frac{\sum_{kq} A^{kq}_{KQ}\, \rho_{kq}(s_a)}{\hat{J}_B \sum_{kq} A^{kq}_{oo}\, \rho_{kq}(s_a)} \tag{17}$$

In the expression $\rho_{KQ}(J_B)$ is the statistical tensor for particle B (spin J_B) and is defined in the usual way[32] in terms of the density matrix elements $\rho_{M_B M'_B}(J_B)$ for the spin state of B:

$$\rho_{KQ}(J_B) = \sum_{M_B M'_B} \langle KQ | J_B J_B M_B - M'_B \rangle (-)^{J_B - M'_B}\, \rho_{M_B M'_B}(J_B) \tag{18}$$

The polarization state of the incident beam is specified by $\rho_{kq}(s_a)$ which likewise is defined as in Eq. (18). The transfer coefficients A^{kq}_{KQ} can be defined explicitly in terms of the reaction matrix elements. In a representation where $\underset{\sim}{j} = \underset{\sim}{\ell} + \underset{\sim}{s}_a$ in the initial channel, $\underset{\sim}{J} = \underset{\sim}{j} + \underset{\sim}{J}_A$, *etc.*, the reduced reaction matrix elements can be written as

$$R = \langle (\ell_1 s_a) j_1 J_A J \| R \| (\ell_2 s_b) j_2 J_B J \rangle$$

$$R' = \langle(\ell_1' s_a)j_1' J_A J' \| R \| (\ell_2' s_b) j_2' J_B J'\rangle \tag{19}$$

and

$$A_{KQ}^{kq} = \sum (-)^{\ell_1' + \ell_2'} \hat{\ell}_1 \hat{\ell}_1' \hat{\ell}_2 \hat{\ell}_2' \hat{j}_1 \hat{j}_1' \hat{j}_2 \hat{j}_2' \hat{J}^2 \hat{J}'^2 \frac{\hat{b}}{\hat{A}} \hat{k}_1 \hat{\kappa} \hat{k} \hat{K} \hat{k}_2^2$$

$$\times \langle \kappa Q | k_2 K 0 Q\rangle \langle k_1 0 | \ell_1 \ell_1' 00\rangle \langle k_2 0 | \ell_2 \ell_2' 00\rangle \langle \kappa q | k_1 k 0 q\rangle \tag{20}$$

$$\times \begin{Bmatrix} j_1 & J_A & J \\ j_1' & J_A & J' \\ \kappa & 0 & \kappa \end{Bmatrix} \begin{Bmatrix} j_2 & J_B & J \\ j_2' & J_B & J' \\ k_2 & K & \kappa \end{Bmatrix} \begin{Bmatrix} \ell_1 & s_a & j_1 \\ \ell_1' & s_a & j_1' \\ k_1 & k & \kappa \end{Bmatrix} \begin{Bmatrix} \ell_2 & s_b & j_2 \\ \ell_2' & s_b & j_2' \\ k_2 & 0 & k_2 \end{Bmatrix}$$

$$\times \quad RR'^* \mathcal{D}_{Qq}^{\kappa}(R^{-1})$$

where $R^{-1} = (0, -\theta, -\phi)$ specifies the reaction angle θ and the angle ϕ between the normal to the reaction plane and the y-axis is used to define $\rho_{kq}(s_a)$. Both $\rho_{kq}(s_a)$ and $\rho_{KQ}(J_B)$ are defined for the z-axis along $\underset{\sim}{k}_a$ and $\underset{\sim}{k}_b$ respectively. The y-axis for defining $\rho_{KQ}(J_B)$ is along the normal to the reaction plane.

Although the A_{KQ}^{kq} depend in general on the detailed nature of R and R'*, there are certain simplifying symmetries which are independent of the particular reaction matrix elements. From Eq. (20) it can be shown that

$$A_{KQ}^{kq} = (-)^{k + K + Q + q} A_{K-Q}^{k-q}$$

and (21)

$$A_{KQ}^{kq*} = (-)^{k + K} A_{KQ}^{kq}$$

It is just these relations that lead to some interesting applications of particle-gamma correlations with an incident polarized beam.

It should be mentioned that Eq. (17) is normalized such that $T_r[\rho(J_B)] = 1$, or equivalently $\rho_{00}(J_B) = 1/\hat{J}_B$. The denominator of Eq. (17) is responsible for this normalization. The denominator itself corresponds to the normal ("singles") polarized angular distribution $W_p(\theta,\phi)$:

$$W_p(\theta,\phi) = \hat{J}_B \sum_{kq} A_{00}^{kq}\, \rho_{kq}(s_a) \tag{22}$$

It follows that the unpolarized angular distribution is obtained with k = q = 0:

$$W_u(\theta) = \left(\hat{J}_B/\hat{s}_a\right) A_{00}^{00} \tag{23}$$

I dwell on the question of normalization, since often this is not clearly specified in the literature.

An interesting application of these general results concerns the transfer of polarization to the final nucleus B when the emerging particle 'b' is not detected. In this case one must specify $\rho_{KQ}(J_B)$ in some fixed lab frame and integrate over all reaction angles (θ,ϕ). The result for the polarization transfer is

$$\overline{\rho}_{KQ}(J_B) = [\overline{W}_u]^{-1} \sum_k C_{KQ}^{kQ}\, \rho_{kQ}(s_a) \tag{24}$$

where $\overline{W}_u = 2\pi \int W_u(\theta)\sin\theta d\theta$ and C_{KQ}^{kQ} is a transfer coefficient with the properties

$$C_{KQ}^{kQ} = (-)^{k+K}\, C_{K-Q}^{k-Q} = (-)^{k+K}\, C_{kQ}^{kQ*} \tag{25}$$

In general, these transfer coefficients are non-vanishing, but for some reactions they may be small. If the incident beam is vector polarized (k = 1) perpendicular to the beam direction ($\rho_{11}(s_a) = \rho_{1-1}(s_a) = iP_y/2$) then equations (24) and (25) show that the final nuclei B will be vector polarized in the same direction. This offers the possibility of measuring g-factors of beta unstable nuclei by studying the perturbations of the beta particle asymmetry.[33] Even rank polarization can also be transferred to the final nuclei B and if these subsequently decay by gamma emission, an asymmetry should be observed in the gamma ray distribution. Perturbing this asymmetry with a magnetic field again offers the possibility of measuring g-factors. As a test on the idea of polarization transfer, at Stanford we excited the 4.44 MeV level of C^{12} by inelastic scattering, using an incident polarized proton beam at an energy of 8.5 MeV. The gamma rays exhibited a large polarization asymmetry (∿0.5). Stripping reactions, which are peaked in the forward direction, are generally expected to produce large polarization transfers.

B. E2/M1 Mixing

If the emerging particle 'b' is detected at 0° or 180°, Eq. (20) shows that A_{KQ}^{kq} vanishes unless q = Q. In this case

$$A_{Kq}^{kq} = (-)^{k+K} A_{K-q}^{k-q} \tag{26}$$

If the sum of the spins $s_a + s_b + J_A \leq 1$ the polarized angular correlation function for the gamma ray emitted by the final nucleus B is

$$W_p(\theta_\gamma) = W_u(\theta_\gamma) + \underset{\sim}{P}\cdot\underset{\sim}{n} \sum_K A_K(\delta) P_K{}'(\theta_\gamma) \tag{27}$$

where $W_u(\theta_\gamma) = 1 + a_2P_2(\cos\theta) + a_4P_4(\cos\theta)$ is the unpolarized correlation function, $\underset{\sim}{P}$ is the polarization of the incident beam, $\underset{\sim}{n}$ is the normal to the reaction plane, and

$$A_K(\delta) = C \frac{\langle K1|J_B J_B 10\rangle}{\sqrt{K(K+1)}} \tag{28}$$

$$\times \left\{ \frac{\delta^2 F_K(LL) + 2\delta F_K(L,L+1) + F_K(L+1,L+1)}{1 + \delta^2} \right\}$$

The functions $F_k(L,L')$ are the usual particle-gamma correlation functions and L,L' are the multipolarities of the gamma radiation. The quantity C is an overall normalization factor, dependent on the detailed reaction mechanism, but not on K or the multipole mixing ratio δ. (Note: for the purpose of the present discussion δ is defined as M1/E2 whereas the normal definition is E2/M1.) This factorization of the detailed reaction mechanism is a consequence of the spin restrictions and the symmetry expressed by Eq. (26), and has been considered in the literature in a general way.[31,34] Less well recognized is the fact that a similar result is obtained if the spin restrictions are relaxed to $s_a + s_b + J_A \leq 3/2$. Thus, correlations such as $(\vec{p},p'\gamma)$ off a spin 1/2 or spin 0 target, $(\vec{d},p\gamma)$ off a spin 0 target, $(\vec{p},\alpha\gamma)$ off a spin 1/2 or spin 1 target can all be used to find multipole mixing ratios.

At Stanford we have investigated the reaction $Mg^{24}(0^+)(\vec{p},p'_{180}\gamma)Mg^{24}$, $3^+(5.22\text{ MeV}) \rightarrow 2^+(1.368\text{ MeV})$. From a theoretical standpoint the 3^+ level is of interest because the decay to the 2^+ level has been found to be essentially pure E2 radiation, using γ-γ correlation techniques.[35] Even if one does not assume the Davydov-Filippov model[36] (which predicts a zero M1 component) the M1 to E2 intensity ratio is expected[37] to only be 3×10^{-4} in view of the value of the M1/E2 mixture in the $2^+ \rightarrow 2^+$ transition in Mg^{24}. In contrast to the γ-γ correlation using unpolarized incident protons, the polarized particle-gamma correlation is very sensitive to small admixtures of M1 radiation, as shown in Figure 16. The angular dependence of the asymmetry was first measured, as shown in Figure 17, to establish the general validity of the method. The χ^2 as a function of δ(=M1/E2) was determined. There are only two degrees of freedom and the minimum value of χ^2 is 15, corresponding to a confidence limit of 10% and a value of $\delta = +0.07$. A more accurate measurement was then made to precisely determine at the angle at which the asymmetry factor 'y' vanished. For small values of δ this is near 50° or 130°. The asymmetry is necessarily

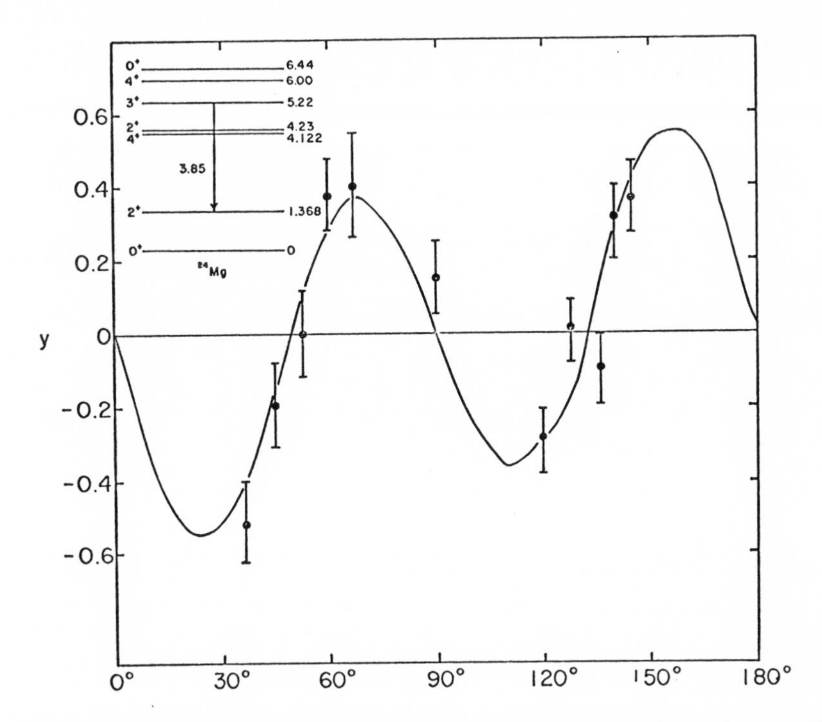

Fig. 16. Stanford measurements of the angular dependence of the polarized correlation function $y(\theta_\gamma)$ for the reaction $Mg^{24}(\vec{p},p'_{180}\gamma)Mg^{24}$, $(3^+ \rightarrow 2^+)$.

zero at 90°. It is clear from Figure 16 that the location of the null angle is very sensitive to the value of δ, if δ is small. The results obtained are shown in Figure 17 and yield a value of δ in the range $+0.1 > \delta > -0$. In this final measurement, the angular geometry was accurately calibrated by observing the corresponding asymmetry in the reaction $C^{12}(\vec{p},p'_{180}\gamma)C^{12}$, $2^+(4.44\text{ MeV}) \rightarrow 0^+(\text{g.s.})$, for which $\delta = 0$ and the angular dependence of the asymmetry is known exactly. The final method adopted for measuring δ does not require the beam polarization to be known accurately.

C. Polarized Spin-Flip Correlations

In a $(p,p'\gamma)$ reaction the spin-flip geometry corresponds to detecting the gamma ray along the normal to the reaction plane, as discussed by Schmidt.[28] The correlations for this type of geometry are conveniently expressed in terms of partial differential cross-sections $\sigma^{++}(\theta)$, $\sigma^{--}(\theta)$, $\sigma^{+-}(\theta)$ and $\sigma^{-+}(\theta)$. In this notation, $\sigma^{+-}(\theta)$, for instance, is the partial differential cross section for scattering from an initial state with incident proton spin up to a final state with

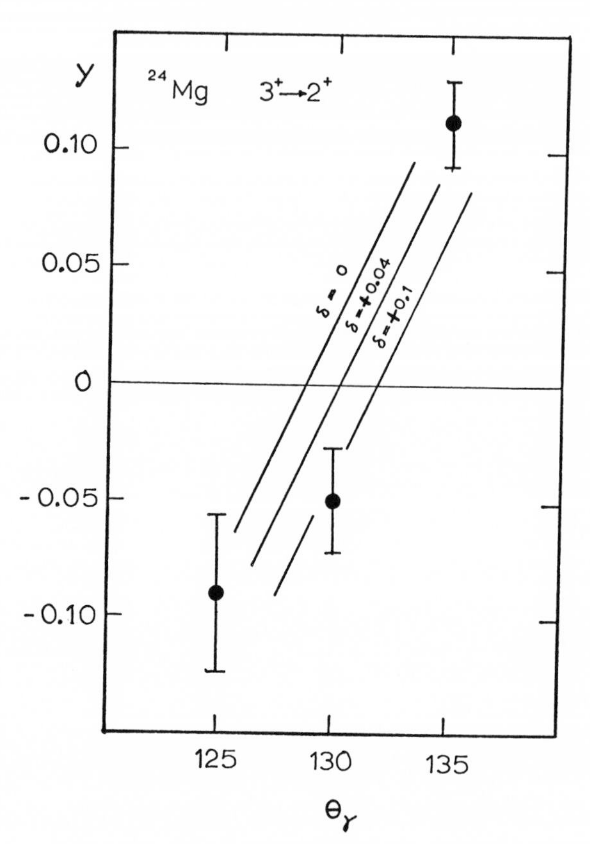

Fig. 17. Stanford measurements of the polarized correlation function $y(\theta_\gamma)$ for the reaction $Mg^{24}(\vec{p},p'_{180}\gamma)Mg^{24}$, $(3^+ \to 2^+)$ near 130°. The solid lines are fits corresponding to different values of δ = M1/E2.

proton spin down. The proton spin states are defined in this notation with respect to a z-axis which is normal to the reaction plane. Thus, by definition, the angular distribution $\sigma(\theta)$, the analyzing power $A_z(\theta)$, and the inelastic proton polarization $P_z(\theta)$ produced by an unpolarized incident beam, are expressed as:

$$
\begin{aligned}
2\sigma(\theta) &= \sigma^{++} + \sigma^{+-} + \sigma^{-+} + \sigma^{--} \\
2\sigma(\theta)A_z(\theta) &= \sigma^{++} + \sigma^{+-} - \sigma^{-+} - \sigma^{--} \qquad (29) \\
2\sigma(\theta)P_z(\theta) &= \sigma^{++} + \sigma^{-+} - \sigma^{+-} - \sigma^{--}
\end{aligned}
$$

The above equations are the "singles" correlations. For the coincident correlations we restrict our attention to gamma rays emerging along the z-axis (*i.e.* normal to the reaction plane) and arising from $2^+ \rightarrow 0^+$ (g.s.) or $1^+ \rightarrow 0^+$ (g.s.) transitions. To emit gamma rays along the z-axis it follows from the properties of gamma ray angular distributions that the magnetic substates of the excited nucleus are restricted to the values $M_z = \pm 1$. By the Bohr theorem[38] only spin-flip scattering can populate the $M_z = \pm 1$ substates. That is to say, only σ^{+-} and σ^{-+} contribute. Thus, if the incident proton beam is unpolarized, the proton-gamma coincident correlation is

$$W_u(\theta) = (\sigma^{+-} + \sigma^{-+})/2 = \sigma(\theta)S(\theta)/2 \tag{30}$$

and this is the quantity measured in the spin-flip measurements discussed by Schmidt.[28] Boyd *et al.*[30] have discussed the case when the incident proton beam is polarized. If the incident polarization is P_B, it follows that the proton-gamma correlation is

$$\begin{aligned} W_p(\theta) &= \frac{1 + P_B}{2}\sigma^{+-} + \frac{1 - P_B}{2}\sigma^{-+} \\ &= W_u(\theta)\left(1 + \frac{\Delta S}{S}P_B\right) \end{aligned} \tag{31}$$

where

$$2\sigma(\theta)\Delta S = \sigma^{+-} - \sigma^{-+} \tag{32}$$

and S is as defined in Eq. (30). It is clear that if both $W_u(\theta)$ and $W_p(\theta)$ are measured σ^{+-} and σ^{-+} may be separately found. As discussed by Boyd *et al.*,[30] when the reaction proceeds through an isobaric analog resonance (IAR), the individual measurement of σ^{+-} and σ^{-+} may provide valuable spectroscopic information.

It should be mentioned that

$$2\Delta S = A_z - P_z \tag{33}$$

is implied by Eqs. (29) and (32). Therefore the equivalent of measuring ΔS in a coincidence experiment is to separately measure P_Z and A_Z. With presently available polarized beams the measurement of A_Z is straightforward. However, the measurement of P_Z requires a second scattering and compared with a coincidence measurement of ΔS this is probably an inferior approach. It should be noted that while ΔS is being

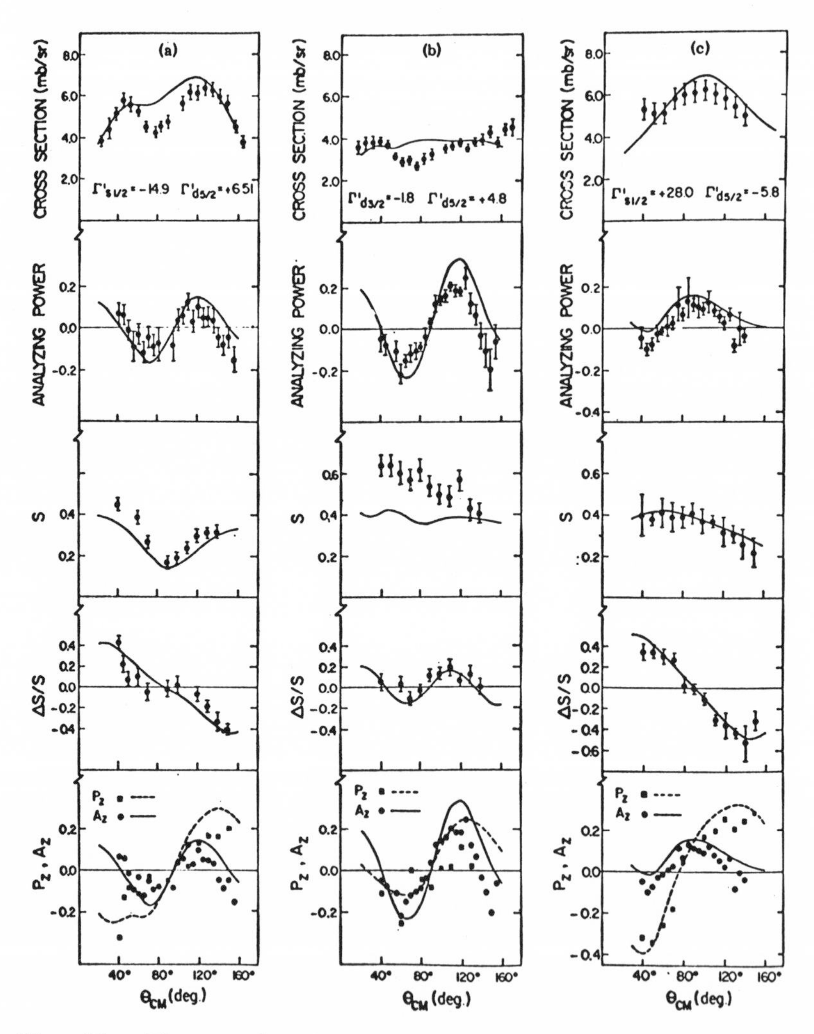

Fig. 18. The angular dependence of the quantities defined in Eqs. (29), (30), (32) and (33) at the (a) 7.00 MeV, (b) 7.08 MeV, and (c) 7.53 MeV resonances in Y^{89}. The partial widths are expressed in keV. The solid curves are preliminary fits as described by Boyd *et al.*[29]

measured by the coincidence method, and can simultaneously measure A_z and S.

A joint experiment[29] was conducted by Rutgers and Stanford Universities to measure ΔS, s and A_z for the following three IAR's in Y^{89}: 7.00 MeV ($5/2^+$), 7.08 MeV ($3/2^+$) and 7.53 MeV ($3/2^+$). These IAR's were observed with the reaction $Sr^{88}(\vec{p},p'\gamma)Sr^{88}$, $2^+(1.836) \rightarrow 0^+$(g.s.). These states are of theoretical interest because the corresponding parent states in Sr^{89} have been described by Spencer *et al.*[39] as weak coupled single-particle states to the 0^+ and 2^+ states in Sr^{88}. In the analog wave function, the coupling to the 2^+ core is characterized by the inelastic proton partial widths $\Gamma^{p'}_{\ell j'}$, and these are contained in the correlation functions for the proton decay of the analog state to the 2^+ state in Sr^{88}. The measurement of ΔS provides a valuable additional constraint for the values that are assigned to the $\Gamma^{p'}_{\ell j}$. The data and preliminary fits taken from Ref. 29 are shown in Figure 18. It may be noted that the varied behavior of ΔS over the three resonances does indeed provide a meaningful signature to the resonances.

The isobaric analog states of Ti^{51}, which have already been investigated[40] by unpolarized elastic and inelastic scattering, provide an interesting application for a polarized spin-flip measurement.[41]

REFERENCES

1. W. Bertozzi *et al.*, *Congr. Intern. Phys. Nucl.*, Vol. II, P. Gugenberger, Ed., (Editions du Centre National de la Recherche Scientifique, Paris, 1964), p. 1026.
2. G. W. Cole, Jr. and F. W. K. Firk, *Proceedings of the Third International Symposium on Polarization Phenomena in Nuclear Reactions*, H. H. Barschall and W. Haeberli, Eds., (University of Wisconsin Press, Madison, 1971), p. 696.
3. T. J. Yule and W. Haeberli, Phys. Rev. Lett. 19, 756 (1967).
4. T. J. Yule and W. Haeberli, Nucl. Phys. A117, 1 (1968).
5. W. Haeberli, *Proceedings of the Third International symposium on Polarization Phenomena in Nuclear Reactions*, H. H. Barschall and W. Haeberli, Eds., (University of Wisconsin Press, Madison, 1971), p. 235.
6. G. Graw, *Proceedings of the Third International Symposium on Polarization Phenomena in Nuclear Reactions*, H. H. Barschall and W. Haeberli, Eds., (University of Wisconsin Press, Madison, 1971), p. 179.
7. G. R. Plattner, *Proceedings of the Third International Symposium on Polarization Phenomena in Nuclear Reactions*, H. H. Barschall and W. Haeberli, Eds., (University of

Wisconsin Press, Madison, 1971), p. 107.
8. D. H. Wilkinson, Physica 22, 1039 (1956).
9. J. P. Elliot and B. H. Flowers, Proc. Roy. Soc. 242, 57 (1957).
10. G. E. Brown, *Unified Theory of Nuclear Models and Forces*, North Holland Publishing Company, Amsterdam, 1967).
11. G. E. Brown and M. Bolsterli, Phys. Rev. Lett. 3, 472 (1959).
12. W. L. Wang and C. M. Shakin, Phys. Rev. C5, 1898 (1972).
13.N.N. W. Tanner, G. C. Thomas and E. D. Earle, Nucl. Phys. 52, 45 (1964).
14. J. T. Caldwell *et al.*, Phys. Rev. Lett. 15, 976 (1965).
15. J. LeTournex, Mat. Fys. Medd. Dan. Vid. Selsk. 34, No. 11 (1965), and R. Ligensa and W. Greiner, Ann. Phys. 51, 28 (1969).
16. R. E. Segel *et al.*, Nucl. Phys. A93, 31 (1967).
17. P. P. Singh *et al.*, Nucl. Phys. 65, 577 (1965).
18. S. A. Farris and J. M. Eisenberg, Nucl. Phys. 88, 241 (1966).
19. L. Meyer-Schützmeister *et al.*, Nucl. Phys. A108, 180 (1968).
20. W. J. O'Connell, Ph.D. thesis, Stanford University (1969), unpublished.
21. B. Buck and A. D. Hill, Nucl. Phys. A95, 271 (1967).
22. J. Raynall, M. A. Melkanoff and T. Sawada, Nucl. Phys. A101, 369 (1967).
23. M. Maragoni and A. M. Sarius, Nucl. Phys. A132, 649 (1969).
24. M. Maragoni and A. M. Sarius, Nucl. Phys. A166, 397 (1971).
25. S. S. Hanna *et al.*, Phys. Lett. 40B, 631 (1972).
26. H. F. Glavish *et al.*, *International Conference on Few Particle Problems in the Nuclear Interaction*, Los Angeles, 1972 (in press).
27. A. E. Litherland and A. J. Ferguson, Can. J. Phys. 39, 788 (1961).
28. F. H. Schmidt *et al.*, Nucl. Phys. 52, 353 (1964).
29. R. N. Boyd *et al.*, Phys. Rev. Lett. 29, 955 (1972).
30. R. N. Boyd *et al.*, Phys. Rev. Lett. 27, 1590 (1971).
31. A. A. Debenham and G. R. Satchler, Particles and Nuclei 3, 117 (1972).
32. D. M. Brink and G. R. Satchler, *Angular Momentum*, (Oxford University Press, London, 1968).
33. A. B. McDonald, Lockheed Palo Alto Research Laboratory, Palo Alto, California, private communication.
34. J. D. McCullen and R. G. Seyler, Nucl. Phys. A139, 203 (1969).
35. C. Broude and H. E. Gove, Ann. of Phys. 23, 71 (1963).
36. A. S. Davydov and G. F. Filippov, Nucl. Phys. 8, 237 (1958).

37. G. E. Gove, *Proceedings of the International Conference on Nuclear Structure*, D.A. Bromley and E. W. Vogt, Eds., (University of Toronto Press, Toronto, 1960), p. 438.
38. A. Bohr, Nucl. Phys. 10, 486 (1959).
39. J. E. Spencer *et al.*, Phys. Rev. C3, 1179 (1971).
40. E. R. Cosman, D. C. Slater and J. E. Spencer, Phys. Rev. 182, 1131 (1969).
41. D. C. Slater, Stanford University, private communication.

DISCUSSION

ENDT: When you discussed the Mg^{24} gamma angular correlation measurement I suppose you mentioned δ but what you actually meant was $1/\delta$. Is that correct?

GLAVISH: This depends on how you define δ.

ENDT: Well, I define it the normal way.

GLAVISH: My definition is M1 over E2; I tried to emphasize that throughout the talk. The reason I used M1/E2 is one of convenience when M1/E2 is small. If one deals with E2/M1 for such a situation, then the numbers are very large and it is cumbersome to assign error bars.

MALIK: I'd like to point out that the extension from a direct capture to the capture through intermediate states is not trivial. The problem is as follows: one can put in a complete set of intermediate states but if one wishes to put in only one state then there are two types of problems. Problem 1 is that the interaction between a continuum and a quasi-bound state for a complex many body system is definitely not the free two-body interaction, but that is immaterial because in all these fits and even in the calculation of Shakin and Wang that has been taken as a free parameter anyway. Problem 2 is probably more important and that is one reason why it was difficult to get an absolute magnitude correctly. This second point is that if only one intermediate state is inserted and an electromagnetic operator takes that state to the ground state, in the normal sense of perturbation theory which one uses in order to calculate the electromagnetic transition, the upper and lower state must be eigenstates of the same type of Hamiltonian. Now, of course we do not know the exact wave functions but a good approximation would be to try to produce the upper and the ground state within the same basis set. If one doesn't do that, one can run into the problem of normalization; *e.g.*, the basis set used to construct the intermediate state with coefficient one should not be used again to construct the ground state. Since the ground state in O^{16} has strong correlations, which was evident in the old days, *e.g.*, from the nice work of Professor Davidson, one should be very careful in constructing intermediate states using the same basis. This third point is of course the most complicated one. There is a problem of overlapping resonances if the resonances are not isolated. It was also not clear to me, what you mean when you say that the theorist should not give the cross sections, angular distributions, and polarizations but they should give the tidbits of the T matrix.

GLAVISH: Yes, well let me try to answer your question in some sort of order. First of all, I believe the formula that Wang and Shakin arrived at is correct. In the case of the giant states in nuclei, present theoretical calculations are aimed at producing good fits but there is no way of relating any new experimental data to these calculations, and the calculations themselves are not really aimed at finding a model to collectively describe a whole lot of experimental data. Where polarized beams have been applied to nuclear spectroscopy success has come about in the following way; one tries to formulate a reaction model. When you do an experiment you take this reaction model which you supposedly have faith in, and by relating this to your experimental data you extract nuclear spectroscopy information. Now such a well tried model does not exist for the capture reaction. The first step towards obtaining a simple physical model that can relate experimental results to nuclear structure in the GDR (giant dipole resonance), is, I believe, the model of Wang and Shakin. In other calculations there does not seem to have been an objective in mind of trying to give some sort of physical picture, a model, something that an experimenter can test time and time again in different ways to see if it works well or to see how it has to be patched up. Now as long as you believe that the model is valid, and that's part of the question I agree, but if you do believe it's valid, then the formula can be accepted, even if it cannot be properly understood or properly justified on theoretical grounds. Then one has to carry out many experiments to test the formula. Of course this is harder for the giant dipole state of nuclei then in other studies, since there is a more limited number of experiments that can reasonably be performed. Unfortunately, at present, not all the experimental information is utilized; often only fits to the total cross section are considered.

One difficulty of coupled channel calculations is that there's no way to look into the theory to see what might be going on. With the Wang and Shakin calculation one can at least look at the matrix elements that they write down. They are simple and are described in terms of bound state wave functions and simple phenomenological potentials. The more complicated features of a GDR are taken into account in a simple parametric way to facilitate comparison with experiment. It would be of great value to the subject of giant dipole states of nuclei, if theoreticians would report the T-matrix elements (amplitudes and phases) they arrive at, in addition to a few selected fits to the experimental data. Then as new experimental data comes to hand, such as the polarization data I have reported here, an immediate comparison with theory could be made.

V.B. BAND STRUCTURE OF Ne^{20}

T. K. Alexander
Chalk River Nuclear Laboratories
Atomic Energy of Canada Limited
Chalk River, Ontario, Canada

I. INTRODUCTION

The nucleus Ne^{20} has been known to exhibit rotational-like motion for some years.[1] The low lying energy levels have been classified into bands with members having energy spacings proportional to $J(J + 1)$ and the same parity. Levels of one band have E2 transition probabilities many times the single particle value and the observation of such strong E2 transitions has been a major experimental tool in assigning a particular level to a particular band. Recent experiments[2,3] have shown that the E2 γ ray transition probabilities between all members of the ground state band up to $J = 8$ are in agreement with shell model calculations[4,5,6,7,8]. This is particularly interesting since the $8^+ \rightarrow 6^+$ transition probability compared to that of the $2^+ \rightarrow 0^+$ is considerably weaker than the rigid rotator prediction.

The α particle reduced widths for the ground state band show a trend which is also very interesting. In particular, the ratio of the reduced widths of the $J^\pi = 6^+$ and 8^+ levels was found to be ~ 8, whereas the SU(3) shell model[9] predicts a constant value for the spectroscopic factor for levels in the ground state band. These new data have stimulated several theoretical studies which contain new ideas about the structure of Ne^{20}.

In this paper I would like to show some of the results outlined above and also present some new data which give information on the higher $K^\pi = 0^+$ "bands" in Ne^{20} and their mixing with the ground state band. Some of these data have been obtained at Chalk River and Oxford, and in the last part of the paper I will describe some of these experiments.

II. THE $K = 0^+$ GROUND STATE AND $K = 2^-$ BANDS

Figure 1 shows the $K = 0^+$ ground state band and the $K = 2^-$ band up to the 7^- level in Ne^{20}. On the left-hand side of Figure 1, the particle threshold energies are shown since these are significant in determining the reactions used to study the levels.

It is seen that the high spin levels are mainly above the $O^{16} + He^4$ threshold energy, so that the natural parity levels with $\pi = (-)^J$ decay predominantly by α particle emission to the ground state of O^{16}. Thus the γ decay properties of these high spin natural parity levels can only be studied

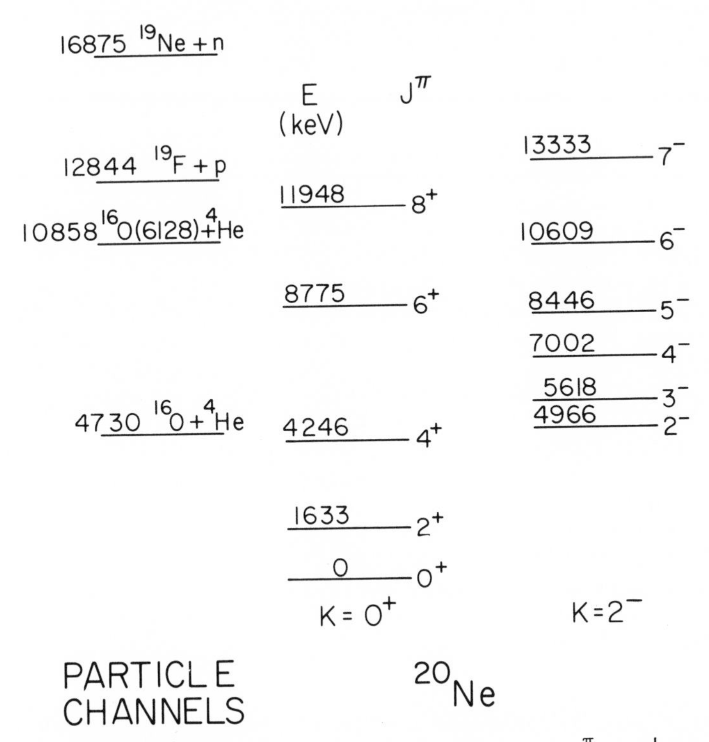

Fig. 1. The Ne^{20} ground state band up to the $J^{\pi} = 8^{+}$ level and the $K^{\pi} = 2^{-}$ band up to the $J^{\pi} = 7^{-}$ level. Particle threshold energies are also shown.

by observing resonances in the $O^{16}(\alpha,\gamma)Ne^{20}$ reaction and measuring their properties. The 6^{+} and 8^{+} levels have been studied[11,3] in the capture reaction $He^{4}(O^{16},\gamma)Ne^{20}$. The α particle widths have been determined by 3 to 5 keV resolution experiments on the elastic scattering of α particles from O^{16} and yield particularly interesting results about the nature of the ground state band.[10]

For the K = 2^{-} band, the γ ray decay of the unnatural parity levels cannot be studied by the capture reaction since the levels cannot be formed. But these levels have been studied by α-γ coincidence experiments and the $C^{12}(C^{12},\alpha\gamma)Ne^{20}$ reaction.[2] These experiments have determined the level energies, the mean lifetimes, and γ ray branching ratios. Some of the significant branches are quite weak and thus difficult to measure. The 5^{-} level of course can be studied by the α particle capture reaction, and experiments have been done by Rogers *et al.*[12] Alpha particle elastic scattering experiments have been carried out with good energy resolution to obtain the α particle widths of the 5^{-} and 7^{-} levels.[10,13]

To illustrate the $O^{16}(\alpha,\alpha)O^{16}$ elastic scattering data,

Figures 2 and 3 show the yield curves[10] in the region of the $J^{\pi} = 6^{+}$ resonance at $E_{\alpha}^{lab} = 5058$ keV and the $J^{\pi} = 8^{+}$ resonance at 9026 keV respectively. The $\ell = 6$ resonance is fitted with $\Gamma_{CM} = 110 \pm 25$ eV and the $\ell = 8$ resonance with $\Gamma_{CM} = 35 \pm 10$ eV.

The differentially pumped He^4 gas target at Chalk River is shown schematically in Figure 4 as it was used for measuring the $\omega\gamma$ of the 8^{+} level.[3] Yield curves of the 4.529 MeV and 2.613 MeV γ rays from the 6^{+}, 8775 keV level excited in the He^4 $(O^{16},\gamma)Ne^{20}$ reaction are shown in Figure 5. The full curves are calculated assuming the beam energy spread is 9 keV fwhm, the energy loss is 1 keV/cm in the gas and $\omega\gamma(6^{+}) = 1.42$ eV. The average value from the two sets of data, $\omega\gamma = 1.37$ eV, agrees well with the average of Diamond *et al.*, and Rogers *et al.*,[11] $\omega\gamma = 1.36$ eV. Figure 6 shows the yield curve of 3.17 MeV γ rays observed for the $J^{\pi} = 8^{+}$ level in the $He^4(O^{16},\gamma)Ne^{20}$ reaction and gives $\omega\gamma = 104 \pm 35$ meV.[3]

The results of the γ ray studies are summarized in Figures 7 and 8, while Table I shows the α particle widths. Figure 7 shows the measured transition probabilities in Weisskopf units (W.u.) between members of the $K = 2^{-}$ band and the predictions of the SU(3) shell model[4] and the rotational

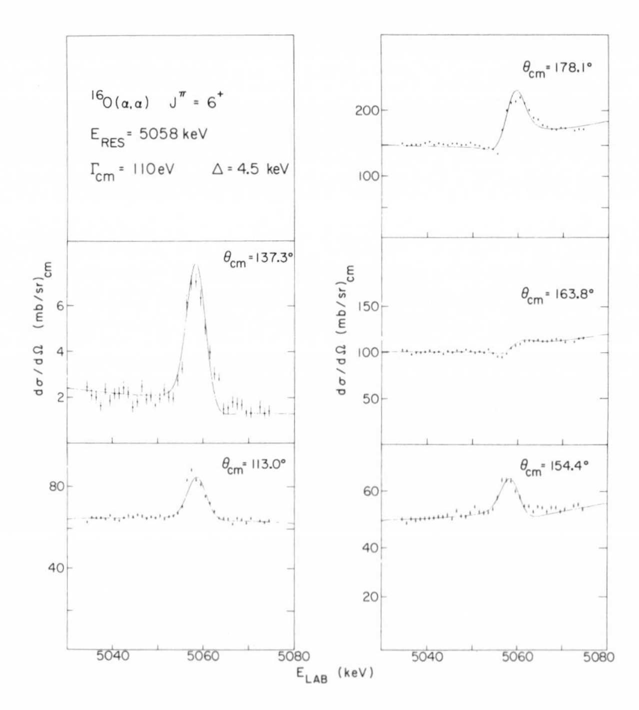

Fig. 2. Yield curve for elastic scattering of α particles from O^{16} at the $J^{\pi} = 6^{+}$ level in Ne^{20}.

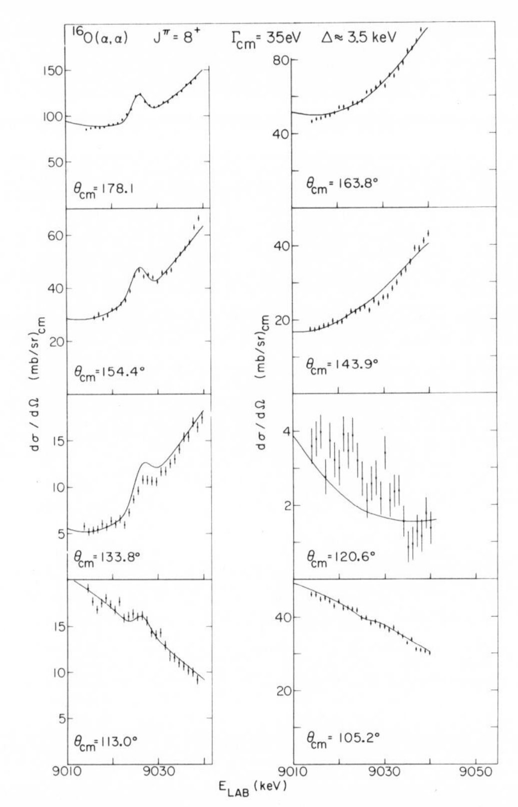

Fig. 3. Yield curve for elastic scattering of α particles from O^{16} at the $J^{\pi} = 8^{+}$ level in Ne^{20}.

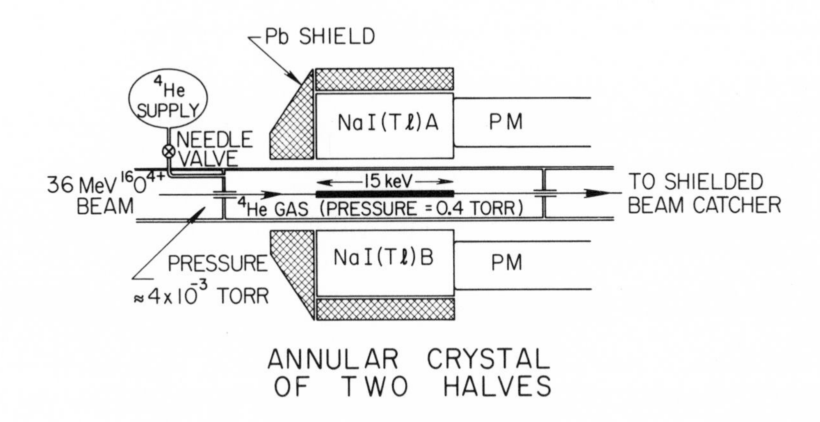

Fig. 4. The Chalk River differentially-pumped He^{4} gas target.

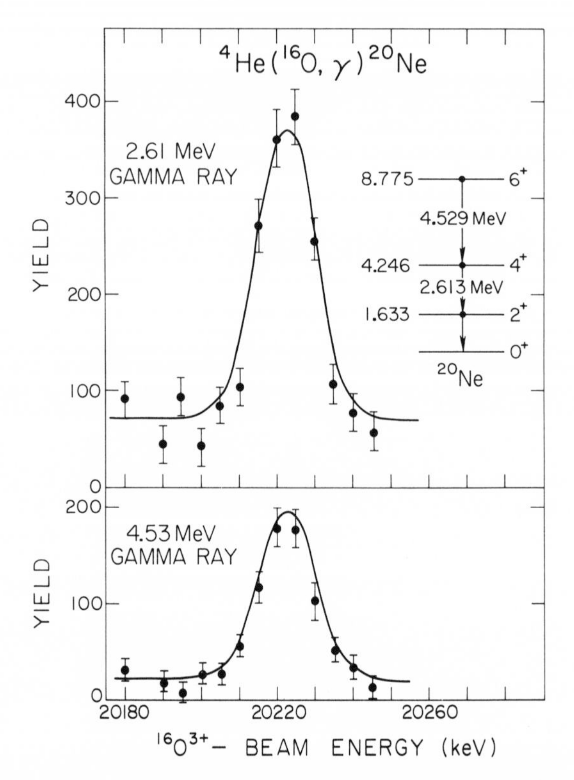

Fig. 5. Gamma ray yield curves showing the $He^4(O^{16},\gamma)Ne^{20}$ resonance at the 6^+ level.

model. It is seen that the transition probabilities between levels of the $K = 2^-$ band are in good agreement with both the shell model and the simple rotational model predictions, particularly for the $J \rightarrow J - 1$ transitions, for which the two models are not significantly different. The $J \rightarrow J - 2$ transitions predictions, however, are different especially for the 7^- level. The 7^- level will be extremely difficult to determine because of its high excitation energy which is well above the energy where the 6.129 MeV level in O^{16} can be excited (see Figure 1). The data points are taken from Häusser *et al.*[2] except for that for the 5^- level which is from Rogers *et al.*[12]

Figure 8 shows the experimental E2 transitions probabilities for the $K = 0^+$ ground state band and the predictions of four calculations using different assumptions. The rotational and SU(3) model predictions[4] have been normalized to the experimental data at the $2^+ \rightarrow 0^+$ transition. The quantity β is the effective charge on the neutrons and protons, *i.e.*, $e_n = \beta e$ and $e_p = (1 + \beta)e$. It is apparent that the observed trend in Γ_γ as J increases agrees much better with the SU(3)

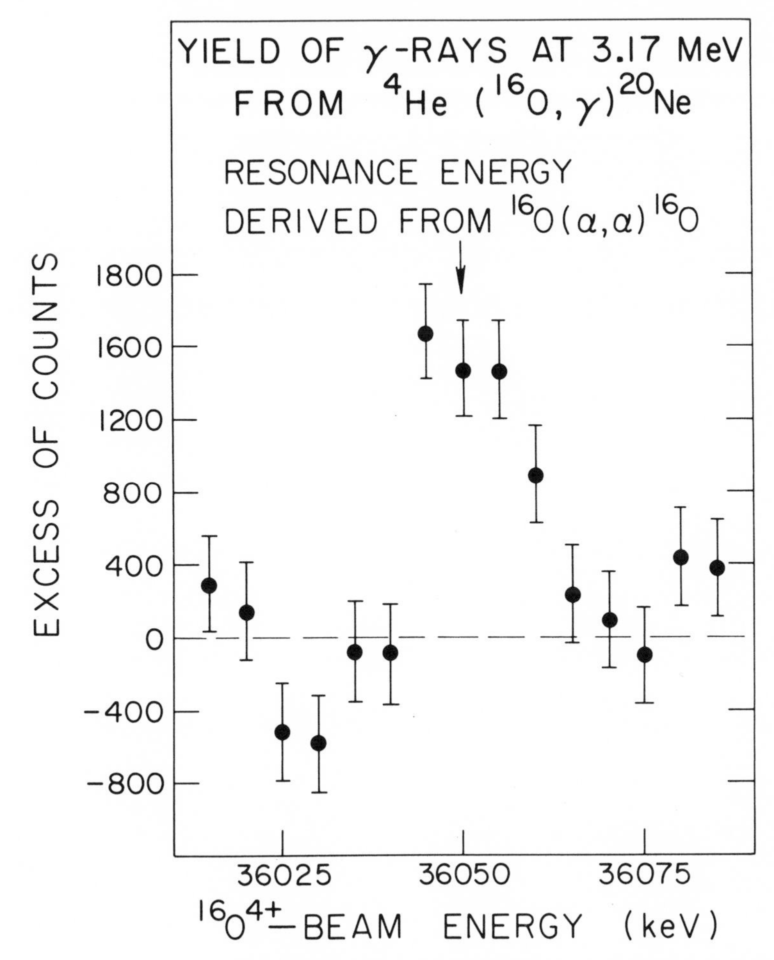

Fig. 6. Gamma ray yield curve showing the $He^4(O^{16},\gamma)Ne^{20}$ resonance at the 8^+ level.

shell model than with the simple rotational model. The SU(3) shell model prediction is typical of the microscopic shell model calculation assuming an "inert" O^{16} core.[5,7,8] This assumption leads to a termination of the ground state band at J = 8. This limitation does not exist if valence particles in more than one major shell are considered.[6]

Also shown in Figure 8 are recent calculations using the cluster model[14] and the projected Hartree-Fock theory.[15] These latter calculations give a more unified description since they not only account for the E2 transition probabilities but also interpret the α particle reduced widths shown in Table I.

In Table I, the measurements[2,10,13] of the α particle

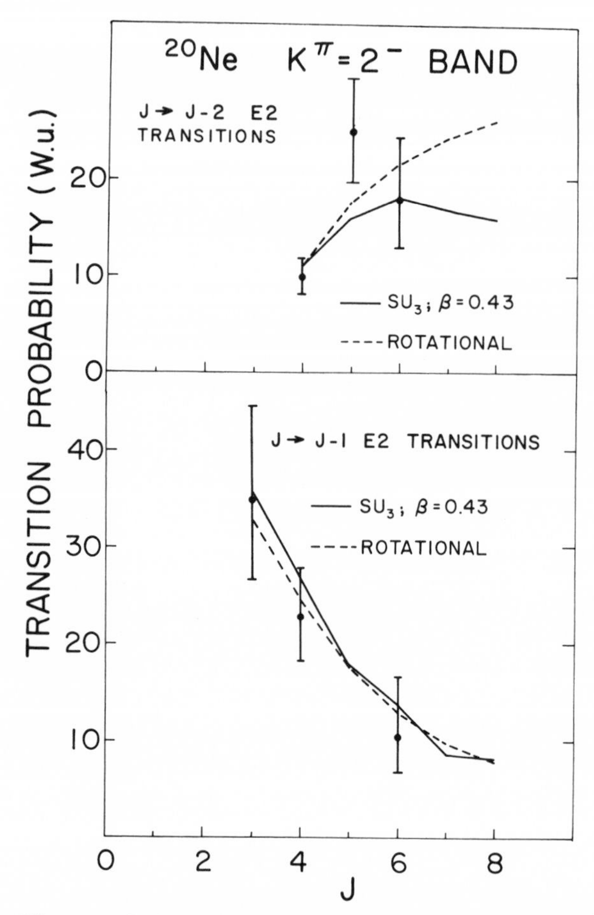

Fig. 7. E2 transitions between the members of the $K^\pi = 2^-$ band in Ne^{20}.

widths, Γ_α, are summarized for both the $K = 2^-$ and $K = 0^+$ bands. In the last column, the α particle reduced widths are shown and illustrate a feature which has stimulated several theoretical interpretations. For both bands, the reduced width evaluated for a fixed radius and a square well potential decreases as the spin increases.

Arima and Yoshida[9] have calculated alpha widths within the framework of the SU(3) shell model using more realistic potentials and have obtained reasonable agreement with experiment when the interaction radius is taken to be $\sim$3.8 fm. If they allow the radius to decrease further by 0.2 fm in going from the 6^+ to the 8^+ level, then excellent agreement is obtained and they suggest that the nucleus shrinks with an increase of ℓ. Vogt[16] has concluded that the 8^+, 11948

TABLE I

Alpha particle Reduced Widths of Some Levels

E_{α}^{LAB} (keV)	E_x (keV)	J^{π}	Γ_{α} (eV)	θ_{α}^{2} [a] (%)
1110	5618	3^-	$(3.1 \pm 0.7) \times 10^{-3}$ [b]	6.0 ± 1.5
4652	8450	5^-	15 ± 4 [c]	0.29 ± 0.08
10759	13333	7^-	80 ± 30 [d]	0.09 ± 0.03
5058	8775	6^+	110 ± 25 [d]	9.1 ± 1.7
9026	11948	8^+	35 ± 10 [d]	1.1 ±± 0.3

(a) $\theta_{\alpha}^{2} = \Gamma/\Gamma_o$, where $\Gamma_o = \frac{2kR}{F_{\ell}^{2} + G_{\ell}^{2}} \frac{\hbar}{\mu R^2}$ was evaluated for a fixed radius R = 5.1 fm. (b) Ref. 2, (c) Ref. 13, (d) Ref. 10

keV level has been misinterpreted, and suggests that this level is the 8^+ member of a 8p-4h (8 particle-4 hole) band based on the 0^+, 7.19 MeV level.

Before discussing this suggestion further, I would like to show Figure 9 which is a variation-after-projection, Hartree-Fock calculation by Lee and Cusson.[15] The calculation shows that the deformation, β, of Ne^{20} decreases with increasing spin in the ground state band up to spin 8. This is consistent with the reduction in the α particle width observed since their calculation also predicts a decrease in the rms distance between the center of mass of O^{16} and α in a clustering state of Ne^{20}. Their model also leads to theoretical B(E2) values which reproduce the measured values (shown at the right of Figure 9) without the help of an effective charge.

Having reviewed the data and interpretations of the properties of the two lowest-lying bands in Ne^{20}, I would like to turn now to some recent data[17] obtained on the two excited $K^{\pi} = 0^+$ bands near 7 MeV in Ne^{20}. But first let us look at the predictions of the quartet model. Figure 10 shows the systematics of particle-hole excitations in even-even, N = Z nuclei from A = 8 to 44 calculated by Harvey[18] using a generalized quartet model. The energies of these states are th the mean energies between the intrinsic states of the excited bands and the ground state band. The energy is ex-

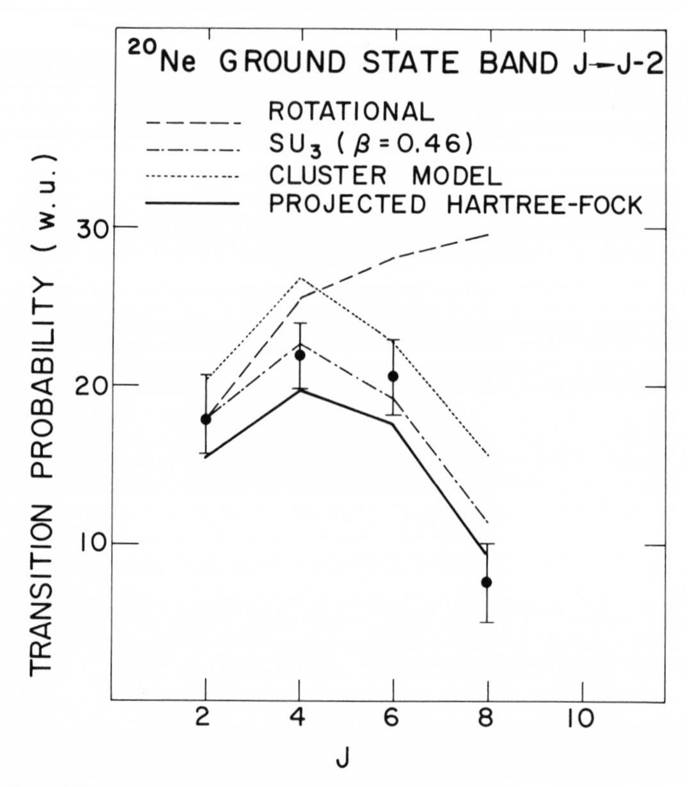

Fig. 8. E2 transitions between the members of the $K^\pi = 0^+$ ground state band of Ne^{20}. The theoretical calculations are described in the text.

pressed in terms of the observed binding energies of neighboring nuclei and the residual interaction between n = 1, 2, 3 ... particle excitations across major shells having a "quartet" symmetry. In the quartet model[19] the residual interaction was a quartet-quartet interaction whereas Harvey has introduced one additional parameter. In addition Harvey does not assume constancy of the particle-hole matrix elements with mass number.

The 4(p → sd) quartet state is shown in heavy lines and reaches a minimum at Ne^{20}. Above A = 24, the lowest quartet is the 4(sd → pf) state. The next higher states are the 1(p → sd) and 1(sd → pf) states which are at the same energy at Ne^{20}. The $K^\pi = 2^-$ band discussed previously is related to the 1(p → sd) state and in addition a $K^\pi = 0^-$ band based on the 1^-, 5.93 MeV level is also found in Ne^{20}.

This schematic correspondence is shown in more detail in Figure 11 where the bands of Ne^{20} (taken from Ref. 20) and the states calculated by Harvey are shown. A common

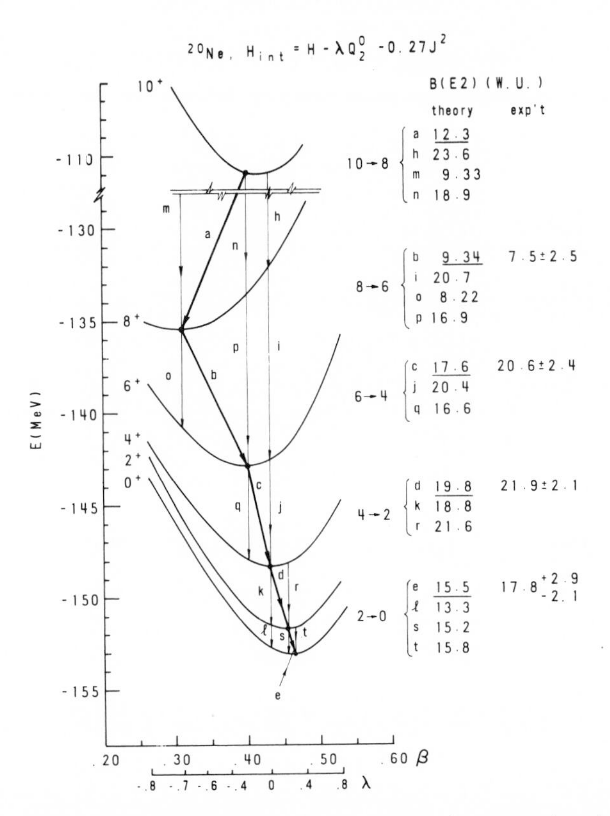

Fig. 9. Projected energies of the ground state band members in Ne^{20} versus the mass quadrupole cranking parameter λ defined in the Hamiltonian $H_I = H - 0.27J^2 - \lambda Q_2{}^0$ after Lee and Cusson (Ref. 15). The abscissa, β, is the quadrupole deformation parameter (see Ref. 15).

property of the $K^\pi = 0^-$ band members pointed out by Kuehner and Almqvist[21] is their large reduced alpha-particle widths. This band has $(\lambda,\mu) = (9,0)$ in the SU(3) classification scheme which has a considerable admixture of the $1(sd \rightarrow pf)$ excitation. The bands I would like to discuss in more detail now are the two $K = 0^+$ bands based on the 6722 keV and 7191 keV levels. The former is believed to arise from an $(sd)^4$ configuration and the latter is believed to have a large admixture of the $4(p \rightarrow sd)$ quartet state. Hartree-Fock calculations by Benson and Flowers[7] predict a $K^\pi = 0^+$ band of the $(sd)^4$ configuration near these levels at 7 MeV.

The assignment of levels to particular bands becomes increasingly speculative with increasing excitation energy because the assignments rely entirely on measured spin values and excitation energies. The assignment of the higher levels

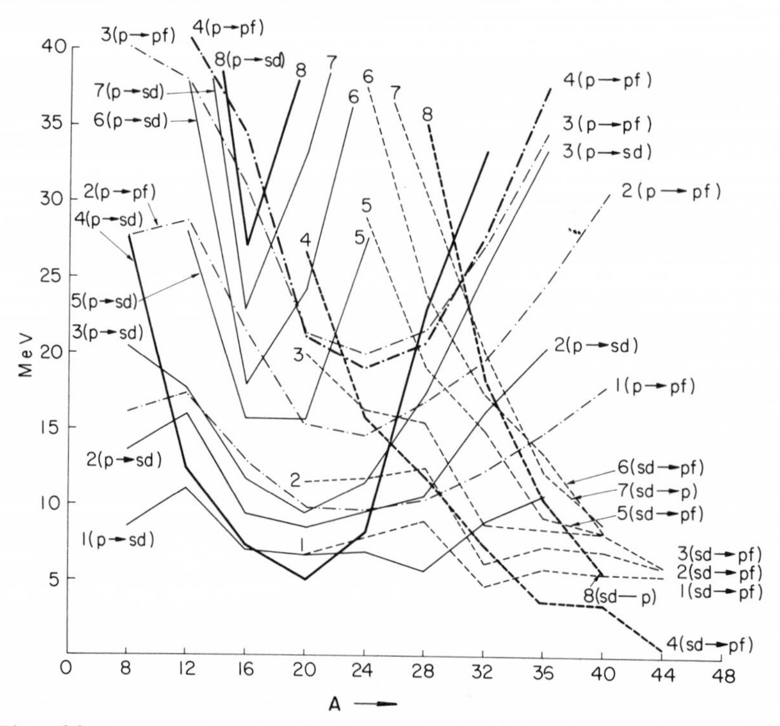

Fig. 10. Systematics of particle-hole excitations after Harvey (Ref. 18).

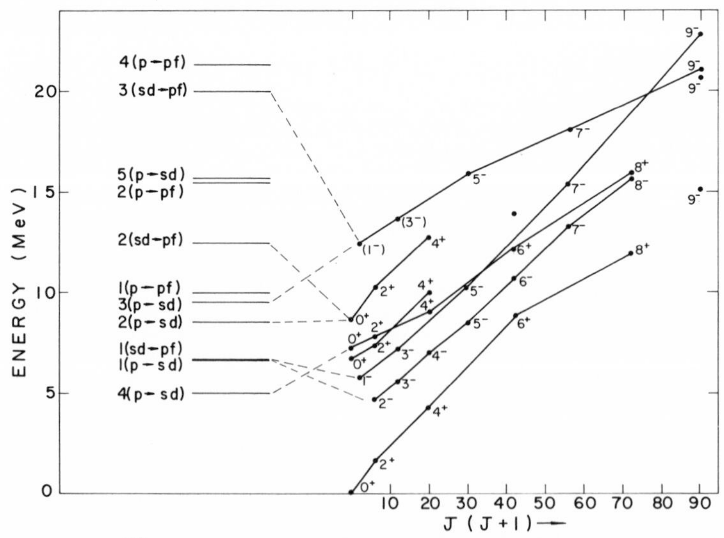

Fig. 11. Ne^{20} bands and particle-hole energies calculated by Harvey (see Figure 10). Lines join levels thought to be members of a band.

used here is thus not unique among authors and should be regarded as tentative.

III. THE K = 0^+ BANDS NEAR 7 MeV

As mentioned previously, Vogt[16] has concluded that the 8^+, 11,948 keV level has been misinterpreted, and suggests that this level is the 8^+ member of an $(sd)^8\ p^{-4}$ band, *i.e.*, the quartet band, based on the 0^+, 7.19 MeV level. The suggestion that the 0^+, 2^+ and 4^+ levels near 7 MeV form bands had been made by Kuehner and Almqvist[21] and Hunt *et al.*[22] Vogt's reassignment of the 8^+ level is based on the α particle reduced widths shown in Figure 12, where sets of levels with similar reduced widths are grouped together. The α particle widths, Γ_{CM} of these levels had been measured some time ago by Cameron,[23] McDermott *et al.*,[24] and Pearson and Spear.[25]

Recent experiments by Middleton *et al.*,[26] on the angular distributions and excitation functions of the $C^{12}(C^{12},\alpha)Ne^{20}$ reaction suggest that the 0^+, 7.19 MeV and 2^+, 7.83 MeV levels are members of the quartet configuration corresponding to two alpha particles outside a C^{12} core. Nagatani *et al.*,[27] have studied the $C^{12}(N^{14},Li^6)Ne^{20}$ reaction and found that the 4^+, 9.03 MeV level (see Figure 12) is populated more strongly

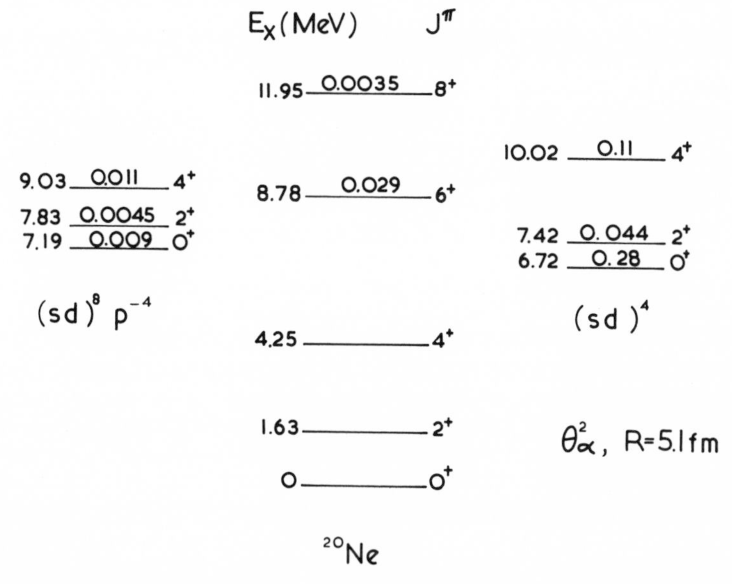

Fig. 12. An energy level diagram of Ne^{20} showing the $K^\pi = 0^+$ ground state band and the levels associated with the two excited $K^\pi = 0^+$ "bands". Levels other than those in the ground state band have been grouped together according to their alpha reduced widths, θ_α^2, after Vogt (Ref. 10). The value of θ_α^2 is shown for each level.

than the 4^+, 10.02 MeV level as expected if the former belongs to a 8p-4h configuration and the latter to a $(sd)^4$ configuration. Their evidence on the 0^+ and 2^+ levels is less conclusive.

Evidence that the 0^+, 6722 keV level has a large component of the $(sd)^4$ configuration comes from the results of the $F^{19}(He^3,d)Ne^{20}$ reaction[28] when compared to shell model calculations.[8] The large α particle width of the 6722 keV level is believed to arise from mixing with a state of fp-shell character.[29]

It is evident that the classification of these α unstable levels into bands is a difficult exercise. By radiative capture, the γ ray decay modes can be studied, but the observation of enhanced in-band E2 transitions to establish members of a given band is difficult since the transition energies are small. However, the energetic cross-band decays to the ground state band can be observed and some measurements have been reported in the literature. Pearson and Spear[25] have measured the radiative widths of the 9.03 MeV and 10.02 MeV, $J^\pi = 4^+$ levels; and van der Leun *et al.*,[30] and more recently Toevs,[31] have measured the radiative width of the 0^+, 6.72 MeV level. All of these measurements yielded E2 transitions of $\sim$5 W.u. strength, indicating quite strong cross-band transitions and implying mixing into the ground state band. Until recently, shell model calculations[32] have not been able to reproduce these large transition strengths.

It was the aim of the $O^{16}(\alpha,\gamma)Ne^{20}$ reaction experiments[17] I would like to describe now to measure the γ ray transition probabilities of the 0^+ level at 7.19 MeV, the 2^+ levels at 7.42 and 7.83 MeV and to search for a possible 6^+ level required by Vogt's suggestion in the region from 10.0 to 10.8 MeV. In addition, the radiative width of the 4^+ level at 9.03 MeV was remeasured to confirm its enhanced E2 decay. Measurements were made on the underlined resonances in Table II, which summarizes the resonances of interest. Those listed as "reference resonances" were used to obtain thick target measurements of the resonance strength, $\omega\gamma$, since their strength has been previously determined by Pearson and Spear[25] by Diamond *et al.*,[11] and by Rogers *et al.*[11] The experimental methods employed are described briefly in Section A and the results are presented resonance by resonance in Section B. A discussion of the results and comparison with theoretical calculations are contained in Section C.

A. Experimental Methods

1. Locating Resonances

The 8 MV injector Van de Graaff accelerator at Oxford has been used to study resonances in the $O^{16}(\alpha,\gamma)Ne^{20}$ reaction. The water cooled targets can stand up to 10-15 μA

TABLE II

Resonances in $O^{16}(\alpha,\gamma)Ne^{20}$, Q_o = 4730 keV

J^{π}	$E_R(\alpha,\gamma)$ (keV)	E_x (a) (keV)	$E_r(\alpha,\alpha)$ (keV)	Γ_{CM} (keV)	Note
0^+	2490 ± 8	6722 ± 8	2490	19	(b)(d)
0^+	3074 ± 5	7191 ± 3	3090	4	(b)
2^+	3363 ± 5	7420 ± 1	3380	8	(b)
2^+	3872 ± 5	7826 ± 3	3885	2.4	(b)
4^+	5368 ± 5	9029 ± 3	5432	3.2	(c)
$(4)^+$	6610 ± 30	10020 ± 30		155 ± 30	(e)
		REFERENCE RESONANCES			
2^+	6919 ± 5	10271 ± 3		$\lesssim$ 2	(e)
6^+	5052 ± 5	8775 ± 2.6	5058	0.11 ± 0.025	(f)

Underlined values from the present work. (a) Corrected for the recoil energy of the Ne^{20} nucleus; (b) (α,α) data from Ref. 23; (c) (α,α) data from Ref. 24; (d) (α,γ) data from Ref. 30 and 31; (e) (α,γ) data from Ref. 25, $\omega\gamma$ = 22.2 ± 2.4 eV; (f) (α,α) data from Ref. 9; (α,γ) data from Ref. 11, $\omega\gamma$ = 1.36 ± 0.14 eV.

beams of 5 MeV α particles for about 24 hours without serious deterioration. The targets consisted of $Ta_x O^{16}_y$ films formed by sputtering[33] Ta in low pressure O^{16} gas depleted to 42 ppm of O^{18} and 80 ppm of O^{17} on to high purity gold backings 0.13 mm thick. The oxide thickness was chosen to be from 15 keV to 70 keV for 4 MeV alpha particles depending on the resonance under investigation.

For observation of the capture γ rays, a 9.9% efficient Nuclear Diode and a 7% RCA Ge(Li) detector were used in close geometry (source to detector front face $\lesssim$ 5 cm) at 30° and 60° or at 0° and 90° with respect to the beam. About 2 mm of lead was placed between the detectors and target chamber to absorb very low energy radiation. Both detectors had an energy resolution of less than 4 keV at 3854 keV. The pulse height spectra from these detectors were accumulated either in 4096-channel Laben analyzers or in the on-line PDP-7 computer.

To locate resonances quickly, a beam energy modulator[34] connected to the injector accelerator was used to modulate the beam energy in a cyclic manner and generate yield curves automatically over an energy range up to 200 keV wide. On-resonance and off-resonance data are accumulated at the same time, a particularly useful feature when background subtraction is critical.

A 12.7 cm by 15.2 cm, or a 7.6 by 7.6 cm, NaI(Tl) detector was also used with the beam modulation system for quickly finding resonances and determining target thicknesses. For example the yield curve inset into Figure 13 was obtained in this way. The underlined resonance energies, $E_R(\alpha,\gamma)$, listed in Table II were obtained from the yield curves and the accelerator energy calibration previously determined by the $Li^7(p,n)$, $F^{19}(p,n)$ and $Al^{27}(p,n)$ reaction thresholds. In general, the excitation energies calculated from $E_R(\alpha,\gamma)$ and Q_o = 4730 keV agree with those obtained from γ ray energy measurements within 5 keV.

The excitation energy of a resonance was determined by measurements of the decay γ ray energies. These energies are shown in the third column of Table II and are mainly based on the calibration obtained from the 6129.3 ± 0.4 keV[35]

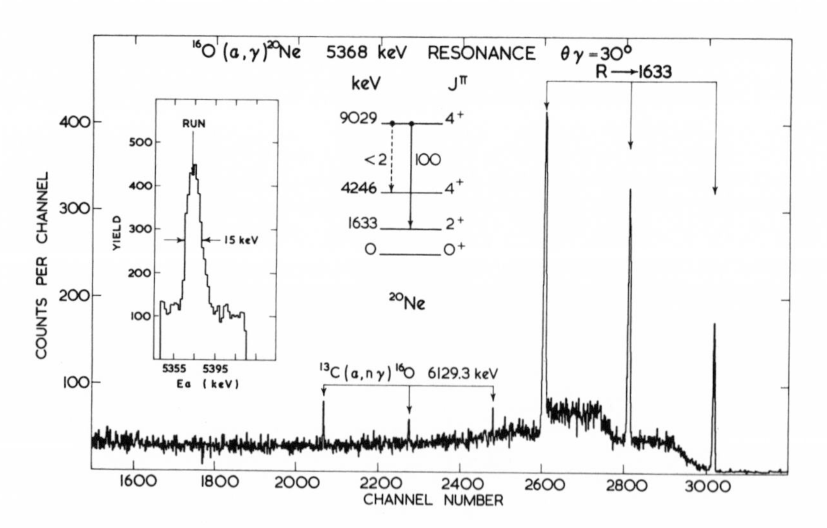

Fig. 13. A partial Ge(Li) detector spectrum obtained at θ_γ = 30° on the 5368 keV, J^π = 4^+ resonance in the $O^{16}(\alpha,\gamma)Ne^{20}$ reaction. The inset shows the yield curve observed with a 7.6 cm by 7.6 cm NaI(Tl) detector for the target used to accumulate the Ge(Li) spectrum.

contaminant line (and its escape peaks) from the $C^{13}(\alpha,n\gamma)O^{18}$ reaction and from the background RdTh line at 2614.47 ± 0.10 keV.[35]

2. Radiative Width Measurements

The radiative widths of the resonances were measured by comparison of their thick target yields with that from either the 2^+, T = 1, 6919 keV or the 6^+, 5052 keV resonance in $O^{16}(\alpha,\gamma)Ne^{20}$. The 'reference' resonance yield was measured before and after the resonance under investigation to check the target durability. The values of $\omega\gamma \equiv (2J+1)\Gamma_\gamma\Gamma_\alpha/\Gamma$ for these 'reference' resonances are shown at the bottom of Table II and were obtained from the literature.[11,25] The value $\omega\gamma$ = 27.8 ± 3 eV quoted by Pearson and Spear[25] is assumed to refer to the laboratory system and has been multiplied by 16/20 since here quoted results refer to the center of mass system.

The ratio of the radiative widths, $\omega\gamma$, of two resonances in the same reaction measured with the same thick target and geometry is

$$\frac{\omega\gamma_1}{\omega\gamma_2} = \frac{E_1}{E_2} \cdot \frac{\varepsilon(E_1)}{\varepsilon(E_2)} \cdot \frac{Y_1(\theta)}{Y_2(\theta)} \cdot \frac{e_2(E_\gamma)}{e_1(E_\gamma)} \cdot \frac{f_1(\theta)}{f_2(\theta)} \cdot \frac{\tan^{-1}(\xi_2/\Gamma_2)}{\tan^{-1}(\xi_1/\Gamma_1)}$$

where E is the resonance energy, $\varepsilon(E)$ is the stopping power taken from the tables of Northcliffe and Schilling,[36] $Y(\theta)$ is the observed intensity at angle θ per incident particle and corrected for dead time, $e(E_\gamma)$ is the experimentally determined relative γ ray efficiency of the detector and $f(\theta)$ is the calculated (or measured) angular distribution factor. The factor $\tan^{-1}(\xi/\Gamma)$ corrects the observed yield to correspond to that from an infinitely thick target.[35] In general, the target thickness was at least five times the natural resonance width to ensure the observed yield was close to the maximum. The $60^\circ/30^\circ$ detector geometry was usually used to obtain yield measurements that were relatively insensitive to the γ ray angular distributions.

B. Results

1. The 5368 keV, $J^\pi = 4^+$ Resonance

The inset in Fig. 13 shows a yield curve of γ ray pulses greater than 4 MeV in a 7.6 cm by 7.6 cm NaI(Tl) detector near the bombarding energy region of the 4^+ resonance. The resonance occurred at 5368 keV and the target was ∿15 keV thick, *i.e.*, much greater than the resonance width of 3.2 keV (see Table II). The partial Ge(Li) spectrum shown in Figure 13 was accumulated with the 9.9% efficient detector placed at 30° for a total incident $^4He^{1+}$ beam charge of 0.101 Coul. at the beam energy shown in the inset. Similar data were accumulated concurrently in the second Ge(Li) detector at 60°. A comparison of the yields at both angles with the 5052 keV, 6^+ resonance yield gave $\omega\gamma(4^+)/\omega\gamma(6^+) = 2.24 \pm 0.16$ or $\omega\gamma(4^+) = 3.05 \pm 0.38$ eV. The ratio of the yields at 30° and 60° was consistent within experimental errors of 3% with that expected for a $4^+ \rightarrow 2^+$ transition. These results are in good agreement with the measurements of Pearson and Spear.[25] Their quoted value $\omega\gamma(4^+) = 3.4 \pm 0.4$ eV has been multiplied by the factor 16/20 to convert to the c.m. system used in this paper.

An upper limit on the intensity of the unobserved $4^+ \rightarrow 4^+$, 9029 → 4246 keV transition is <2% relative to the $4^+ \rightarrow 2^+$ transition.

The excitation energy of the 4^+ level is 9029 ± 3 keV from measurements of the γ ray energy, corrected for the Doppler shift and recoil energies.

2. The 3872 keV, $J^\pi = 2^+$ Resonance

Measurements of the resonance strength and gamma decay branching ratios of the 2^+ resonance at 3872 keV bombarding energy were carried out using a 16 keV target, which is thick compared with Γ = 2.4 keV. An 8½ hour run in which 0.338

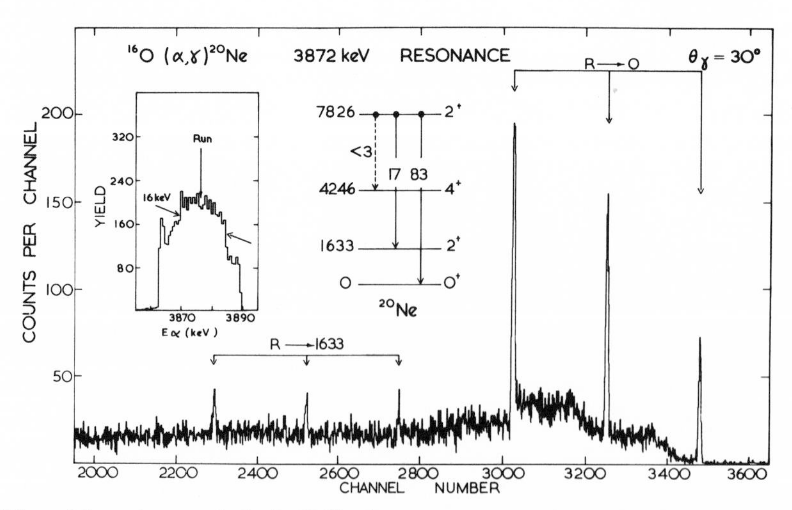

Fig. 14. A partial Ge(Li) detector spectrum obtained on the 3872 keV, $J^\pi = 2^+$ resonance in the $O^{16}(\alpha,\gamma)Ne^{20}$ reaction.

Coul. of He^+ beam was incident resulted in the spectrum shown in Figure 14. The bombarding energy used to obtain this Ge(Li) detector spectrum is shown on the yield curve inset. This level decays (83 ± 1)% to the ground state and (17 ± 1)% to the 1633 keV first excited state. Again spectra were obtained concurrently at 30° and 60° with respect to the beam direction and $\omega\gamma$ values obtained from both spectra. The average value from both angles for $\omega\gamma(2^+ \rightarrow 0^+)$ is 285 ± 34 meV and for $\omega\gamma(2^+ \rightarrow 2^+)$, 58 ± 8 meV. These values were measured relative to the T = 1, $J^\pi = 2^+$ resonance at E_α = 6919 keV (see Table II).

The excitation energy of this 2^+ level is 7826 ± 3 keV from measurements of the γ ray energies, corrected for the Doppler shift and recoil energies.

The complete angular distribution of the $2^+ \rightarrow 2^+$ transition would have been desirable to obtain the E2/M1 mixing ratio. However, the low intensity of this transition prohibited this measurement, particularly because it is lower in energy than the much stronger $2^+ \rightarrow 0^+$ transition.

A branching ratio limit of <3% can be set on the $2^+ \rightarrow 4^+$, 7826 → 4246 keV transition which was not observed.

3. The 3363 keV, $J^\pi = 2^+$ Resonance

The 2^+ resonance at 3363 keV has a broad natural width and decays mainly to the first excited level at 1633 keV. The latter gave the opportunity of measuring the E2/M1 mixing ratio for the $2^+ \rightarrow 2^+$ transition. A preliminary run on this resonance with an 11 keV thick target gave $\Gamma_{lab} = 8 \pm {}^{2}_{3}$ keV in agreement with the natural resonance width, Γ_{lab} = 10 keV

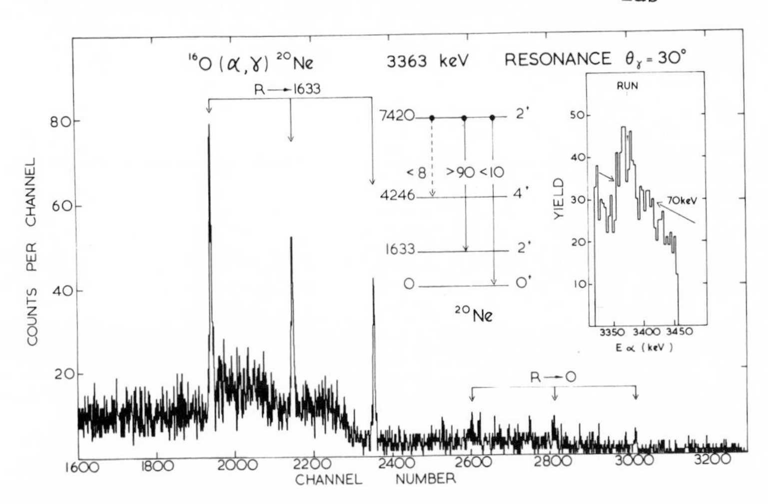

Fig. 15. A partial Ge(Li) detector spectrum obtained on the 3363 keV, $J^\pi = 2^+$ resonance in the $O^{16}(\alpha,\gamma)Ne^{20}$ reaction.

determined by Cameron.[23]

A substantially thicker target was made and found to be 56 keV thick at the 6^+, 5052 keV resonance. This target was then 70 keV thick at 3363 keV and suitable for thick target yield measurements on the 2^+, 3363 keV resonance. Spectra were accumulated with the two Ge(Li) detectors at 30° and 60° and again at angles of 0° and 90°. Figure 15 shows the θ_γ = 30° spectrum which was accumulated over 8 hours for a total charge of 0.289 Coul. Before the 30°/60° run, a yield measurement from the 6^+ resonance yield was made and again at the end of the run before the detectors were rotated. After the detectors were rotated into the 0°/90° position, the 6^+ resonance yield was again measured before and after the longer run on the 2^+ resonance. This procedure gave assurance that the target was not deteriorating, gave direct reference to the 6^+ resonance intensities for each angle, and yielded a four point angular distribution for the $2^+ \rightarrow 2^+$ transition.

The angular distribution is shown in Figure 16(a). Since the spin of the resonance studied was already established, the angular distribution was measured to obtain the E2/M1 mixing ratio. Figure 16(b) shows the value of χ^2 as a function of the E2/M1 mixing ratio, δ, for various initial level spin assumptions, J = 1, 2, 3 or 4. The best fit yields J = 2 with $\delta = -8.4 \pm^{1.0}_{1.5}$. The χ^2 values for J = 1, 3, and 4 are shown for completeness. A weak (9.4 ± 1.4)%, ground state transition (see Figure 15) was also observed from the 3363

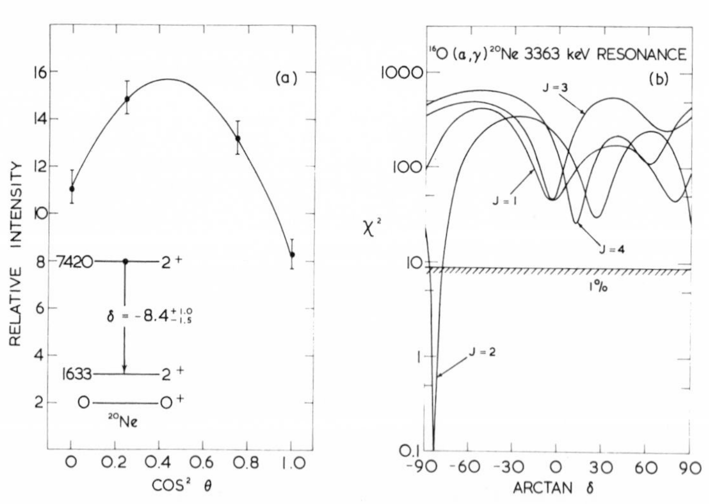

Fig. 16. (a) The angular distribution of the $2^+ \rightarrow 2^+$, 7420 $\rightarrow$ 1633 keV transition in Ne^{20}. The curve is the best fit for J = 2, δ = -8.4 and attenuation coefficients of $P_2(\cos\theta)$ and $P_4(\cos\theta)$ equal to 0.95 and 0.85 respectively. (b) χ^2 fits to the angular distribution to obtain the E2/M1 mixing ratio. The horizontal hatched line is the 1% confidence limit.

keV resonance. This measurement is considered an upper limit since off-resonance spectra of sufficient accuracy were not accumulated.

A comparison of the yields of the $2^+ \rightarrow 2^+$ and $2^+ \rightarrow 0^+$ primary transitions with that from the 6^+ resonance at 5052 keV gave $\omega\gamma(2^+ \rightarrow 2^+) = 146 \pm 19$ meV and $\omega\gamma(2^+ \rightarrow 0^+) \leq 15 \pm 2.6$ meV.

A branching ratio limit of < 7.6% can be set on the $2^+ \rightarrow 4^+$, 7420 → 4246 keV transition which was not observed. The excitation energy in Ne^{20} corresponding to this 2^+ resonance is 7420 ± 3 keV as determined from the γ ray energies.

4. The 3074 keV $J^\pi = 0^+$ Resonance

A preliminary attempt to locate the $J^\pi = 0^+$ resonance known from elastic α particle scattering[23] showed that its radiative width was much weaker than that of the resonances described so far. The resonance energies from the present investigation were consistently about 15 keV lower than the (α,α) data[23] (see Table II) which enabled us to predict quite well where to find the resonance. To increase the detection sensitivity a 12.7 cm by 15.2 cm NaI(Tl) detector was placed at 45° at a distance of 3 cm from the target. Two-parameter data were accumulated in an 8192 channel Laben analyzer allowing 32 channels for beam energy by 256 channels for γ ray energy. A slice in γ ray energy from 4.5 to 5.6 MeV produced the yield curve shown in the inset of Figure 17. Although

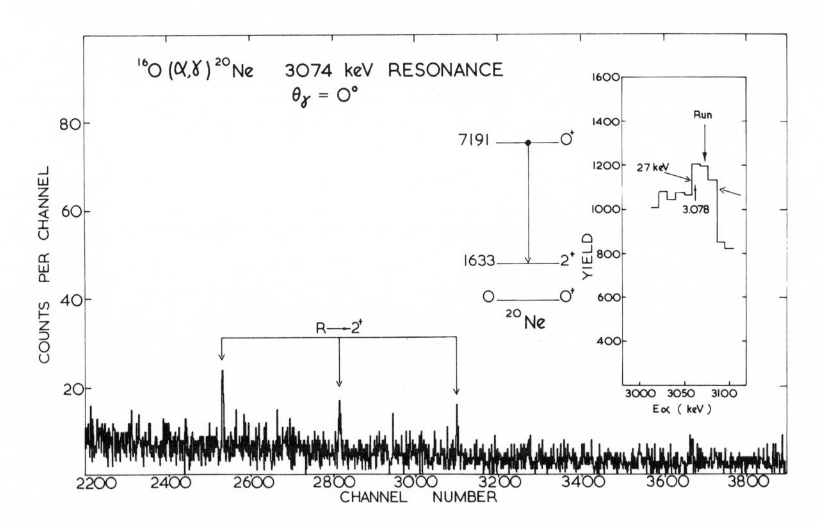

Fig. 17. A partial Ge(Li) detector spectrum obtained on the 3074 keV, $J^\pi = 0^+$ resonance in the $O^{16}(\alpha,\gamma)Ne^{20}$ reaction. The yield curve was obtained with a 12.7 cm by 15.2 cm NaI(Tl) detector as explained in the text.

the background was large, presumably because of neutron interactions in the NaI detector, the up and down sweep yield curves clearly showed a resonance at 3074 ± 5 keV. Analysis of the two dimensional data to produce an on-resonance minus off-resonance spectrum from the NaI detector clearly showed the $0^+ \rightarrow 2^+$ transition.

To obtain a measurement of $\omega\gamma$, the NaI detector was replaced by the 9.9% efficient Ge(Li) detector and a long run at the bombarding energy indicated in Figure 17 was obtained. The spectrum shown in Figure 17 was accumulated during a 28 hour run in which 0.981 Coul. of integrated beam was incident on the target. Comparison of these data with the yield from the 6^+, 5052 keV resonance measured for the same conditions, gave $\omega\gamma(0^+ \rightarrow 2^+) = 4.35 \pm 0.75$ MeV.

The excitation energy in Ne^{20} corresponding to the 0^+ resonance is 7191 ± 3 keV from the gamma-ray energy measurement.

5. Search for a possible 6^+ level in the region 10.2 to 10.8 MeV excitation energy in Ne^{20}

As mentioned earlier, a possible explanation of the small reduced α particle width of the 8^+ level at 11948 keV was Vogt's suggestion[16] that this state should be assigned to the band beginning with the 0^+, 7.19 MeV level and not the ground state band. Assuming this, a 6^+ level should exist at $\sim$10.5 MeV. We have searched for a possible $R \rightarrow 4^+$ (4246 keV level) transition using the beam modulator and a Ge(Li) detector. Data were accumulated on magnetic tape in address-recorded form with the PDP-7 computer and were sorted in detail off-line with the PDP-10 computer.

The apparatus was checked and calibrated on the 2^+, $T = 1$, 6919 keV resonance resulting in the data shown in Figure 18. For this test the scan width was 124 keV and the yield was from a γ ray energy window from 6 MeV to 9 MeV. The bombarding energy region up to 7.6 MeV was then scanned in $\sim$180 keV regions in four additional runs.

Figure 19 shows the result of an analysis of the data in which the counts in 24 keV wide windows centered at the γ ray energy $(E_x - 4246)(1 + v/c)$ keV and the corresponding escape-peak energies are plotted as a function of the incident beam energy. The peak at the extreme left results from the Compton tail of gamma rays from the $T = 1$ resonance. No other resonance in the yield was observed. An upper limit of $\omega\gamma < 0.6$ eV can be deduced from the data of Figure 19. The main problem in obtaining good sensitivity is that the expected energy of the transition is very close to the 6129 keV contamination γ ray from the $C^{13}(\alpha,n\gamma)O^{16}$ reaction. The present data are consistent with those of Pearson and Spear[25] who scanned the same energy region using an NaI(Tl) detector.

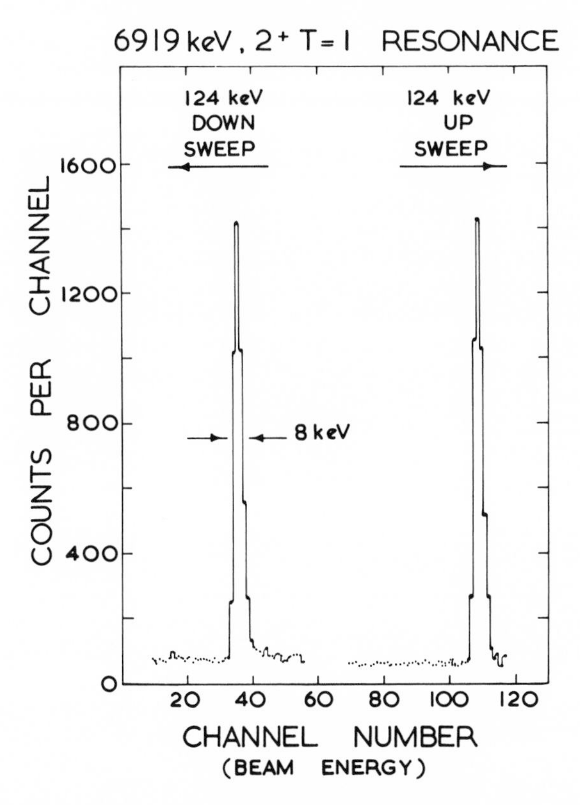

Fig. 18. The yield curves obtained with the beam modulator and a Ge(Li) detector window from 6.0 and 9.0 MeV. The right hand curve was obtained while the beam was increasing in energy, and that on the left was obtained while the beam decreased in energy over a 124 keV interval including the $J^{\pi} = 2^{+}$, T = 1, 6919 keV resonance in $O^{16}(\alpha,\gamma)Ne^{20}$.

C. Discussion

For each of the resonances studied, Γ_{α} is known from the literature as summarized in Table II, and Γ_{γ} could be deduced. The available γ ray E2 transition probabilities in Weisskopf units[37] for the 0^{+}, 2^{+} and 4^{+} levels associated with the 0^{+} band heads at 6722 keV and 7191 keV, are given in Table III. These data are also shown in schematic representation in Figure 20, where the intraband E2 transition probabilities[2,3,11] for the ground state band are also shown for comparison.

In Table III, the measurements of Pearson and Spear[25] and the average value of van der Leun *et al.*,[30] and Toevs[31] are given. The present value of 5.8 ± 0.7 W.u. for the

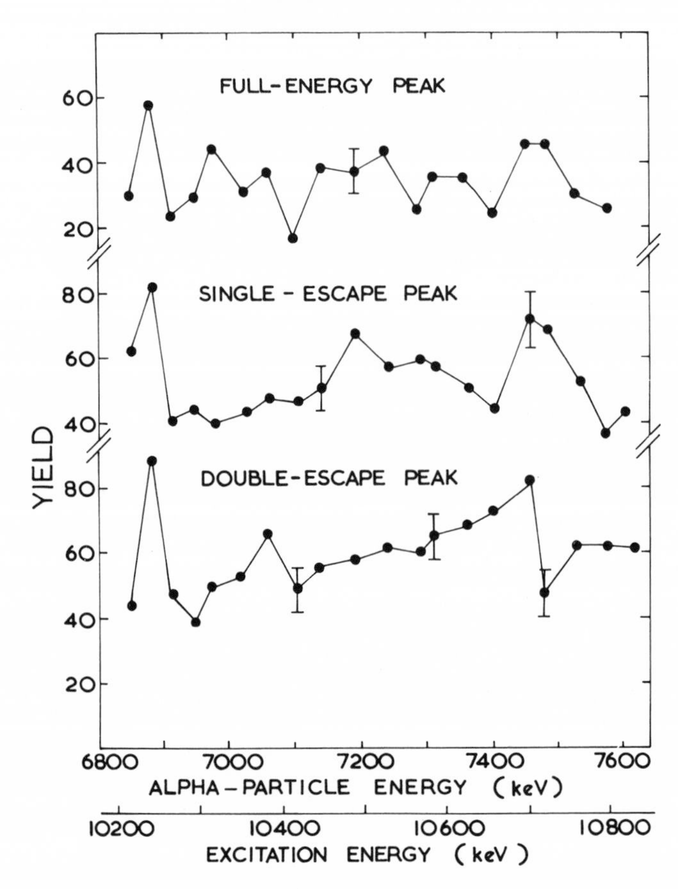

Fig. 19. The yield curves obtained from a Ge(Li) detector during a search for a possible $E_x \rightarrow 4246$ keV transition over the bombarding energy region from 6.8 to 7.6 MeV. The 8 keV thick target illustrated in Figure 7 was used.

$4^+ \rightarrow 2^+$, 9029 → 1633 keV transition is in excellent agreement with the Pearson and Spear value of 5.3 ± 0.7 W.u. There is also agreement of the present result for the $2^+ \rightarrow 0^+$, 7826 → 0 keV transition of 0.73 ± 0.09 W.u. and the value of 0.83 ± 0.13 W.u. published during the present experiments by Mitsunobu and Torizuka[38] using inelastic electron scattering. However, the present result for the $2^+ \rightarrow 0^+$, 7420 → 0 keV transition, $|M|^2 \leq 0.05 \pm 0.01$ W.u., is less than $|M|^2 = 0.13 \pm 0.03$ W.u. obtained by inelastic electron scattering.[38] This may reflect the difficulty of interpreting the inelastic electron scattering data for such weak transitions.

In Figure 20, the grouping of the six levels under discussion into bands is the same as in Figure 12, *i.e.*, based on the similarity of the α particle reduced widths. Other

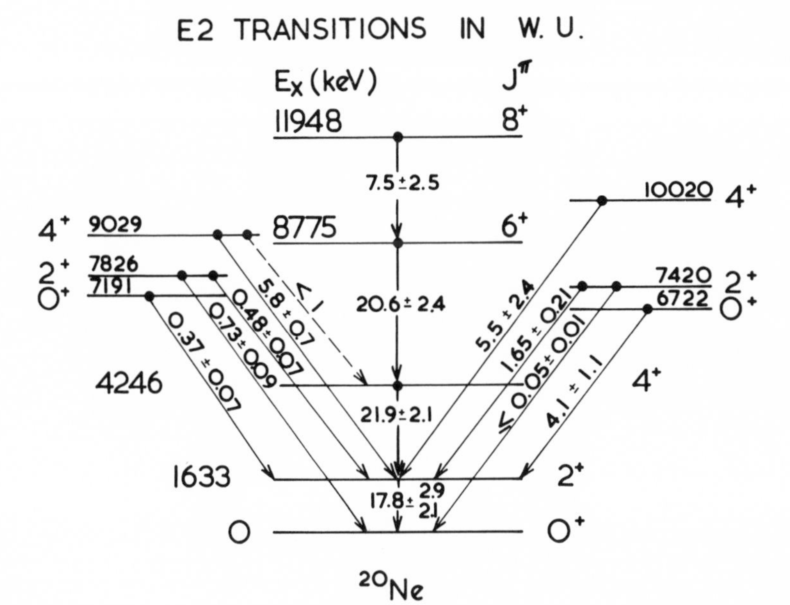

Fig. 20. A summary of the "interband" E2 transitions in Ne^{20}. The ground state intraband E2 transitions are also shown. The $2^+ \to 2^+$, 7826 → 1633 keV transition is assumed to be pure E2. The $4^+ \to 4^+$, 9029 → 4246 keV transition was not observed and is shown with a dashed line. Upper limits for the decay strengths from the 2^+ levels to the 4^+, 4246 keV level are not shown for simplicity of the diagram. The values are listed in Table III.

authors (see *e.g.* Ref. 27) have the 4^+ levels interchanged. It should be noted that although this classification of these levels is debatable, it is convenient for discussion. The following features can be noted from Figure 20:

a. The $0^+ \to 2^+$ transitions differ again by a factor of ∿10, the 6.72 MeV level being the stronger.

b. The $2^+ \to 0^+$ transitions differ again by a factor of ∿10, but the 2^+ associated with the 6.72 MeV level is the weaker.

c. Although the strengths of decays from the 0^+ and 2^+ levels are generally weak, the $4^+ \to 2^+$ transitions are anomalously strong for "interband" transitions. This becomes clearer by comparing the transition strengths to the rotational model ratios as has been done in the last column of Table III. The anomalous be-

TABLE III

Summary of Interband E2 Transition Probabilities

E_x (keV)	$J^{\pi}_i \rightarrow J^{\pi}_f$	$\|M\|^2$ Experimental (W.u.) Present	Previous	Rotational Model $\|M\|^2$ (W.u.)
6722	$0^+ \rightarrow 2^+$		4.1 ± 1[c]	5.8
7420	$2^+ \rightarrow 0^+$	$\leq 0.05 \pm 0.01$	0.13 ± 0.03[a]	1.16
	$2^+ \rightarrow 2^+$	1.65 ± 0.21		<u>1.65</u>
	$2^+ \rightarrow 4^+$	< 2.6		2.97
10020	$4^+ \rightarrow 2^+$		5.5 ± 2.4[b]	1.65
7191	$0^+ \rightarrow 2^+$	0.37 ± 0.07		3.65
7826	$2^+ \rightarrow 0^+$	0.73 ± 0.09	0.83 ± 0.13[a]	<u>0.73</u>
	$2^+ \rightarrow 2^+$	0.48 ± 0.07		1.04
	$2^+ \rightarrow 4^+$	< 1.3		1.87
9029	$4^+ \rightarrow 2^+$	5.8 ± 0.7	5.3 ± 0.7[b]	1.04
	$4^+ \rightarrow 4^+$	< 1		0.95

(a) Ref. 38; (b) Ref. 25; (c) Ref. 30 and 31.
Rotational model $|M|^2$ are proportional to $\langle J_i 2 0 0 | J_f 0\rangle^2$ and are normalized to the experimental values where underlined.

havior is accentuated by the weakness, $|M|^2 < 1$ W.u., of the $4^+ \rightarrow 4^+$, 9029 → 4246 keV transition.

The upper limit of 0.6 eV for the γ ray width of a possible $6^+ \rightarrow 4^+$ transition from a level in the region of 10.5 MeV corresponds to $|M|^2 < 3$ W.u. Based on the strength of the $2^+{}_3 \rightarrow 0^+{}_1$, 7826 → 0 keV transition, a $6^+ \rightarrow 4^+$ transition should be approximately 1.1 W.u. for a transition between two K = 0 rotational bands. So, by this comparison, the upper limit does not rule out the existence of the 6^+ state suggested by Vogt.[16] However, by the same comparison, the value of $|M|^2 = 5.8 \pm 0.7$ W.u. for the $4^+ \rightarrow 2^+$ transition is anomalous.

Although the rotational model trend fails for the levels identified with the 8p-4h configuration, partial agreement is seen for the levels associated with the $(sd)^4$ configuration as shown in Table III; the exception is the $2^+ \rightarrow 0^+$, 7420 → 0 keV transition. The $4^+ \rightarrow 2^+$, 10020 → 1633 keV transition strength has large experimental errors.
tions of Millener and Strottman[39] and Halbert *et al.*,[8] have also been tabulated. Good agreement is seen for the first three transitions listed in Table IV, in particular the weak-

TABLE IV

Interband Transitions in Ne^{20} (W.u.)

E_i (keV)	$J_i \rightarrow J_f$	Type	Experiment	Theory	
				Millener[a] & Strottman	Halbert[b]
6722	$0_2 \rightarrow 2_1$	E2	4.1 ± 1.1	3.80	4.2
7420	$2_2 \rightarrow 0_1$	E2	≤ 0.05 ± 0.01	0.008	0.03
	$2_2 \rightarrow 2_1$	E2	1.65 ± 0.2	1.30	1.2
		M1	$(1.0 \pm 0.3)\times10^{-4}$	7×10^{-4}	1.8×10^{-5}
	$2_2 \rightarrow 4_1$	E2	< 2.6	-	0.028
10020	$4_3 \rightarrow 2_1$	E2	5.5 ± 2.4	0.28	1.1
7191	$0_3 \rightarrow 2_1$	E2	0.37 ± 0.07	0.55	
7826	$2_3 \rightarrow 0_1$	E2	0.73 ± 0.09	0.07	
	$2_3 \rightarrow 2_1$	E2	0.48 ± 0.07	0.25	
	$2_3 \rightarrow 4_1$	E2	< 1.3	-	
9029	$4_2 \rightarrow 2_1$	E2	5.8 ± 0.7	0.21	
	$4_2 \rightarrow 4_1$	E2	< 1	-	

(a) Ref. 39; (b) Ref. 8.

ness of the $2_2 \rightarrow 0_1$ transition is predicted whereas it is not by the simple rotational model. Also, fair agreement is obtained for the measured M1 strength, $|M(M1)|^2 = (1.0 \pm 0.3)$ $\times 10^{-}$ W.u., of the $2_2 \rightarrow 2_1$ transition for which Halbert *et al.*[8] obtain $|M|^2 = 1.8 \times 10^{-5}$ W.u. and Millener and Strottman[39] obtain 7×10^{-4} W.u. The results of Halbert *et al.* were obtained assuming an sd configuration space using two body matrix elements deduced from "realistic" two body forces.

The results of Millener and Strottman were obtained with an increased basis which included 6p-2h and 8p-4h states. The enlarged basis was truncated and each 6p-2h and 8p-4h state was limited to one SU(3) representation. This has allowed them to obtain estimates of the $(sd)^8p^{-4}$ levels shown in Table IV with an effective charge of 0.6. Except for the $2_3 \rightarrow 01$ and $4_2 \rightarrow 2_1$ cases the agreement is quite good.

A particular problem with the shell model calculations is their failure to give the energy of the 2^+ level low enough to agree with the small 0^+ to 2^+ energy spacing observed experimentally, especially for the 2^+ level associated with the $(sd)^4$ configuration.

The present γ ray data are in reasonable agreement with calculations for the 0^+ and 2^+ levels especially for the levels associated with the $(sd)^4$ configuration. However the 4^+ levels, particularly the 9029 keV level, have transition probabilities considerably larger than the theoretical predictions.

In conclusion, the theoretical interpretation of the ground state band in Ne^{20} has evolved to a degree of sophistication that allows both the γ ray transition probabilities and the α particle reduced widths to be estimated in a description that is very different from the simple rotational picture. It is possible that the excited $K^{\pi} = 0^+$ band based on the 7.12 MeV level is of a different nature than the ground state band or the band based on the 6.72 MeV level. More experimental data on the candidates for the high spin members of this band would be most interesting.

ACKNOWLEDGMENTS

The work reported here was done in collaboration with O. Haüsser, A. B. McDonald, A. J. Ferguson, W. T. Diamond, A. E. Litherland, I. M. Szöghy, G. T. Ewan and D. L. Disdier at Chalk River and with B. Y. Underwood, N. Anyas-Weiss, N. A. Jelley, J. Szücs, S. P. Dolan, M. R. Wormald and K. W. Allen at Oxford. I would also like to thank M. Harvey, H. C. Lee and D. Strottman for interesting conversations and permission to use the results of their work.

REFERENCES

1. A. E. Litherland *et al.*, Phys. Rev. Lett. 7, 98 (1961).
2. O. Häusser *et al.*, Nucl. Phys. A168, 17 (1971).
3. T. K. Alexander *et al.*, Nucl. Phys. A179, 477 (1972).
4. M. Harvey, *Advances in Nuclear Physics*, Vol. 1, M. Baranger and E. Vogt, Eds. (Plenum Press, New York, 1968).
5. A. Arima, S. Cohen, R. Lawson and M. Macfarlane, Nucl. Phys. A108, 94 (1968).
6. P. Goode and S. S. M. Wong, Phys. Lett. 32B, 89 (1970).
7. H. G. Benson and B. H. Flowers, Nucl. Phys. A126, 305 (1969).
8. E. C. Halbert *et al.*, *Advances in Nuclear Physics*, Vol. 4, M. Baranger and E. Vogt, Eds. (Plenum Press, New York, 1971).
9. A. Arima and S. Yoshida, Phys. Lett. 40B, 15 (1972).
10. O. Haüsser *et al.*, Nucl. Phys. A179, 465 (1972).
11. W. T. Diamond, T. K. Alexander and O. Haüsser, Can. J. Phys. 49, 1589 (1971) and D. W. O. Rogers, *et al.*, Can. J. Phys. 49, 1397 (1971).
12. D. W. O. Rogers *et al.*, Phys. Lett. 37B, 65 (1971).
13. O. Häusser *et al.*, Chalk River report PR-P-94, AECL-4257 (1972).
14. H. Horiuchi, Ph.D. Thesis, University of Tokyo (1970) unpublished.
15. H. C. Lee and R. Y. Cusson (to be published).
16. E. Vogt, Phys. Lett. 40B, 345 (1972).
17. T. K. Alexander *et al.*, (to be published).
18. M. Harvey (to be published).
19. A. Arima and V. Gillet, Annals of Phys. 66, 117 (1971).
20. A. D. Panagioutou, H. E. Gove and S. Harar, Phys. Rev. C5, 1995 (1972).
21. J. A. Kuehner and E. Almqvist, Can. J. Phys. 45, 1605 (1967).
22. W. E. Hunt, M. K. Mehta and R. H. Davis, Phys. Rev. 160, 782 (1967).
23. J. R. Cameron, Phys. Rev. 90, 839 (1953).
24. L. C. McDermott *et al.*, Phys. Rev. 118, 175 (1960).
25. J. D. Pearson and R. H. Spear, Nucl. Phys. 54, 434 (1964).
26. R. Middleton, J. D. Garrett and H. T. Fortune, Phys. Rev. Lett. 27, 950 (1971).
27. K. Nagatani *et al.*, Phys. Rev. Lett. 27, 1071 (1971).
28. R. H. Siemssen, L. L. Lee, Jr. and D. Cline, Phys. Rev. 140B, 1258 (1965).
29. H. T. Fortune, R. Middleton and R. R. Betts, Phys. Rev. Lett. 29, 738 (1972).
30. C. Van der Leun, D. M. Sheppard and P. J. M. Smulders, Phys. Lett. 18, 134 (1965).
31. J. W. Toevs, Nucl. Phys. A172, 589 (1971).

32. A. Arima and D. Strottman, Nucl. Phys. A162, 423 (1971).
33. M. R. Wormald, K. W. Allen and B. Y. Underwood (to be published).
34. A. C. Shotter, J. Takacs and P. S. Fisher, Nucl. Instr. and Methods 88, 233 (1970) and M. R. Wormald *et al.* (to be published).
35. J. B. Marion and F. C. Young, *Nuclear Reaction Analysis Graphs and Tables*, (North Holland, Amsterdam, 1968).
36. L. C. Northcliffe and R. F. Schilling, Nucl. Data Tables A7, 233 (1970).
37. D. H. Wilkinson in *Nuclear Spectroscopy B*, F. Ajzenberg-Selove, ed. (Academic Press, New York, 1960).
38. S. Mitsunobu and Y. Torizuka, Phys. Rev. Lett. 28, 920 (1972).
39. J. Millener and D. Strottman, private communication from Oxford University.

PARTICIPANTS

Larry A. Alexander	Ohio State University
Thomas K. Alexander	Chalk River Nuclear Laboratory
R. E. Azuma	University of Toronto
Jerry D. Brandenberger	University of Kentucky
Louis Brown	Carnegie Institution of Washington
Thomas W. Burrows	University of Kentucky
Marcel Coz	University of Kentucky
Gerard M. Crawley	Michigan State University
Ronald Y. Cusson	Duke University
Sperry E. Darden	University of Notre Dame
John P. Davidson	University of Kansas
Martien J.A. de Voigt	University of Rochester
Edward M. Diener	Naval Research Laboratory
Douglas J. Donahue	University of Arizona
Timothy R. Donoghue	Ohio State University
F. Eugene Dunnam	University of Florida
Pieter M. Endt	State University of Utrecht
John D. Fox	Florida State University
Fletcher Gabbard	University of Kentucky
Hilton F. Glavish	Stanford University
Paul Goldhammer	University of Kansas
Philip Goode	Rutgers University
David R. Goosman	Brookhaven National Laboratory
Christopher R. Gould	North Carolina State University
M. Wayne Greene	McMaster University
Gale I. Harris	Wright Patterson Air Force Base
K. Peter Jackson	University of Toronto
George Kanatas	University of Kentucky
Bernard D. Kern	University of Kentucky
Richard D. Koshel	Ohio University
Ralph W. Krone	University of Kansas
William A. Lanford	Michigan State University
Robert D. Lawson	Argonne National Laboratory
H. C. Paul Lee	Chalk River Nuclear Laboratory
Alan D. MacKellar	University of Kentucky
F. Barry Malik	Indiana University
Sven Maripuu	Michigan State University
Floyd D. McDaniel	University of Kentucky
John D. McCullen	University of Arizona
Marcus T. McEllistrem	University of Kentucky
Gary E. Mitchell	North Carolina State University
Arnold Mueller-Arnke	Technischen Hochschule Darmstadt
Francis W. Prosser	University of Kansas
John R. Risser	Rice University
Barry C. Robertson	Queen's University
Rudolph Schrils	University of Kentucky
Hugh L. Scott	University of Georgia
L. Worth Seagondollar	North Carolina State University

John F. Sharpey-Schafer	University of Liverpool
Douglas M. Sheppard	University of Alberta
Jesse E. Sherwood	University of Kentucky
Franklin D. Snyder	University of Kentucky
Charles P. Swann	Bartol Research Foundation
Paul Taras	University of Montreal
Georges M. Temmer	Rutgers University
Thomas T. Thwaites	Pennsylvania State University
David R. Tilley	North Carolina State University
Stein Tryti	University of Oslo
Cor van der Leun	State University of Utrecht
Henri A. Van Rinsvelt	University of Florida
James F. Walker	University of Massachusetts
Ernest K. Warburton	Brookhaven National Laboratory
Jesse L. Weil	University of Kentucky
John W. Woodring	University of Kentucky